Environmental Toxicology

Eco-toxicology, an offshoot of toxicology, is a multidisciplinary science that integrates toxicology and ecology by drawing knowledge and procedures from both fields. It analyses the effects of toxic chemicals or biological agents on living organisms at different levels of organization. Eco-toxicology classifies different contaminants, their characteristics, release and ecological fate, and predicts their effects so that timely action can be taken to prevent or minimize any detrimental effects.

Environmental Toxicology starts with the basics of toxicology, briefly touching on the sources of toxic compounds, classification of toxicants, and factors affecting toxicity, and then elaborates on heavy metal toxicity. The individual chapters on various heavy metals and radioactive metals discuss the sources and routes of exposure, aetiology, pathophysiology, clinical manifestations, and mechanisms of toxicity, toxicological effects, diagnosis, treatment, management, and ecological impact. The book covers the field in its greatest width and provides an insight into pesticide and radiation toxicity, and recent advances in eco-toxicology, with a special focus on the removal of HMs and the latest bioremediation techniques.

This book serves as a reference work for advanced students pursuing degrees in environmental toxicology and across various disciplines, such as biomedical and environmental sciences, toxicology, eco-toxicology, pharmacology, public health, etc. and all interested in learning the concepts of eco-toxicology.

Features:

- A systematic overview of the key concepts of eco-toxicology, its relationship with other disciplines, and recent advances in the area
- Detailed classification of toxicants, types of toxicity, and mechanisms of the action of toxicants
- An in-depth coverage of topics on the mechanism toxicity of HMs, in addition to exclusive sections on pesticide and radiation toxicity
- A fact file in each chapter, highlighting its key points
- Flow charts, tables, diagrams, and illustrations in easily understandable language

Environmental Toxicology

Pankaj Gupta, Amit Chanjta, and Yogesh Mehta

CRC Press
Taylor & Francis Group
Boca Raton London New York

CRC Press is an imprint of the
Taylor & Francis Group, an **informa** business

Cover image credit: © shutterstock
First edition published 2024
by CRC Press
2385 Executive Center Drive, Suite 320, Boca Raton, FL 33431

and by CRC Press
4 Park Square, Milton Park, Abingdon, Oxon, OX14 4RN

CRC Press is an imprint of Taylor & Francis Group, LLC

Library of Congress Cataloging-in-Publication Data
Names: Gupta, Pankaj, author. | Chanjta, Amit, author. | Mehta, Yogesh, author.
Title: Environmental toxicology / authored by Pankaj Gupta, Amit Chanjta, Yogesh Mehta.
Description: First edition. | Boca Raton : CRC Press, 2024. |
Includes bibliographical references and index.
Identifiers: LCCN 2023030259 (print) | LCCN 2023030260 (ebook) |
ISBN 9781032381510 (hbk) | ISBN 9781032381527 (pbk) | ISBN 9781003343721 (ebk)
Subjects: LCSH: Environmental toxicology. | Heavy metals–Toxicology. |
Heavy metals–Physiological effect.
Classification: LCC RA1226 .G873 2024 (print) | LCC RA1226 (ebook) |
DDC 615.9/02–dc23/eng/20231002
LC record available at https://lccn.loc.gov/2023030259
LC ebook record available at https://lccn.loc.gov/2023030260

ISBN: 978-1-032-38151-0 (hbk)
ISBN: 978-1-032-38152-7 (pbk)
ISBN: 978-1-003-34372-1 (ebk)

DOI: 10.1201/9781003343721

Typeset in Times
by Newgen Publishing UK

Contents

Figures

Tables

Preface

Eco-toxicology, being a branch of toxicology, is a multidisciplinary science that integrates toxicology and ecology by drawing the knowledge and procedures from both fields. Principally, it examines the effects of toxic chemicals or biological agents on living organisms at different levels of organization. It not only categorizes different pollutants, their characteristics, release, exposure, and ecological fate, but also predicts their effects so that timely action can be taken to prevent or reduce the detrimental effects.

Scientists and researchers across the globe have been evaluating the effects of toxic materials on the individual and species in the food chain to analyse, predict, and prevent harmful events in the natural environment, by undertaking toxicity testing and risk assessment. To achieve this, monitoring studies are carried out so that the effects of toxins and poisons on the food chain, metabolism, deposition, and transfer of toxic compounds in different food chains can be predicted. This further helps in safeguarding the environment and ecosystems and is a significant contribution to the conservation of resources.

The book *Environmental Toxicology*, with 13 chapters starts with the basics of toxicology, briefly touching the sources of toxic compounds, classification of toxicants, factors affecting toxicity, and elaborates on heavy metal toxicity. The individual chapters on heavy metals, namely lead, arsenic, cadmium, mercury, chromium, and radioactive metals discuss the sources and routes of exposure, aetiology, pathophysiology and clinical manifestations, mechanisms of toxicity, toxicological effects, diagnosis, treatment, management, and ecological impacts of each of these.

The book covers the field in great detail and provides an insight into the occupational health problems, pesticide toxicity, radiation toxicity, and other contemporary issues. The book has chapters on recent advances in the field of eco-toxicology. Therefore, it is likely to enhance the quality, steadiness, and transparency of the teaching/learning process in the field of environmental toxicology. The book aims to provide basic knowledge on the eco-toxicology principles and related sciences. It provides an inclusive overview of various toxicants and their effects on the environment, plants, and animals to meet the requirement of students, researchers, and teachers working in diverse areas.

There is a fact file at the beginning of each chapter, highlighting the key points covered in the chapter which complements the main chapter and addresses the contemporary issues. The book extensively covers a range of toxicants that are of current interest and will be of growing interest for years to come. The content of each chapter is supported with flow charts, tabulations, diagrams, and illustrations. At the end of each chapter a bibliography of important works is provided.

This will be a reference for PG students and researchers in the areas of environmental and pharmacological sciences, ecology, environmental health, etc., in addition to industry and government departments dealing with environmental regulations and directives.

Author Biographies

Pankaj Gupta, a Faculty Member in the School of Environmental Studies, Department of Interdisciplinary Studies, Himachal Pradesh University, Shimla, completed his doctorate in Life Sciences at the Department of Biosciences, Himachal Pradesh University. He has completed more than 12 research projects and eight consultancies. He is a Consultant of Himachal Pradesh Power Corporation Ltd. for conducting Environment & Social Impact Assessment Studies and baseline studies. Dr. Gupta is a member of various academic and professional bodies: he is a Life Member, Society for Environment and Development; Member, Ethnobotanical Society, Lucknow; and Member, International Society for Ethnopharmacology. He contributed to the design of the Course Curriculum for the M.Sc. (Environment Science). He is a content writer of IGNOU for various units of learning material for the M.Sc. (Environmental Sciences). He has published more than 20 books, and more than 45 research papers in journals of repute besides 23 chapters in various national and international books.

Amit Chanjta, Gold Medalist in M.Sc. Environmental Sciences, has worked on the Community-Biodiversity Inventory for developing a database of Bioresources and Ethno-Medicinal Wisdom of Communities Living around the Sacred Groves of District Shimla, Himachal Pradesh. He is working as a Project Fellow in the Department of Interdisciplinary Studies, Himachal Pradesh University, Shimla. He has worked in various assignments for the Himachal Pradesh Power Corporation Ltd., Tribal Development Department, and NABARD. He has actively participated in various conferences and presented his research papers. He has published his work in journals of repute.

Yogesh Mehta, M.Sc. Environmental Sciences, has worked on Traditional Natural Resource Management practices among the mountain communities of Shimla district. He works as a Project Fellow in the Department of Interdisciplinary Studies, Himachal Pradesh University, Shimla. He has worked on diverse themes pertaining to the environment and social impact assessment and impact analysis of government-funded projects, such as HPPCL, Tribal Development Department, and NABARD. He has presented papers in various conferences and published his work in reputable journals.

Abbreviations

AAS	Atomic Absorption Spectrometry
AAUs	Assigned Amount Units
AC	Activated Carbon
ACGIH	American Conference of Governmental Industrial Hygienists
ACh	Acetylcholine
AchE	Acetyl Cholinesterase
ACR	Acute Chronic Ratio
AD	Anno Domini
ADP	Adenosine Diphosphate
AF	Accumulation Factor
AFM	Acute Flaccid Myelitis
Ag	Silver
AIHA	American Industrial Hygiene Association
Al	Aluminium
ALAD	Aminolevulinate Dehydratase
ALT	Alanine Transaminase
AMF	Arbuscular Mycorrhizal Fungi
ARI	Acute Respiratory Infections
As	Arsenic
AST	Aspartate Aminotransferase
ATM	Ataxia Telangiectasia Mutated
ATP	Adenosine Triphosphate
ATR	Ataxia telangiectasia and Rad3-related protein
Au	Gold
AUC	Area Under the Curve
Ba	Barium
BAC	Bioaccumulation Coefficient
BAF	Bioaccumulation Factor
BAL	British Anti Lewisite
BCE	Before the Christian Era
BCF	Bioconcentration Factor
Be	Beryllium
BHA	Butylated Hydroxyanisole
BHT	Butylated Hydroxytoluene
Bi	Bismuth
BIBRA	British Industrial Biological Research Association
BOD	Biochemical Oxygen Demand
BOHS	British Occupational Health Society
BSAF	Biota Sediments Accumulation Factor
BUN	Blood Urea Nitrogen
CAD	Contained Aquatic Disposal
CAT	Catalase

CBs	Carbamates
Cd	Cadmium
CDFs	Confined Disposal Facilities
CDKI	Cyclin-dependent protein kinase inhibitors
CDM	Clean Development Mechanism
CEC	Cation Exchange Capacity
CHD	Coronary Heart Disease
CIIT	Chemical Industry Institute of Toxicology
CNS	Central Nervous System
CNTs	Carbon Nanotubes
CO	Carbon Monoxide
Co	Cobalt
CoA	Coenzyme A
COD	Chemical Oxygen Demand
COPD	Chronic Obstructive Lung Diseases
COPT1	Cu-transporter
CP-1	Chicago Pile 1
Cr	Chromium
CS	Chitosan
CSF	Cerebrospinal fluid
CSIs	Chitin Synthesis Inhibitors
Cu	Copper
CV	Chronic Value
CV	Cardio-vascular
CVD	Cardiovascular Disease
DBCP	Dibromo-3-Chloropropane
DBP	Dibutyl Phthalate
DDE	Deep Dose Equivalent
DDT	Dichlorodiphenyltrichloroethane
DNA	Deoxyribonucleic Acid
DO	Dissolved Oxygen
DON	Deoxynivalenol
DSBs	Double Strand Breaks
e.g.	For Example
EC	Effective Concentration
EC	Electrical Conductivity
ECG	Electrocardiography
ED	Effective Dose
EDTA	Ethylene Diamine Tetra Acetic Acid
EEL	Environmental Exposure Level
EF	Ecological Footprint
EP	Erythrocyte Protoporphyrin
ETC	Electron Transport Chain
ETV	Electrothermal Vaporization
FAAS	Flame Atomic Absorption Spectrometry

FAD	Flavin Adenine Dinucleotide
FAO	Food and Agriculture Organization
Fe	Iron
FSH	Follicle Stimulating Hormone
Ga	Gallium
GABA	Gamma-Aminobutyric Acid
GDP	Gross Domestic Product
Ge	Germanium
GEM	Genetically Modified Microorganism
GFAAS	Graphite Furnace Atomic Absorption Spectrometry
GHG	Greenhouse Gases
GI	Gastrointestinal
GJIC	Gap Junctional Intercellular Communication
GPx	Glutathione Peroxidase
GRIHA	Green Rating for Integrated Habitat Assessment
GSH	Glutathione
GSSG	Glutathione Disulfide
HABs	Harmful Algal Blooms
HC	Hydrocarbons
HCN	Hydrocyanic Acid
Hg	Mercury
HMs	Heavy Metals
HNE	Hydroxy-2-nonenal
HSI	Hepatosomatic Index
HSPs	Heat Shock Proteins
i.e.	That is
IAEA	International Atomic Energy Agency
ICP	Inductively Coupled Plasma
ICP-MS	Inductively Coupled Plasma Mass Spectrometry
ICP-OES	Inductively Coupled Plasma Optical Emission Spectrometry
ICRP	International Commission on Radiation Protection
ICRU	International Commission on Radiation Units and Measurements
IET	International Emissions Trading
IGRs	Insect Growth Regulators
IM	Intramuscular
IMT	Intimal Medial Thickness
In	Indium
IP	Intra-Peritoneal
IPM	Integrated Pest Management
IR	Ionizing Radiation
IV	Intravenous
JI	Joint Implementation
LA	Laser Ablation
LC	Lethal Concentration
LD	Lethal Dose

LDH	Lactate Dehydrogenase
LEED	Leadership in Energy and Environmental Design
LH	Luteinizing Hormone
Li	Lithium
LNAPLs	Light Non-Aqueous Phase Liquids
LOD	Limit of Detection
LOEL	Lowest Observed Effective Dose
LOOH	Lipid Hydroperoxides
LPO	Lipid Peroxidation
MAPK	Mitogen-Activated Protein Kinase
MCL	Maximum Contaminant Level
MDA	Malondialdehyde
MEUF	Micellar Enhanced Ultrafiltration
MFO	Mixed Function Oxidases
Mg	Magnesium
MG	Methylglyoxal
MI	Myocardial Infarction
MiADMSA	Monoisoamyl DMSA
MMA	Monomethylarsonic Acid
Mn	Manganese
MNA	Monitored Natural Attenuation
Mo	Molybdenum
MoU	Memorandum of Understanding
MRE	Magnetic Resonance Elastography
MRI	Magnetic Resonance Imaging
MRSA	Methicillin-resistant Staphylococcus Aureus
MS	Mass Spectrometry
MSWI	Municipal Solid Waste Incineration
MTF	Metal Transfer Factor
MTs	Metallothioneins
NAD	Nicotinamide Adenine Dinucleotide
NCDs	Non-Contagious or Non-Communicable Diseases
NED	No Effect Dose
NEL	No Effect Level
Ni	Nickel
NIV	Nivalenol
NLMOAs	Natural Low Molecular Organic Acids
NMR	Nuclear Magnetic Resonance
NO	Nitric Oxide
NOAEL	No Observed (Adverse) Effect Level
NOEL	No Observed Effect Level
OA	Osteoarthritis
OEL	Occupational Exposure Limit
OPs	Organophosphorus
PAH	Polycyclic Aromatic Hydrocarbons

PAs	**Pyrrolizidine Alkaloids**
Pb	Lead
PC	Phytochelatin
PC	Protein Carbonylation
PCBs	Polychlorinated Biphenyls
PCs	Phytochelatins Compartmentation
PEB	Pro-Environmental Behaviour
PEL	Permissible Exposure Limit
PES	Plasma Emission Spectroscopy
PEUF	Polymer Enhanced Ultrafiltration
PFT	Pulmonary Function Test
PGPR	Plant Growth Promoting Bacteria
PIXE	Spectroscopy and Micro-Particle Induced X-Ray Emission
PM	Particulate Matter
PNS	Peripheral Nervous System
ppb	Parts per billion
PPE	Personal Protective Equipment
ppm	Parts per million
PRB	Permeable Reactive Barrier
Prx	Peroxiredoxins
PS I	Photosystem I
PS II	Photosystem II
Pt	Platinum
PTMs	Post Translational Modifications
PTSD	Post-Traumatic Stress Disorder
PUFA	Poly Unsaturated Fatty Acids
RA	Rheumatoid Arthritis
RBCs	Red Blood Cells
RDS	Respiratory Distress Syndrome
REM	Roentgen Equivalent Man
RIBE	Radiation Induced Bystander Effect
RNS	Reactive Nitrogen Species
RO	Reverse Osmosis
ROS	Reactive Oxygen Species
RT	Radiation Therapy
RVF	Rift Valley Fever
SA	Spark Ablation
Sb	Antinomy
SC	Sub-Cutaneous
SDGs	Sustainable Development Goals
SDI	Species Diversity Index
Se	Selenium
SGOT	Serum Glutamic Oxaloacetic Transaminase
SGPT	Serum Glutamic Pyruvic Transaminase
Sn	Tin

SOD	Superoxide Dismutase
SOM	Soil Organic Matter
SPECT	Single-Photon Emission Computed Tomography
SPM	Suspended Particulate Matter
Sr	Strontium
SSBs	Single Strand Breaks
STEL	Short-Term Exposure Limit
SVOCs	Semi-Volatile Organic Compounds
SXRF	Synchrotron Radiation Based X-Ray Fluorescence
TAC	Total Antioxidant Capacity
TCE	Volatilize Trichloroethylene
TDI	Tolerable Daily Intake
TDS	Total Dissolved Solids
Te	Tellurium
TEDE	Total Effective Dose Equivalent
TF	Transfer Factor
TF	Translocation Factor
Ti	Titanium
TIX	Tetrodotoxin
Tl	Thallium
TODE	Total Organ Dose Equivalent, Maximum Organ
TTC	Threshold of Toxicological Concern
TWA	Time-Weighted Average
U	Uranium
UF	Ultrafiltration
UNEP	United Nations Environment Programme
UV	Ultraviolet
UV/VisAS	Ultraviolet/Visible Absorbance Spectroscopy
V	Vanadium
VEGF	Vascular-Endothelial Growth Factor
VOCs	Volatile Organic Compounds
VSM	Vascular Smooth Muscle
WBCs	White Blood Cells
WHO	World Health Organization
ZEN	Zearalenone
Zn	Zinc
ZNT1	Zn-transporter
α	Alpha
β	Beta
γ	Gamma

1 Introduction

All substances are poisons; there is none that is not a poison. The right dose differentiates a poison and a remedy.

Paracelsus

After going through this chapter, the reader will learn the basic terminology, historical background, benchmarks in the history of toxicology, sub-disciplines of toxicology, eco-toxicology as a sub-discipline of toxicology, the relationship of toxicology with other disciplines and the scope of toxicology

FACT FILE

- The earliest poisons used for hunting, waging war, etc. consisted of plant extracts, animal venoms and minerals, including arsenic, lead, opium and cyanogenic glycosides
- Volcanic gas, mainly oxides of nitrogen and sulphur from the explosion of Mount Vesuvius, resulted in thousands of deaths at Pompeii in 79 AD
- Shen Nung, the "*Father of Chinese Medicine*", lived around 2695 BCE and is said to have tasted and catalogued 365 different herbs. He introduced herbal drugs to Chinese medicine but later died of a toxic overdose
- Cleopatra also experimented with poisons. It is believed that she poisoned her own brother to obtain the throne. She took her own life by allowing an asp to bite her
- Greeks and Romans had a good knowledge of naturally occurring poisons
- Socrates was sentenced to drink poisonous hemlock for allegedly humiliating the youth of Athens
- Mathieu Orfila, a Spanish physician and *father of modern toxicology*, was the first toxicologist to use autopsy material and chemical analysis as legal proof of poisoning in the French court. He established the relationship between the chemical and biological properties of poisons
- Paracelsus set the basics of pharmacology, toxicology and therapeutics

DOI: 10.1201/9781003343721-1

- Dioscorides, a Greek physician, was the first to classify poisons accompanied with descriptions and drawings in the court of the Roman emperor
- In the 1700s, there was a high incidence of scrotal cancer in chimney sweeps on account of exposure to polycyclic aromatic hydrocarbons—the first occupation-related cancer reported from England
- In the 1800s, an outbreak of mercurialism was reported from New Jersey in hatters involved in the felting process of the hatting industry
- In the 1900s, an increased incidence of bladder cancer was reported globally, especially among dye industry workers because of exposure to ß-naphthylamine
- Since the 1920s till present times, millions of people have been at risk from asbestos exposure, causing asbestos-related disease and cancer across the globe
- There were smog-related illnesses and 64 deaths in 1930 in the Meuse valley of Belgium
- Cadmium-contaminated water resulted in *Itai-Itai* disease in Japan in 1954
- In 1948, smog resulted in 20 deaths and thousands were ill due to polluted air at Donora, Pennsylvania
- Organic mercury poisoning was reported in 1950 after the consumption of toxic fish at Minamata Bay, Japan
- Four thousand deaths were reported in London, England, due to photochemical smog in 1952
- Methylmercury-contaminated grain resulted in 459 deaths in 1971 in Iraq
- In the 1970s, a nematocide, 1,2-dibromo-3-chloropropane (DBCP) resulted in infertility among pesticide workers in California
- Mustard gas used in chemical warfare caused hundreds of casualties in Iraq and Iran in 1982
- Toxic smoke, including CO and cyanide, resulted in 87 deaths at the Happy Land Social Club at Bronx, New York, in 1990
- Currently, more than 10,000 chemical entities are used in industries for the manufacturing nearly 500,000 goods
- Environmental toxicology is a newly emerging field of research and its rapid development has been attributed to Rachel Carson, who wrote *Silent Spring* (1962)

1.1 INTRODUCTION

Toxicology is the study of poisons, or broadly, the identification and assessment of undesirable impacts linked with the exposure to physical or chemical agents and similar environmental conditions. Toxicology looks at the characteristics of toxic substances and the mechanism of action of toxic substances, and provides significant input both for medical sciences and epidemiology in analysing the aetiology, possibility of a link between dose and impact, incidence of occupational diseases and related health conditions. In modern society, toxicology has become an important component of environmental and occupational health. The reason being that the inference drawn from toxicological studies is utilized to assess and regulate occupational and non-occupational health hazards. Knowledge about the toxicological impacts of poisons and toxic elements helps human society in many ways, not only in safeguarding man and the environment from the harmful effects of toxicants but also in evolving anti-cancer and other clinical medications.

1.1.1 Terminology

Toxicology is the branch of science that deals with poisonous and toxic substances. A poison is any substance that produces a detrimental effect in a living organism when consumed accidently or intentionally. In principle, it includes the study of harmful effects produced by physical agents, e.g., radiations and noise. Derived from two Greek words, *toxikos* connoting "poisonous" and *logos* meaning "discourse", toxicology may be defined as the branch of biology and chemistry which studies the harmful and injurious effects of chemical, biological and physical agents on biological systems. It examines adverse effects ranging from acute to long term. It is used to assess the probability of hazardous results or adverse effects of toxic substances and predict its impacts on individuals, populations and ecosystems. Adverse effects may appear in several forms, ranging from instant death to indirect changes not realized until months or years after exposure. They may occur at various levels within the body, such as at the organ or cell level, or a specific biochemical level. While studying toxicology, one comes across some common terms, which are described below.

Abortifacient: Substance capable of terminating pregnancy or causing premature abortion

Absolute Lethal Concentration (LC100): Lowest concentration of a substance in an environmental medium which can kill 100 per cent of test organisms or species

Absolute Lethal Dose (LD100): Lowest amount of a substance capable of killing 100 per cent of test animals

Absorbed Dose/Internal Dose: Amount of a substance taken by an organism or by organs or tissues

Acceptable Risk: Probability of suffering disease or injury identified by an individual

Accidental Exposure: Inadvertent contact with a substance or change in the physical environment as a consequence of an accident

Acclimatization: Methods by which a group of individuals acquires the capacity to terminate a substance, or develops a tolerance to it

Acute Effects: Effects which manifest after controlled exposure and just (hours, days) after exposure and may be retrievable or irretrievable

Acute Exposure: An exposure of short duration, generally less than 24 hours, often believed to be a single exposure (or dose) but may comprise of recurrent exposures within a short time period

Acute Toxicity: An adverse effect revealed in a relatively short time interval ranging from almost instantaneously to several days after exposure/dosing

Acute Toxicity Test/ Short-Term Toxicity Test: Investigation wherein organisms are observed for a short time period and there is a single exposure to the test agent in the experiment

Adaptation: Change in an organism, in response to changing environmental conditions or the process by which the physiological condition of an organism becomes stable after a change

Adverse Effect: Change in biochemical, physiological, growth, development morphology, behaviour or lifespan of an organism resulting in injury to functional capacity or damage to the capacity to recompense for extra stress

Aetiology: Science dealing with the cause or origin of disease

Age sensitivity: Measurable and qualitative age-dependence of an effect

Allergic Reaction: Anaphylactic or delayed response to a toxicant produced by a changed state of the normal immune response

Antidote: Substance able to specifically counteract or decrease the effect of a probable toxic substance in an organism by a specific chemical or pharmacological action

Antigen: Substance capable of prompting the immune system to generate specific antibodies or cells

Antihistamine: Substance that blocks or steadies the action of histamine

Antimetabolite: Substance, structurally similar to a metabolite, which competes with it or replaces it, and suspends or decreases its normal use

Antipyretic: Substance capable of reducing fever

Anti-resistant: Substance used as a stabilizer during the preparation of a pesticide to decrease the resistance of insects to the pesticide

Asphyxia: Condition attributable to insufficient intake of oxygen

Asphyxiant: Substance able to block the transport or use of oxygen by living organisms

Assimilation: Uptake and absorption of a material by a living organism

Astringent: Substance capable of causing contraction, generally local after topical application

Autoimmune Disease: Pathological condition in which an organism produces antibodies or specific cells

Bio-accessibility: Likelihood of a substance coming in contact with a living organism and interacting with it

Bio-accumulation: Progressive increase in the amount of a substance in an organism or part of an organism when the rate of intake surpasses organism's ability to remove the substance from the body

Bio-activation: Metabolic transformation of a xenobiotic to a more toxic derivative

Bio-degradation: Breakdown of a substance catalysed by enzymes *in vitro* or *in vivo*

Bio-magnification: Sequence of processes in an ecosystem by which higher concentrations are attained in organisms at higher trophic levels

Carcinogen: Chemical, physical or biological agent which increases the incidence of malignant neoplasms, resulting in cancer

Carcinogenesis: Process of inducing cancer

Carcinogenicity: Process of initiation of malignant neoplasms, and cancer, by chemical, physical or biological agents

Cardiotoxic: Chemicals harmful to the heart cells

Chronic Effects: Effects revealed after prolonged exposure and/or that continue even after exposure has stopped

Chronic Exposure: Exposures, repeated or continuous over a period of time, generally more than three months

Chronic Toxicity: Permanent or long-lasting adverse effect presented after exposure to a toxicant

Co-carcinogen: Chemical, physical or biological factor that increases the effects of a carcinogen

Corrosive: Substance capable of producing a surface-destructive effect on contact

Critical Concentration: Concentration of a substance at and above which detrimental functional alterations occur in a cell or organ

Critical Dose: Dose of a substance at and above which detrimental functional changes (reversible or irreversible) occur in a cell or organ

Critical Effect: First adverse effect that appears when a critical organ achieves the threshold concentration or dose

Cyanotoxin: Toxin produced by cyanobacteria or blue–green algae

Cytotoxic: Toxin that damages cell structure or function

Decontamination: Method of translating a potentially toxic substance in the natural environment or laboratory areas to avoid detrimental effects by counterbalancing or removal

Delayed or Latent Effect: Result after a latent period or termination of exposure to a toxic substance or other harmful ecological factor

Delayed or Latent Toxicity: An adverse or undesirable effect occurring after a long period of initiation and/or termination of exposure to the toxicant

Delivered/Effective/Target Organ Dose: Quantity of toxicant reaching an organ that is extremely transformed by the toxicant

Deposition: Route by which a substance reaches a particular organ or tissue and sediments out

Deterministic Effect/Deterministic Process: Phenomenon devoted to a specific result controlled by vital physical codes

Deterministic: Health effects, the gravity of which varies with the dose

Detoxification/Detoxication: Process, or processes, of chemical modification that make a toxic molecule less toxic

Detriment: Measure of the probable harm or loss linked to a harmful occurrence

Dosage: Dose divided by product of mass of organism and time of dose

Dose Threshold: Dose level below which no detectable effect occurs

Dose: Amount of a xenobiotic entering an organism (in units such as mg/kg body weight) at specific time intervals. It is the total amount of a substance administered to/taken up/absorbed by an organism, organ or tissue

Dose–Effect Curve: Graph representing the relationship between the dose and the extent of biological change produced

Dose–Effect Relationship: Relationship between the dose and consequential degree of a constantly graded biological change, either in an individual or in a population

Dose–Response Curve: Graph representing the link between the dose and proportion of individuals in a population responding to a biological effect

Dose–Response Relationship: Relationship between the dose and percentage of individuals exhibiting an exact effect

Drug/Pharmaceutical: A substance when taken/absorbed by a living organism that may alter one or more of its functions.

ED_{50} (Effective Dose): A dose producing a specific effect other than lethality in 50 per cent of animals

Effective Concentration (EC): Concentration capable of producing a distinct degree of response

Effluent: Solid, liquid or gas released from a source into an external environment

Endotoxin: Toxin of a cell wall of some bacteria released after collapse of the bacterial cell

Environmental Contaminant: A component or material which doesn't occur in nature, but is introduced by anthropogenic activity into the environment, upsetting its composition

Environmental Damage: Adverse impacts on the natural environment

Environmental Exposure Level (EEL): Amount of a substance to which an organism or other environmental constituent is exposed in its natural surroundings

Environmental Fate: Destiny of a chemical or biological pollutant after being released into the natural environment

Environmental Health Impact Assessment: Estimation of health implications likely to change the environment

Environmental Health: Human wellbeing and environmental impacts of chemical or biological pollutant

Environmental Pollutant: Substance present in nature, in greater than natural abundance attributable to human activity, which has a detrimental effect on environment and, therefrom, on living organisms including humans

Environmental Toxicant: Any substance and/or its metabolites in the environment which exerts toxic effect over organisms, undesirably upsetting their structures and functions

Epidemiology: Study of distribution and factors accountable for health-related status

Exposure Assessment: Method of quantifying or estimating concentration, duration and frequency of exposures to an agent present in the environment

Exposure: Way by which an organism comes in contact with a toxic substance that may be through air, water, soil, food, medication, etc.

Exposure Dose: Air concentration of pollutant breathed during a certain time period

Fetotoxicity: Toxicity to the foetus

Food Additive: Any consumable, not usually consumed as a food by itself and not normally used as an ingredient of a given food

Food Allergy: Hypersensitivity reaction to substances in food to which an individual has earlier been sensitized

Food Intolerance: Physiologically based adverse reaction to a specific food or any ingredient

Half-life: Time required for the concentration of a reactant in a given reaction to reach a value that is the arithmetic mean of its initial and final (equilibrium) values

Harmful Occupational Factor: Component of the working environment that results in ill-health or reduces the working ability

Harmful Substance/Noxious Substance: A substance, which after coming in contact with an organism, produces unpleasant health impacts either at the time of exposure or later in life of present and future generations

Harmful/Adverse Effect: Damaging effect to a population, species, individual organism, organ, tissue or cell

Hazard Assessment: Measurement of factors governing probable effects of a hazard, such as dose–effect/dose–response relationships, variations in target susceptibility and mechanism of toxicity

Hazard: Set of intrinsic properties of a substance, mixture of substances, or a process capable of causing harmful effects on organisms or the environment, depending on the extent of exposure

Hit-and-Run Effect: Toxicity following a single exposure to a substance

Human Equivalent Dose: Human dose of an agent that is thought to induce the same magnitude of a toxic effect that the known animal dose has induced

Human Exposure Threshold: Generic value of exposure to a substance, or a group of substances, below which there is probability of no substantial risk to human health

Hyperreactivity: Reactions of an individual to an agent.

Hypersensitivity: Condition in which an individual reacts to allergic effects after exposure to a certain substance

Idiosyncratic Reaction: Response to a toxicant occurring at exposure levels much lower than those usually essential to produce the same effect in most individuals within a population

Immunosuppression: Reduction in the functional capacity of the immune response

Immunotoxin: Toxin detrimental to the immune system

Incidence: Number of illness events during a given period of time in a specific population

Indirect Exposure: Exposure to a substance in a medium or vehicle other than the one originally receiving the substance

Individual Susceptibility: Includes changes in types of reaction to hazardous substances, among a group of people. Each individual is unique and thus there may be disparities in response to exposure

Inflammation: The body's reaction to injury or infectious, allergic or chemical irritation, which usually manifests as redness, swelling, heat and pain

Internal/Absorbed Dose: Actual quantity of a toxicant absorbed by an organism and dispersed throughout the body

Intoxication/Poisoning: Pathological process with clinical signs and symptoms produced by a substance of exogenous or endogenous origin

Irritant: Material capable of producing inflammation after instant, persistent or recurrent contact with skin, mucous membranes or other biological material

Latency time: Duration between first exposure and appearance of a noticeable effect or response

Latent Period: Interval between exposure to a toxic substance and appearance of a disease or other detrimental effects

LD_{50} (Effective Dose): Dose causing 50 per cent lethality in an animal population. The higher the LD_{50}, the lower is the acute toxicity

Lethal Concentration (LC): Concentration of a substance in an environmental medium that results in death sometime after exposure

Lethal Dose (LD): Quantity of a substance or physical agent that results in death when absorbed by the body/tissue

Lethal Synthesis: Formation of a highly toxic metabolic compound leading to death of affected cells

Lethal: Deadly/disastrous, capable of causing death

Local Effect: Change occurring at the site of contact between an organism and a toxicant

Local Toxicity: An undesirable effect that is exhibited at a toxicant's site of contact, e.g., burning sensation in eyes, upper respiratory tract irritation, etc., after exposure to a particular chemical

Lowest Observed Effective Dose (LOEL): Lowest observed effective dose on a dose–response curve, or the lowest dose that produces an effect

Malignancy: Cells showing both uncontrolled growth and an inclination to attack and destroy other tissues

Mutagen: Agent able to induce heritable changes (mutations) of the genotype in a cell on account of changes in or loss of genetic material

Mutagenesis: Induction of heritable changes (mutations) of the genotype

Mutagenicity: Ability of a physical, chemical or biological agent to induce or produce heritable alterations in the genotype in a cell due to changes or loss of genes or chromosomes

Mutation: Relatively stable heritable change in genetic material that may be a chemical variation/modification of an individual gene, changing its function

Negligible Risk: Likelihood of detrimental effects occurring at an insignificant rate

Nephrotoxic: Chemically injurious to kidney cells

Neurotoxic: Chemically detrimental to the nervous system

No-Effect Dose (NED) Sub-threshold Dose: Quantity of a substance having no effect on an organism

No-Effect Level (NEL): Maximum dose of a substance that produces inconspicuous changes under distinct conditions of exposure

No Observed (Adverse) Effect Level (NOAEL): The highest dose that does not produce a toxic effect

Non-Occupational Exposure: Environmental exposure outside the workplace to substances that are otherwise linked with specific work environment

Occupational Environment: Workplace-adjacent settings/circumstances

Occupational Exposure: Contact with substances, radiation or other conditions at the workplace

Occupational Exposure Limit (OEL): Regulatory level of exposure to substances, radiation, etc., quantified appropriately by relevant government legislation

Organ Dose: Amount of a substance or physical agent absorbed by an organ

Para-occupational Exposure: Exposure of a worker's family to substances carried from the workplace to the home or contact of visitors to workplace substances

Phototoxicity: Antagonistic effects produced by exposure to light energy

Poison: Substance which is taken into or formed within the organism, and harms the health of an individual or results in death

Poisoning/Intoxication: Morbid condition produced by a poison

Pollutant: Any undesirable solid, liquid or gaseous matter in environmental medium that worsens the natural environment by altering the growth of species, affecting the food chain, is harmful, or affects health, comfort, amenities or properties of a population

Regulatory Dose: Term used to define the probable dose resultant from human contact to a substance at a level where it is controlled in the environment

Retained or Absorbed Dose: Amount of toxin or chemical present in the body at a certain time during or after exposure

Reversible Toxicity: An adverse effect that can be inverted once the exposure is withdrawn

Risk Assessment: Estimation process of possible categorizing the harmful health effects of exposure

Risk Estimation/Evaluation: A qualitative or quantitative link between risks and benefits

Risk Identification: Identifying probable hazard and factors vital to evaluate the possibility of contact of organisms or people to that threat and of harm subsequent to such exposure

Risk Management: Decision-making method which involves the considerations of political, socio-economic and engineering factors with suitable risk assessments

Risk: Likelihood of a specific adverse effect to occur. Often expressed as the percentage of cases in a given population during a specific time period

Route of Exposure: Agencies by which a toxic agent gains access to an organism after ingestion, inhalation, topical or by other routes, viz. intravenous, subcutaneous, intramuscular or intraperitoneal routes

Safety Factor: Formal, random number with which one divides NOEL or LOEL derived from animal experiments to obtain a tentative permissible dose for humans

Safety: Degree or possibility that a specific exposure or dose will not produce a toxic effect

Sensitive Sub-Population: Group of people likely to be at more risk owing to exposure to harmful substances including young, chronically ill, aged, pregnant women and women of childbearing age

STEL (Short-Term Exposure Limit): Highest concentration allowable for a short time period, in general for 15–30 minutes

Sub-Acute Exposure: Similar to acute exposure except that the duration of exposure is longer, ranging from a few days to one month

Sub-Chronic Exposure: Recurrent exposure that spreads over an intermediate time range

Systemic Effect: Effect that is either comprehensive or occurs at a site away from the entry point of a substance

Systemic Toxicity: An unfavourable effect that can be seen all over an organism or in an organ with selective distance from the entry point of the toxicant

Target Dose: Amount of substance bound to a critical molecule, and is more closely linked with the toxic effect

Target Organ: Primary or most sensitive organ affected after exposure

Teratogen: Substance which, if taken prenatally by a mother, induces permanent structural malformations or defects in offspring

Teratogenicity: Manifestation of non-heritable structural abnormalities or defects in offspring

Teratology: Study of malformations or serious abnormalities from normal development in organism.

Threshold Dose: A dose or exposure intensity below which harmful or adverse effects of a substance are not perceived in a population

Tissue Dose: Amount of substance in a specific tissue

Tolerable Daily Intake (TDI): Estimation of the amount of a potentially harmful substance in food or drinking water that can be ingested daily over a lifetime without appreciable health risk

Tolerable Risk: Probability of suffering disease or injury that can, for the time being, be tolerated, taking into account the linked benefits

Tolerance: Adaptive state categorized by abridged effects of an exact dose of a substance or ability of an organism to survive in the presence of a toxic substance

Topical Effect: Result of application of a substance on the surface of the body which occurs at the point of application

Toxaemia: A condition in which blood contains toxins produced by body cells

Toxic: Substance capable of producing injury to living organisms after physico-chemical interaction

Toxic Agent: A chemical, physical or biological substance capable of producing an undesirable biological effect

Toxic Dose/Super-Threshold Dose: Amount of a substance that produces intoxication without lethal outcome

Toxic Organisms: Organisms capable of excreting chemicals producing toxic effects

Toxic Substance: Material having toxic properties

Toxicant: Substance that produces a detrimental or adverse effect on coming into contact with a living organism at a sufficiently high concentration

Toxicity: Inherent capability of a chemical agent to badly affect an organism

Toxicity Rating: Random grading of doses or exposure levels producing toxic effects. Grading can be "super-toxic," "highly toxic," "moderately toxic" and so on

Toxicity Test: Experimental investigation of the harmful effects of exposure of a living organism to a toxic substance for a definite duration

Toxico-dynamics: Method of interface of possibly toxic substances with target sites, and biochemical and physiological results leading to adverse effects

Toxico-Genetics: Study of the impact of hereditary factors on results of possibly toxic substances on individual organisms

Toxico-Genomics: Scientific sub-discipline linking toxicology with genomics to regulate how an organism's genetic make-up affects its response to a toxic substance

Toxico-Kinetics: Overall process of absorption/uptake of toxic substances by the body, their distribution and metabolites in tissues and organs, besides their metabolism

Toxicologist: A scientist dealing with detrimental effects of toxic substances and cellular, biochemical and molecular mechanisms of the effects

Toxicology: Scientific discipline involving the study of actual or potential danger presented by detrimental effects of substances on living organisms and ecosystems, of the relationship of such harmful effects to exposure, and of mechanisms of action, diagnosis, prevention and treatment of intoxications

Toxicometry: Combination of investigative methods and techniques for making a quantitative valuation of toxicity and hazards of potentially toxic substances

Toxicovigilance: Active process of identification, study and evaluation of various toxic effects in a community to reduce or control the substance(s) which produces these effects

Toxification: Metabolic transformation of a possibly toxic substance to a product that is more toxic

Toxin: Toxicant produced by a living organism, i.e., floral, faunal or microbial entity

Toxinology: Area of science dealing with microbial, floral and faunal venoms, poisons and toxins

Tumorigenic: Tumour-causing agents

Uptake: Entry of a substance in the body, organ, tissue, cell or body fluids through a membrane or by other means

Vehicle: Substance(s) used to make active ingredients for administration or use

Xenobiotic: A term used for "*foreign substances*" to an organism, including drugs, industrial chemicals, naturally occurring poisons and environmental pollutants

1.2 HISTORICAL BACKGROUND

The history of toxicology is believed to be as old as the history of human civilization, ever since early mankind learned to differentiate between consumable and non-consumable. While exploring his environment, in search of food, man must have observed the curative or detrimental effects of plants and minerals, and that the stings of some insects and reptiles which resulted in disease or death. Primitive man must have noticed that animals naturally avoid consuming some plants. This may have created interest, and it must have taken only a little research to discover that some plant extracts could be fatal. Since early humans were hunters, the next logical step was to apply the extracts of poisonous plants on tips of arrows. Toxicology studies the adverse effects of chemical, biological or physical agents on living beings, ranging from instant death to subtle alterations that may be unrecognized for months or years. Understanding of how toxic substances harm the body has evolved with the knowledge of medicines and pharmaceuticals. the evolution of toxicology started when the early cave dwellers used toxic plants and animals for hunting or in warfare. By 1500 BCE, written recordings indicate that hemlock, opium, arrow poisons and certain metals were used to poison rivals or for state executions. Notable poisoning victims include Socrates, Cleopatra and Claudius. With the passage of time, humans learnt the art of using toxic or poisonous substances for hunting, against rivals in combat, and for killing. One of the most remarkable narratives is the tale of King Mithridates of Pontus (120–63 BCE) who was so afraid of becoming a victim of political poisoning that he invented a concoction using a number of herbs for his own consumption. Later, he accepted that by consuming small quantities of potential poisons, he might safeguard himself from any would-be poison. Actually, his strategies worked so well that he gained a name for himself as the person who could not be assassinated. Fatefully, it is said that when his condition was such that he wanted to kill himself, he was unable to do so even by consuming poison.

Arrow poisons developed by primitive man during the Mesolithic age are still in use among many indigenous communities around the world. Reptile poisons or venoms from toads and salamanders were in use for a long time. Pharmacology, the science dealing with the preparation, usage and impacts of drugs, has advanced from the understanding of poisons, and some of the active ingredients derived from these are used for therapeutic purposes in Western medicine. One finds numerous instances of killing by professional poisoners in Indian mythology. The origin of toxicology in the Indian context can be traced to the Vedic period. In Indian mythology and tradition, the origin of poisons is attributed to Lord Brahma, the creator of the universe. It is said that when Lord Brahma was insulted by an *asura* (devil), he created poison to kill him. Lord Brahma was successful in killing the demon with his weapon, but its evil spread over the entire world. So much so that, to reduce its harmful effects, Lord Brahma had to distribute it through the vegetable, animal and mineral kingdoms, and also create its antidote. Brahma thus distributed poisons into three categories: animal, mineral and vegetable. Lord Brahma is known as a pioneer in classifying the poisons. One of the main Indian experts in the science of toxicology was *Kashyapa*, a physician, believed to be a curer of snakebites. Ancient scriptures such as *Atharva Ved* have details of poisons and remedial measures. *Charak* and *Susrut Samhita* have detailed descriptions of modes of poisoning, and the signs, symptoms and treatments of poisons.

These ancient texts describe secrets of poisoning natural water sources to kill animals. *Agad tantra*, a branch of Ayurved, mainly deals with toxins of mineral, animal or plant origins and the methods of getting rid of poisons and the use of antidotes. It deals with the diagnosis and treatment of any one bitten by poisonous insects or venomous reptiles or affected by any natural, artificial or compound poison. To safeguard kings from being poisoned, food preparation was carried out under supervision in a well-guarded place. The king was advised to share fresh dishes after offering them to a fire and then to birds. According to a legend, Chanakya, who served as the minister of Chandragupta Maurya, started adding small amounts of poison in Chandragupta's food so that he became immune to it. This was done to prevent the king from being poisoned by enemies. One day, the pregnant queen shared food with the king and died as a result. Chanakya decided that the baby should not die; hence, he removed the baby from the queen's belly. The colour, texture, odour and appearance of eatables were used to draw inferences about the toxic properties of food. Another popular Indian legend related to poisons is about Lord Shiv, whose neck turned blue when he sipped the deadly poison produced by the churning of the ocean. For this reason, Lord Shiv is also known as *Neelkanth*, the one with a blue throat.

1.2.1 Benchmarks in the History of Toxicology

Early Period: Most evidence of the existence of toxicology in early history has been lost and that which has survived deals mainly with medicine. Some evidence, however, deals more precisely with its lethal action or with the use of poisonous substances for legal trials, suicide or political assassination. Despite the dearth of early evidence, and given the need for people to avoid poisonous animals and plants, toxicology is ranked as one of the ancient sciences. The word "*poison*" is historically linked with conspiracy and known as a coward's weapon. Poison is a substance that is capable of ending life or causing illness, through biochemical action, when introduced into, or absorbed by, a living system in small quantities. Since time immemorial, magic, myths and tales of poison have been linked with hunting, crime, punishment, politics, medicine and development of antidotes. Plants containing substances similar to digitalis and belladonna alkaloids were known to people. Hippocrates mentioned a number of poisons and clinical toxicology principles regarding bioavailability in therapy and overdoses. The *Book of Job* described poison arrows. There is mention of several poisons and their use in the ancient literature of Greece. Theophrastus (370–286 BCE), a student of Aristotle, discussed poisonous plants in his multi-volume book *De Historia Plantarum*.

Dioscorides, a Greek physician, was a pioneer in classifying poisons accompanied with descriptions and drawings in the court of the Roman emperor. His cataloguing into plant, animal and mineral poisons remained a standard for centuries. Dioscorides also experimented in therapy, identifying the use of emetics in poisoning and the use of caustic agents and cupping glasses in snakebites. Socrates' (470–399 BCE) cup of hemlock extract was assessed to be the proper dose. Cleopatra's (69–30 BCE) familiarity with basic toxicology allowed her to use a cobra to kill herself. The Romans too made substantial use of poisons. According to one legend, King Mithridates VI of Pontus carried out acute toxicity experiments on ill-fated convicts that led to his ultimate claim that he had found an antidote for every venomous reptile and poisonous substance. The king was so scared of poisons that he used to ingest a mixture of thirty-six ingredients as protection against assassination. On the occasion of his impending capture by opponents, his attempts to kill himself with poison failed because of his popular antidote concoction, and he was forced to use a sword held by a servant. Poisonings in Rome were prevalent during the 4th century BCE. It was during this period that a plot by women to eliminate men from whose killing they might gain was revealed. Similar poisoning incidences continued until Sulla issued the *Lex Cornelia*, the first law against poisoning.

Middle Age: Before the Renaissance, the writings of Maimonides (1135–1204) included a discourse *Poisons and Their Antidotes*, dealing with poisonings due to insects, snakes and mad dogs. It is believed that experimenters of this period, in search of a universal antidote, learned to distil a 60 per cent ethanol beverage. In the early Renaissance, the Italians, with typical rationality, brought the art of poisoning to its peak. Poisoning was a prominent part of politics and there is mention of the political use of poisons.

Later Period: A significant figure in the history of science and medicine in the later period was Paracelsus (1493–1541). Paracelsus was a physician who set the basics of pharmacology, toxicology and therapeutics. He stated that toxicity is related to chemical structure and described the clinical manifestations of two common poisoning agents, namely, arsenic and mercury. In his words, "*All substances are poisons; there is none which is not a poison. The right dose differentiates a poison and a remedy.*" He determined that specific chemicals are responsible for the toxicity of a plant or animal poison and he documented that the body's response to these chemicals depends on the dose.

Pharmacology had its beginnings during the middle period and early Renaissance. Simultaneously, the study of toxicity and the dose–response relationship of drugs started. Occupational hazards linked with metal works were recognized during the 15th century. Early publications by Ellenbog cautioned about the toxicity of mercury and lead.

In 1556, Agricola published a paper on mining diseases. However, the major work on occupational diseases, *On the Miners' Sickness and Other Diseases of Miners* (1567), was published by Paracelsus, which talked about the aetiology of miners' disease, with cure and prevention methods. Occupational toxicology obtained further input with Bernardino Ramazzini's work entitled *Discourse on the Diseases of Workers*, which set principles for occupational medicine during the 19th century.

Percival Pott (1775) recognized the role of soot in scrotal cancer among chimney sweeps, which enhanced the related medical practices. The 19th century started an environment of industrial and political revolution. In 1800, organic chemistry was in its embryonic stage, but by 1825, phosgene ($COCl_2$) and mustard gas were used in World War I. By 1880, over 1000 organic compounds had been manufactured, including chloroform, carbon tetrachloride, diethyl ether, carbonic acid, petroleum and coal gasification by-products. The toxicity of benzene was established at the turn of the 20th century. Estimation of the toxicological potential of newly synthesized chemicals became the basis of toxicology as experimental toxicology accompanied the advancement of organic chemistry and developed quickly during the 19th century.

Magendie (1783–1885), Orfila (1787–1853) and Bernard (1813–1878) undertook influential research in experimental toxicology and laid the foundation for pharmacology and experimental therapeutics and occupational toxicology. Magendie, a physician and experimental physiologist, studied the mechanisms of action of emetine, strychnine and arrow poisons. Orfila, a Spanish physician, was the first toxicologist to use autopsy material and chemical analysis as legal proof of poisoning in the French courts. He published on the toxicity of natural agents in 1815. Orfila, known as the founder of toxicology, was a Spanish physician who first established the relationship between the chemical and biological properties of poisons. Forensic toxicology has its roots in his work , as he was the first scientist to use chemical analysis as proof of poisoning in a court of law. During the 19th century, with the rapid advent of organic chemistry, many new chemical compounds were synthesized. This gave new impetus to the science of toxicology as it was necessary to test the newly developed compounds for their toxicological properties. Claude Bernard continued studying arrow poisons and the mechanism of carbon monoxide poisoning. Bernard's discourse, *An Introduction to the Study of Experimental Medicine*, added to the expansion of toxicology.

Schmiedeberg made contributions to the field of toxicology by training students in the field of pharmacology and toxicology. His research was dedicated to analysis of the synthesis of hippuric acid in the liver and the detoxification mechanism of the liver.

Lewin, formerly skilled in medicine and natural sciences, was trained in toxicology at the Pharmacological Institute of Berlin. He published a work on the toxicity of methanol, glycerol, acrolein and chloroform.

Arsenic-containing cosmetics were believed to be responsible for deaths during the 20th century.

Finally, it was during the 20th century that toxicology developed as a multidisciplinary science. Owing to the significant development of chemicals used for warfare such as pesticides, bullets and drugs, World War II deeply impacted the discipline. It was mainly during the second half of the century that science developed rapidly and reshaped many disciplines, and toxicology was no exception.

Contemporary Toxicology: World War II marked the increased production of drugs, pesticides, munitions, synthetic fibres and industrial chemicals. During the second half of the 19th century, the world witnessed the expansion of science that started the contemporary era of genetics, medicine, synthetic chemistry, physics and biology. With the arrival of anaesthetics and disinfectants and the development of experimental pharmacology in the late 1850s, toxicology made a forward leap. By the late 19th century, the use of organic chemicals became widespread, and benzene, toluene and xylene were produced on a commercial scale. Benzene was used as a drug to treat leukaemia in the early 1900s. During this period, there were several instances of poisoning from these drugs.

Toxicology expanded in response to legislation. The National Safety Council was founded in 1911, and the Division of Industrial Hygiene was established by the US Public Health Service in 1914. The publication of scientific journals and creation of societies were ignited by the expansion of this new field. The *Journal of Industrial Hygiene* was started in 1918. Chemical manufacturers in the US established internal toxicology research laboratories to facilitate verdicts on worker health and product safety. During the 1890s and early 1900s, the French scientists Becquerel and Curies reported radioactivity, which was opened up for exploration in the field of physics, biology and medicine. Philip B. Hawk, a young associate of Bernard L. Oser, devloped many toxicologic assays. Oser's later inputs to food and regulatory toxicology were amazing. The 1920s witnessed many activities, e.g., the use of arsenicals for treating diseases like syphilis, which resulted in the analysis of acute and chronic toxicity. The ban on alcoholic beverages in the US opened avenues for early studies of neuro-toxicology. Mueller's discovery of dichlorodiphenyltrichloroethane (DDT) and several other organo-halides, during the late 1920s, resulted in wider use of insecticidal agents.

Efforts to synthesize steroid-like chemicals were led by E. C. Dodds and his associates. Dodds' work on the bioactivity of estrogenic compounds resulted in the synthesis of diethylstilbestrol (DES), hexestrol and other stilbenes. Golberg's managed the establishment of the laboratories of British Industrial Biological Research Association (BIBRA) and Chemical Industry Institute of Toxicology (CIIT). During the World War II, pharmaceutical and chemical industries were established in Germany and the United States to manufacture readymade antibiotics and warfare agents. One of the first journals devoted to experimental toxicology, *Archiv fur Toxikologie*, began publication in Europe in 1930. Herbert Hoover established the National Institutes of Health (NIH) in the US. The discovery of sulfanilamide turned out to be a major event in combating bacterial diseases. The sulfanilamide disaster played a significant role in the development of toxicology, resulting in work by Eugene Maximillian Geiling in pharmacology. Scientists connected with Lehman and Geiling become the frontrunners of toxicology. The main research areas during the World War II focused on the pharmacology of organophosphate chemicals, anti-malarial drugs and radionuclides. Hence, the crisis of the World War II triggered the development of toxicology. The role of the Chicago and Rochester groups in the history of the toxicology of metals is self-evident. Use of uranium in the atomic bomb and research into the role of metals changed the entire scenario. These areas transformed modern biology, chemistry, therapeutics and toxicology. Stafford Warren developed a programme with colleagues. Another important event during the World War II period was the discovery of organophosphate cholinesterase inhibitors which was a driving force in the study of toxicology. Geiling's group, especially Kenneth Dubois, was a frontrunner in the areas of toxicology and pharmacology. Elizabeth and James Miller, a husband–wife team, contributed to the development of toxicology under the mentorship of Professor Rusch. Adrian Albert's "*Selective Toxicity*" published in 1951 presented a brief documentation of the principles of the site-specific action of chemicals.

1.3 SUB-DISCIPLINES OF TOXICOLOGY

Toxicology, like medicine, is both a science and an art. The science of toxicology has observational and data-gathering stages, whereas the art of toxicology uses data to predict the consequences of exposure in human and animal populations. Toxicology has the following sub-disciplines or sub-specialties:

Analytical Toxicology: Detects toxicants through examination of body fluids, contents of gastrointestinal tract, excretion or skin

Applied Toxicology: Linked with the use and impacts of toxic chemicals outside the laboratory and attempts to regulate the physiologic effects or safety of administered toxic chemicals

Aquatic Toxicology: Solves practical problems related to the standards of water quality assessment and standardization of human-induced water pollution

Behavioural Toxicology: Studies the alterations in the behaviours of organisms after exposure to particular toxicant.

Biochemical Toxicology: Studies the changes at cellular and sub-cellular levels as a result of the toxic action of chemicals.

Clinical Toxicology: Studies the short-term or long-term toxic effects of various drugs on the body, in addition to the treatment and prevention of drug toxicity including accidental and intentional abuse of chemical substances, therapeutic agents, unwanted excessive and non-therapeutic drug effects, harmful interaction of xenobiotics, etc. It is closely related to forensic toxicology

Comparative Toxicology: Studies the relative toxicities of xenobiotics to various organisms

Computational Toxicology: Helps in developing mathematical and computer-based models to understand and predict the unwanted health effects caused by environmental pollutants and pharmaceuticals

Descriptive Toxicology: Gathers toxicological information from animal experimentation to establish the quantity of a chemical resulting in illness or death

Economic Toxicology: Related to the economics of a nation and deals with the harmful effects of chemicals which are deliberately administered for achieving a specific effect in a biological system

Entomotoxicology: Examines the toxins in arthropods that feed on decaying flesh of dead animals, including human flesh

Environmental Toxicology: Linked with the study of harmful effects of various chemical, biological and physical agents on living organisms. It incorporates pollution studies, residue analysis, industrial hygiene and occupational health. In general, pollution can be defined as "*matter in the wrong place*" or any substance released into the environment which degrades it. Eco-toxicology, a sub-discipline of environmental toxicology differs from it in the way that environmental toxicology includes the influences of stressors across all degrees of biological organization from the molecular level to entire communities and ecosystems, while environmental toxicology emphasise on the impact at an individual or lower level. Thus, it explores the impact of toxic agents on individual organisms, organs, tissues, cells, organelles and biochemical reactions. Rachel Carson is described as the *Mother of Environmental Toxicology*. Her book *Silent Spring*, published in 1962, was a ringing alarm bell which highlighted the effects of uncontrolled pesticides and made a remarkable contribution in the field of toxicology. The book was based on a series of reports by Lucille Farrier Stickel on the environmental impacts of DDT. Living beings are exposed to numerous toxicants at one stage of life. Some organisms are more sensitive than others and, likewise, toxicity also varies with the organism's position in the food web. The storage of toxicants in fatty tissues results in bioaccumulation, which may eventually introduce a trophic cascade and biomagnification of toxicants. Biodegradation liberates carbon dioxide and water as by-products into the environment, which is limited in areas influenced by environmental toxicants. The detrimental effects of chemical and biological agents such as toxicants from pollutants, insecticides, pesticides, etc. affect organisms and their communities by decreasing the diversity and abundance of species. Of the various sub-disciplines of eco-toxicology, some of the most important ones are listed below.

Mechanistic Toxicology: Studies how chemical or physical agents interact with living organisms and result in toxicity. It relates to the molecular basis of disruption of biological targets by chemicals

Aquatic Eco-toxicology: Studies the properties and behaviours of pollutants on aquatic organisms in aquatic ecosystems

Ethnic or Geographical Toxicology: Analyses of climatic, nutritional and genetic factors, and other environmental impacts in response to a xenobiotic

Food Toxicology: Evaluates the nature, properties, effects and detection of toxic substances in food including microbial toxins, radioactive elements, heavy metals, or packing materials used in food processing and their manifestation as disease

Forensic Toxicology: Establishes the cause-and-effect relationship between exposure and medico-legal aspects of poison/drug or chemical and toxic or lethal effects resulting from exposure

Genetic Toxicology: Studies the deleterious effects of xenobiotics, if any, on genetic material of individuals

In silico Toxicology/Computational Toxicology: Creates mathematical models based on existing knowledge and utilizes them for making calculations

Nutritional Toxicology: Deals with the effects of diet in causing toxicity and the mechanism of these effects

Occupational/Industrial Toxicology: Deals with health effects due to exposure to chemicals at the workplace including the occupational diseases that are due to industrial chemicals which account for thousands of deaths annually.

Preventive Toxicology: Deals with the evaluation and recommendations of suitable measures for protection of the environment and living resources

Regulatory Toxicology: Deals with imposing legal restrictions on the entry of chemicals into the environment for which legislations are framed, passed and enforced by the government

Regulatory Toxicology: Develops regulations for reducing or controlling exposure to dangerous chemicals and helps government in framing regulations and putting them into practice

Soil Eco-toxicology: Debates on the sources, destiny and transportation of harmful chemicals in soil. It highlights the destiny and probable impacts of pesticides on soil ecosystems and provides the basic procedures for performing toxicity assessments

Systemic Toxicology: Deals with the harmful effects of chemicals on a particular system of organisms, viz., respiratory system, immune system, nervous system, etc.
Terrestrial Eco-toxicology: Evaluates the mechanisms for how environmental pollutants affect land-dependent organisms and their environment
Toxinology: Studies the toxins/poisons produced by microbes
Veterinary Toxicology: Deals with the diagnosis and treatment of poisoning in animals, mainly livestock
Wildlife Toxicology: Studies the effects of toxicants on wildlife, in a similar manner to veterinary toxicology that is related to domestic animals

1.3.1 Ecotoxicology as a Sub-Discipline of Toxicology

Eco-toxicology as a branch of toxicology deals with the effects of toxic chemicals on biological systems, mainly the population, community, ecosystem and biosphere. Being a multidisciplinary science, it:

- integrates toxicology and ecology and draws knowledge and procedures from them to analyse the effects of toxic chemical or biological agents on living organisms
- combines ecology, toxicology, physiology, analytical chemistry, molecular biology and arithmetic
- studies the impacts of pollutants on living beings and ecosystems
- identifies the main classes of organic and inorganic contaminants, their properties, release and ecological fate, and transmission in air, water and food chains

The final aim of this approach is to predict the effects of toxicants so that timely action can be taken to prevent or lessen any detrimental effects. Scientists and researchers dealing with eco-toxicology are known as eco-toxicologists, and are involved in assessing the instant effects of a toxic substance not only on the individual but on species in food webs and also in predicting its effects on flora, fauna and natural resources. They try to analyse, predict and prevent the detrimental events in the natural environment by:

- undertaking toxicity testing and risk assessment on chemicals
- undertaking monitoring studies on the effects of toxins and poisons on the food chain
- tracing the metabolism, deposition and transfer of natural and manmade toxic compounds in different food chains
- ascertaining population modifications after exposure to toxicants, mainly genetic changes
- safeguarding the environment and ecosystems for future generations
- making a significant contribution to maintain food resources across the world

1.4 RELATIONSHIPS WITH OTHER DISCIPLINES

Toxicology is an all-encompassing science and human activity drawing from, and adding to, a comprehensive range of other sciences and human activities. At one end of the spectrum are those disciplines that contribute their methods and theoretical ideas to serve the requirements of toxicologists, either in research or in its use to promote human welfare. Chemistry, biochemistry, pathology, physiology, epidemiology, immunology, environmental science and bio-statistics have long been significant, while molecular biology has contributed to the advancement of toxicology (see Figures 1.1 and 1.2). At other end of the spectrum are other disciplines to which toxicology contributes, e.g., medicine as forensic medicine, clinical toxicology, pharmacy and pharmacology, public health, etc. Toxicology contributes in a significant manner to veterinary medicine, and to agriculture through the development and safe use of agricultural chemicals.

The role of toxicology in environmental studies has gradually become increasingly important in recent years. Undoubtedly, toxicology is primarily an applied science, devoted to the improvement of quality of life and environmental protection. Toxicology has advanced immeasurably in recent decades, both in the number of toxicologists and in the knowledge gained. This has brought about a change from a principally descriptive science to one which makes use of a wide range of procedures to study the factors involved.

Toxicology, as a branch of pharmacology, deals with the therapeutic impacts of exogenous substances and the biochemical consequences involved in those effects. As the therapeutic dose range of drugs is quite small, and most of these drugs are toxic at higher doses, pharmacology may be considered as a branch of toxicology. It is related to the applied sub-divisions of biology, namely medicine and agriculture. In the former, clinical diagnosis and treatment of poisoning and controlling the toxic side effects are areas of importance, while in the latter the development of agricultural biocides is of great significance.

The detection and management of the off-target effects of toxic compounds are of great concern that is indispensable to their constant use. Toxicology is believed to be an area of fundamental biology as the acclimatization of living beings to toxic

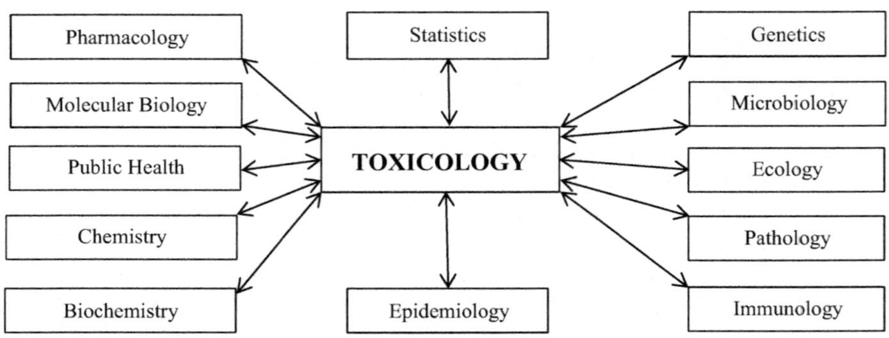

FIGURE 1.1 Relationship of toxicology with other branches of science.

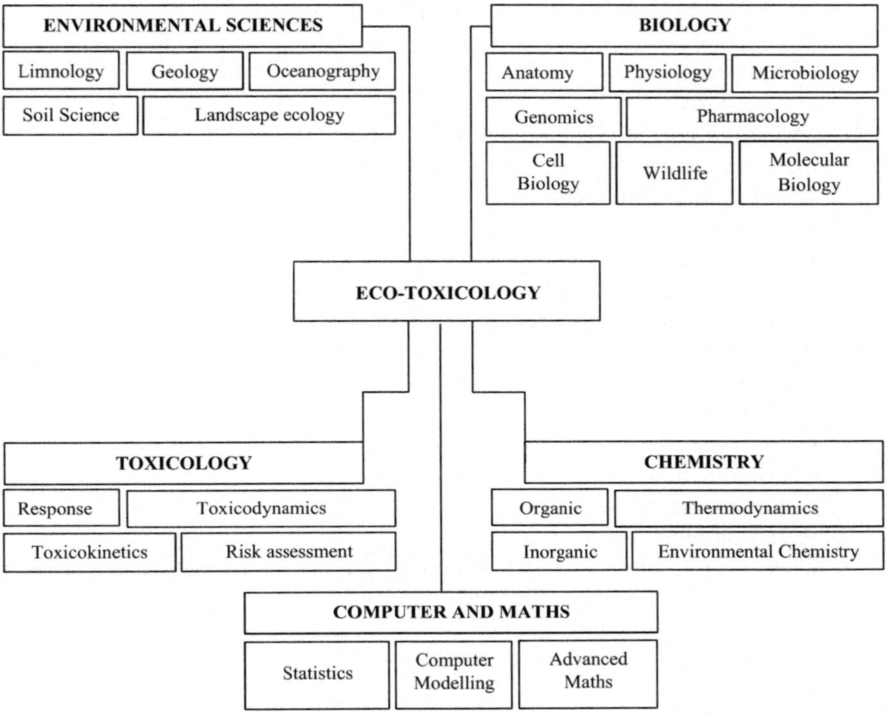

FIGURE 1.2 Relationship of eco-toxicology with other disciplines.

environments has significant effects on ecology and evolution. Chemistry and chemical biology are closely linked to the development of the new methodology of toxicology. Chemistry contributes by providing analytical tools for the evaluation of toxic compounds, mainly for forensic toxicology and residue analysis, and those of biochemistry provide the methods to examine the metabolism and mode of action of toxic compounds.

On the other hand, studies of the chemistry of toxic compounds have added to fundamental organic chemistry, and studies of enzymes involved in detoxification and toxic action have contributed to the basic knowledge of biochemistry. Toxicology facilitates the risk assessment of human, animal and environment exposure to industrial chemicals, plant safeguarding products, biocides, medical drugs and devices, and consumer goods, including food and cosmetics.

1.5 SCOPE OF TOXICOLOGY

The scope of toxicology can be discussed based on the following:

A. **Mechanisms of Toxic Action:** Based on the mechanism of action, the range of toxicology is confined to:

- Biochemical, behavioural and nutritional toxicology
- Carcinogenesis and teratogenesis
- Mutagenesis
- Organ toxicity

B. **Measurement of Toxicants and Toxicity:** Based on the dimension of toxicants and toxicity, the scope of toxicology is limited to:
- Analytic toxicology
- Toxicity testing
- Toxicologic pathology
- Structure–activity study
- Biomathematics, statistics and epidemiology

C. **Applied Toxicology:** Based on the applications, the scope of toxicology is restricted to:
- Clinical toxicology
- Veterinary toxicology
- Forensic toxicology
- Environmental toxicology
- Industrial toxicology

D. **Classes of Chemicals Used:** Based on the classes of chemicals used, the scope of toxicology is confined to:
- Agricultural chemicals
- Clinical drugs and drugs of abuse
- Food additives
- Industrial chemicals
- Naturally occurring substances, including phytotoxins, mycotoxins, inorganic minerals
- Combustion products
- "Toxin" refers to toxic substances that are produced naturally
- "Toxicant" refers to substances that have an anthropogenic origin

E. **Regulatory Toxicology:** This pertains to:
- Legal aspects (formulation of laws and their execution)
- Risk assessment (description of risks, potential risks and risk–benefit equations)

F. **Development of Antidotes:** Antidotes play a major role in curing poisoning. The correct use of antidotes may significantly enhance the removal of and stabilization of the toxic actions of poisons. In certain situations, they may considerably shorten the period of therapy, and, in some instances, save the living organism from any aftereffects.

BIBLIOGRAPHY

Carson, R. (1962) *Silent Spring*. Boston: Houghton Mifflin.

Casarett, L. J., Klaassen, C. D., Amdur, M. O., and Doull, J. (1996) *Casarett and Doull's Toxicology: The Basic Science of Poisons*. New York: McGraw-Hill.

Chapman, P. M. (2002) Integrating toxicology and ecology: Putting the "eco" into ecotoxicology. *Marine Pollution Bulletin* 44(1): 7–15.

DuBois, K. and Geiling E. M. K. (1959) *Textbook of Toxicology*. New York: Oxford University Press.

Hodgson, E., Mailman, R. B., and Chambers, J. E. (1998) *Dictionary of Toxicology*. London: Macmillan.

Kallet, A. and Schlink, F. J. (1933) *100,000,000 Guinea Pigs: Dangers in Everyday Foods, Drugs and Cosmetics*. New York: Vanguard.

Newman, M. C. and Unger, M. A. (2003) *Fundamentals of Ecotoxicology*. Boca Raton: Lewis Publishers.

Olmsted, J. M. D. (1944) *Francois Magendie: Pioneer in Experimental Physiology and Scientific Medicine in XIX Century France*. New York: Schuman.

Rand, G. M. and Petrocelli, S. R. (1985) *Fundamentals of Aquatic Toxicology: Methods and Applications*. Washington: Hemisphere Publishing.

2 Basics of Eco-Toxicology

Ecotoxicology, a multidisciplinary discipline integrates toxicology and ecology.
Anonymous

After going through this chapter, the reader will learn: sources of toxic compounds, classification of toxicants, types of toxicity, factors affecting toxicity, environmental persistence, and toxicity testing

FACT FILE

- Toxicology is the scientific discipline that intersects with other branches of science besides studying the harmful effects of chemical substances on living organisms
- Dioscorides, a Greek physician, first classified plants according to their toxic and therapeutic effects
- Jean Stas was the first to separate plant poisons from human tissue
- Eco-toxicology studies the impact of toxic chemicals on biological systems, mainly the population, community, ecosystem, and biosphere. It differs from environmental toxicology as it combines the impact of stressors across different levels of biological organization from the molecular level to communities and ecosystems, while environmental toxicology analyses toxicity to humans and often emphasize on effects at the organism level and below
- René Truhaut, a toxicologist, used the term "*ecotoxicology*" for the first time in 1969 during the Stockholm conference
- Rachel Carson is regarded as the *mother of environmental toxicology*, after she published her book, *Silent Spring*, in 1962
- An expert committee convened by World Health Organization and Food and Agriculture Organization of the UN, known as JECFA, is the international body accountable for assessing the health risk caused by natural toxins including mycotoxins
- Natural toxins cause a variety of harmful health effects and pose serious health threats to both humans and livestock. Some of the naturally occurring toxins are enormously potent

DOI: 10.1201/9781003343721-2

- Adverse health effects of natural toxins include acute poisoning which manifests in the form of allergic reactions or acute stomach-ache and diarrhoea, and it can also be fatal
- Longer-lasting health concerns of toxins include the immune, reproductive, and nervous systems, and also cancer
- Mycotoxins are naturally occurring toxins produced by some moulds (fungi), which have a variety of adverse health implications, posing a serious threat both to human and livestock health, ranging from acute poisoning to long-standing adverse effects

2.1 INTRODUCTION

Toxicology studies the harmful consequential effects of biological and physical phenomena ranging from acute to long-term on individuals, populations, and ecosystems, which may manifest in diverse forms, ranging from immediate death to indirect changes that are not realized until months or years later. These effects manifest at various levels, such as at the organ, cell, biochemical, or molecular levels. Likewise, toxicity is a sequence of events starting with exposure, continuing through distribution and metabolism, and ending with interaction with cellular macromolecules (i.e. DNA or protein). The harmful effects of any toxic substance depend on the following factors:

- **Quantity/Amount/Dose:** Both large single exposures (acute) and continuous small exposures (chronic)
- **Route of Exposure:** Oral (ingestion), respiratory (inhalation), and dermal (absorption)
- **Other Factors:** Type of species, age, gender, well-being, physical environment, and individual features

2.2 SOURCES OF TOXIC COMPOUNDS

1. **Air, Water, and Food Pollutants**
    - **Air**: CO, nitrogen and sulphur oxides, hydrocarbons, and particulates
    - **Water**: Agricultural chemicals including pesticides, herbicides, fungicides, nematicides, rodenticides, and fertilizers
    - **Halogenated hydrocarbons**: Chloroform, dichloroethane, tetrachloride
    - **Chlorinated aromatics**
    - **Detergents**: Alkyl benzene sulphonates
    - **Food contaminants**: Bacterial toxins, mycotoxins, plant alkaloids, animal toxins, pesticide residues, and residues of animal food additives
2. **Chemical additives in food** (includes preservatives: antibacterial, antifungal, or antioxidant used to change the physical characteristics, taste, colour, and odour)
3. **Workplace Chemicals** (includes inorganic compounds, namely metals and fluorides, CO, etc.; organic compounds, namely, aliphatic hydrocarbons,

aromatic hydrocarbons, halogenated hydrocarbons, alcohols, esters, organometallics and amino compounds

4. **Drugs of Abuse** Includes CNS-depressants (ethanol), CNS stimulants (cocaine, nicotine, and caffeine), opioids (heroin, morphine), and hallucinogens (LSD – lysergic acid diethylamide)
5. **Therapeutic Drugs** Therapeutic drugs also cause toxic effects but the danger to the individual depends on the nature of the toxic response, dose necessary to produce the toxic response, and relationship between the therapeutic dose and toxic dose, e.g., anticancer drugs may at times act as carcinogens
6. **Pesticides** Used to control pests such as rodents, insects, or plants
7. **Solvents** Many solvents and organic liquids are inflammable and highly volatile, which enter the body through inhalation, ingestion, or through dermal routes, and may irritate the eyes, respiratory tract or have detrimental effects on the liver, kidneys, heart, blood vessels, bone marrow, and nervous system
8. **Polycyclic Aromatic Hydrocarbons (PAH)** Includes organic compounds containing carbon and hydrogen having multiple aromatic rings and discharged after incomplete combustion of organic materials, coal, oil, gasoline, trash, tobacco, wood, and naturally from forest fires and volcanoes, in addition to high-temperature cooking). Long-standing exposure to PAHs can cause cataract, kidney, liver damage, and jaundice
9. **Cosmetics**: Ingredients of cosmetic and personal care products contain chemicals that may be dangerous to health. Some of these are:
   - Formaldehyde used as a preservative in cosmetics (carcinogenic, skin irritant, allergic reactions and rashes)
   - Butylated hydroxyanisole (BHA) and butylated hydroxytoluene (BHT) used as antioxidants, preservatives, additives, perfumed ingredients (cause skin irritation, hormone disruption, etc.)
   - Diethanolamine (DEA) used as a pH adjuster and foaming agent in face makeup and hair products (causes skin irritation and organ system toxicity)
   - Phthalates dibutyl phthalate (DBP), dimethyl phthalate (DMP), and diethyl phthalate (DEP) used as plasticizers, solvents, and perfume components of nail polish, hair sprays, perfumes, etc. (damages the male reproductive system)
   - Perfume (causes skin irritation, allergic reactions, cancer, or reproductive toxicity)
10. **Naturally Occurring Toxins:** Toxins in the environment, derived from animals, water, and food, are a universal problem and are more prevalent in tropical and less developed regions, and present a greater potential hazard to humans. Natural toxins being naturally produced by living organisms are not harmful to the organisms themselves but these may be poisonous to other living beings, including humans, if consumed. These chemical compounds have varied structures and differ in biological function and toxicity. Natural toxins are formed in food as defence mechanisms against predators, insects,

or microorganisms, or as a consequence of infestation with microorganisms, such as mould, in response to climate stress. Natural toxins can cause harmful health effects to both humans and livestock, causing acute poisoning ranging from allergic reactions to severe stomach-ache and diarrhoea, and even death, while longstanding health effects affecting on the immune, reproductive, or nervous systems, and can cause cancer. Plants produce some toxins as a natural defence mechanism against predators, insects, or microorganisms, or as a consequence of infestation with microbes, such as moulds, or in response to climate stress, such as drought or extreme humidity. Microscopic algae and plankton in water bodies produce natural toxins which are toxic to humans but not to fish or shellfish, and are dependent on toxin-producing organisms. People consuming fish or shellfish containing toxins are likely to develop illness. Some of the toxin-producing organisms are:

A. Natural Toxins of Plant Origin: Of the more than 300,000 different plant species around the world, at least 2,000 species are considered poisonous. Cases of poisoning are reported after consumption of wild non-edible mushrooms, berries, or other plants. Worldwide, hundreds of plant species are commonly eaten, however many of them can be toxic to the body if consumed in excess or if not treated properly before consumption. Plants of the same genera may show similar or dissimilar toxicities. The quantity and distribution of toxins in a plant depend on the type of species and the topographical settings where it is grown. Furthermore, the incidence of toxic effects depends on the toxin level present in the plant as well as the individual's vulnerability. Generally, the plant organs important for survival and reproduction, such as flowers and seeds, concentrate defence compounds. These compounds may be more quickly produced or stored at certain stages of critical growth, i.e., in buds, young tissue, or seedlings (see Table 2.1).

B. Aquatic Bio-Toxins: Aquatic biotoxins like algal toxins during blooms are produced by naturally occurring algae in oceans and fresh water. Mussels, scallops, and oysters may contain these toxins. Algal toxins in humans, mammals, or fish can cause:

- Diarrhoea
- Paralysis
- Tingling
- Vomiting

Algal toxins are retained by shellfish and fish, and pollute drinking water. They don't have taste or odour, and cannot be removed by cooking or freezing. Ciguatera fish poisoning (CFP) is caused by eating fish contaminated with dinoflagellates producing ciguatoxins. Some fish, such as barracuda, black grouper, dog snapper, and king mackerel, harbour ciguatoxins. The symptoms of ciguatera poisoning include nausea, vomiting, and neurological signs, such as a tingling sensation on the fingers and toes.

TABLE 2.1
Natural Phyto- and Phycotoxins and Their Modes of Toxicity

Toxin	Organism (example)	Mode of toxicity
Amanitin	*Amanita phylloides*	Hepatotoxin
Ricin	*Ricinus communis*	Cytotoxin
Senecionine	*Jacobaea vulgaris*	Hepatotoxic, potential carcinogen
Coniin	*Conium maculatum*	Neurotoxin
Solanine	*Solanum tuberosum*	Cell membrane disruption
Formononetin	*Trifolium pratense*	Estrogenic activity
Linamarin	*Manihot esculenta*	Cyanide poisoning
Ptaquiloside	*Pteridium aquilinum*	Carcinogen
Microcystin LR	*Microcystis aeruginosa*	Hepatotoxin
Saxitoxin	*Dolichospermum circinale*	Neurotoxin
Agaric acid	Food containing mushrooms, alcoholic beverages	Mucosal membrane irritant
Hydrocyanic acid	Nougat, marzipan or its substitutes, or similar products Canned stone fruits Alcoholic beverages Confectionery Stone fruit juices	
Hypericine	Alcoholic beverages	
Saffrole	Meat preparations and meat products, including poultry and game	

C. Cyanogenic Glycosides occur in almost 2000 plant species including cassava, sorghum, stone fruits, bamboo roots, and almonds; and the potential toxicity of a cyanogenic plant depends mainly on the likelihood that its consumption produces a concentration of cyanide that is toxic to exposed beings. The clinical symptoms of acute cyanide intoxication include convulsions, diarrhoea, dizziness, headache, low blood pressure, mental confusion, terminal coma, etc. Death can occur due to cyanide poisoning when the cyanide level surpasses the limit an individual is able to decontaminate.

D. Furocoumarins includes stress toxins discharged in response to stress and are present in many plants such as celery roots, citrus plants, some medicinal plants, parsnips, etc., and can cause gastrointestinal problems and skin reactions on UVA exposure.

E. Lectins produced by different varieties of beans, with kidney beans having the highest concentrations, and mainly red kidney beans can cause severe stomach-ache, vomiting, and diarrhoea. Soaking in water for at least 12 hours, followed by boiling for at least 10 minutes in water, destroys lectins.

F. Oxalic acid: Stems of rhubarb, a plant used for preparing pies and other desserts, contain oxalic acid, which is particularly high in the leaves, making it toxic for consumption. Its toxicity can cause neurological, cardiac, respiratory, and gastrointestinal (GI) effects.

G. Cucurbitacins include naturally occurring toxins in courgetti [zucchini] are used for preparing savoury dishes but can cause severe GI problems.

H. Mycotoxins natural toxins produced by some moulds, which often grow on foodstuffs such as cereals, dry fruits, nuts, and spices. Moulds grow before or after harvest, or during storage, on/in the food itself often under moist and humid conditions. Hundreds of mycotoxins have been identified, the most common being aflatoxins, ochratoxin A, patulin, fumonisins, zearalenone, and nivalenol/deoxynivalenol. These are chemically stable and persist after food processing. The impact of foodborne mycotoxins is acute, with symptoms of severe illness and even death appearing quickly after consuming tainted food products. The long-standing effects of chronic mycotoxin exposure on health include cancer and immune deficiency disorders. Exposure to mycotoxins can be direct on consuming infected food or indirect from animals fed contaminated feed. Aflatoxins are among the most poisonous mycotoxins produced by certain moulds (*Aspergillus flavus* and *Aspergillus parasiticus*) and grow in soil, decaying vegetation, hay and grains. Crops that are frequently affected by *Aspergillus* spp. include:

- Cereals (corn, sorghum, wheat, and rice)
- Oilseeds (soybean, peanut, sunflower, and cotton seeds)
- Spices (chilli peppers, black pepper, coriander, turmeric, and ginger)
- Tree nuts (pistachio, almond, walnut, coconut, and Brazil nut)

Toxins can be found in the milk of animals consuming contaminated feed, in the form of aflatoxin M1. Large doses of aflatoxins can lead to aflatoxicosis or acute poisoning which can be life threatening, usually due to liver damage. Aflatoxins are known to be genotoxic, i.e., they can damage DNA, causing cancer in animals. Ochratoxin A, produced by species of *Aspergillus* and *Penicillium* during storage of crops, is a food-contaminating mycotoxin, known to produce toxic effects in animals, with the most sensitive and notable effect being kidney damage. These toxins may also affect foetal development and the immune system. Patulin, a mycotoxin produced by a range of moulds, mainly *Aspergillus*, *Penicillium*, and *Byssochlamys*, is found in rotting apples and apple products. Patulin can also occur in mouldy fruits, grains, and other foods. Acute symptoms of patulin toxicity in animals include liver, spleen, and kidney damage, and toxicity in the immune system. In humans, it causes nausea, gastrointestinal disturbances, and vomiting. *Fusarium* fungi produce a range of toxins, including trichothecenes such as deoxynivalenol (DON), nivalenol (NIV), T-2 and HT-2 toxins, zearalenone (ZEN), and fumonisins. The formation of moulds and toxins occurs on a variety of cereal crops. Different *Fusarium* toxins are linked with certain types of cereal. Trichothecenes can be acutely toxic to humans, causing rapid irritation to the skin or intestinal mucosa, resulting in diarrhoea. ZEN is known to cause hormonal and oestrogenic effects, and infertility with a high intake, mainly in pigs. Fumonisins are linked to the incidence of oesophageal cancer in humans, and liver and kidney toxicity in animals.

I. Solanines and Chaconine are produced by plants belonging to the solanaceae family, including tomatoes, potatoes, and aubergines [eggplants] in response to stresses such as bruising, UV light, microorganisms, and attacks of insect pests and herbivores. Though the levels are usually low, higher concentrations are found in potato sprouts, bitter-tasting peel and green parts, and green tomatoes. To reduce the formation of solanines and chaconine, potatoes should be stored in a dark, cool, and dry place, and consumption of green or sprouting parts should be avoided.

J. Poisonous Mushrooms: Wild mushrooms containing toxins, e.g., muscimol and muscarine, can cause vomiting, diarrhoea, confusion, visual disturbances, salivation, and hallucinations. The onset of symptoms occurs 6–24 hours or more after consuming mushrooms. Fatal poisoning is usually linked with severe delayed symptoms affecting the liver, kidney, and nervous systems. Cooking or peeling do not incapacitate these toxins.

K. Pyrrolizidine Alkaloids (PAs): These are produced by many plants belonging to the Boraginaceae, Asteraceae, and Fabaceae families. Most are weeds that grow in fields and contaminate food crops, causing a range of adverse health effects. PAs are detected in herbal teas, honey, herbs and spices, and cereal-based products.

L. Natural Toxins of Animal Origin: More than 400 species of venomous snakes, nearly 1500 species of toxic marine animals, and numerous poisonous or venomous arthropods are distributed throughout the world. Almost every phylum of the animal kingdom is characterized by species that produce toxins, known as zootoxins, which may be in the form of poisons or venoms. A natural toxin of animal origin may be an outcome of metabolic activity or a chemical passed along the food chain. While poisoning after consuming terrestrial animals is comparatively rare, poisoning due to marine toxins occurs around the world. Marine toxins produced by toxic microalgae are stored in shellfish, crustaceans, and finfish after their consumption. Tetrodotoxin, a potent marine neurotoxin, is believed to be produced by some bacteria. It is found in over 90 species of pufferfish and may be lethal even in small quantities. Seafood poisoning, generally reported in coral reef fish, is due to the presence of ciguatoxin found in more than 300 species of fish. Some fish that harbour ciguatera toxin include black grouper, black fin snapper, barracuda, dog snapper, king mackerel, etc. Symptoms of ciguatera poisoning include nausea, vomiting, and neurologic symptoms.

Currently, there is no cure for ciguatera poisoning and the symptoms diminish gradually; treatment is symptomatic. There are nearly 1,200 poisonous and venomous animals in the world. The glands of some animals not considered poisonous or venomous, if consumed, can cause food poisoning, such as the gall-bladder of grass carp which contains cyprinol-related chemicals. Mercury occurs naturally in sea water, and accumulates over the years in the bodies of large ocean-dwelling fish, such as sharks, through direct absorption and feeding on smaller fish. Histamine produced by bacterial spoilage of scombroid fish also causes seafood poisoning. Scombroid poisoning occurs because of inadequate storage conditions and poor temperature control of fish. If fish are not preserved by freezing during transportation, histamine collects in the fish and is responsible for poisoning. Symptoms of scombroid poisoning start quickly, often 30 minutes to 1 hour after consuming the toxin, and include abdominal cramps, blurred vision, burning and tingling of the lips and mouth, diarrhoea, dizziness, etc. In severe cases, breathing difficulties can occur, especially in asthmatics. The oil of some fish is not metabolized by humans after consumption. This type of toxicity is self-limiting, as the diarrhoea discontinues when the fish oil is excreted from the body. Diarrhoea usually does not cause mortality, because it don't cause any dehydration. Toxins produced as secondary metabolic products are poisons which accumulate in the tissues of hosts or predators after consuming toxin-bearing prey. Not all exposures to toxic animals cause toxicoses, as disparity in toxicity of zootoxins occurs owing to host and environmental factors such as age, sex, nutritional status, season, geographic location, and the toxin constituents. Likewise, target animals may vary in their vulnerability to zootoxins due to factors such as age, size, species, location, and severity of bites or stings. Some animals that produce zootoxin include the following:

a. Marine Zootoxins

Dinoflagellates or algae produce marine zootoxins, e.g., domoic acid, gonyautoxins, and ciguatera toxin which accumulate in the tissues of marine animals. Toxins may be formed by the fish itself or by the marine plankton or algae consumed by the fish. Fish are a main source of proteins for many people of coastal areas and for animals in the form of fish meal in their feed. Ichthyotoxicosis fish poisoning is more common in humans than other animals. Some common ichthyotoxins are the following:

1. Domoic Acid/Heterocyclic Amino Acid (Neurotoxin)

Producer: Marine algae

Effect: Neurological and behavioural effects due to the mimicry of neurotransmitter glutamate in the nervous system

- **Gastrointestinal**
  - ✓ Nausea
  - ✓ Diarrhoea
  - ✓ Abdominal discomfort
- **Cardiovascular**
  - ✓ Neurologic dysfunction
  - ✓ Headache
  - ✓ Disorientation
  - ✓ Loss of short-term memory

2. **Ciguatoxins (Neurotoxin)**

Producer: Dinoflagellate *Gambierdiscus toxicus* (ingestion of fish, Moray eel) that accumulate ciguatoxins produced by dinoflagellates

Effect: Acts on voltage-sensitive sodium channels on excitatory membranes of nerves and muscles resulting in autonomic dysfunction, bradycardia, cramping, diarrhoea, excessive secretion of intestinal fluid, fatigue, hypotension, muscle weakness, myalgia, etc.

3. **Tetrodotoxin/TIX** (widely distributed among marine as well as terrestrial animals)

Producer: Marine bacteria

Effect: Selective voltage-dependent sodium channel blocker non-protein toxin

- Accumulates in the tissues of a poisonous Pacific fish, the pufferfish
- Neurological
- Gastrointestinal symptoms

4. **Saxitoxins**

Producer: *Alexandrium* spp. and related dinoflagellates

Effect: Binding directly to voltage-dependent sodium channels in nerve and muscle cell membranes interrupts nerve signal transmission and leads to paralysis, in addition to GI symptoms, ataxia, dysphagia, and hypertension.

b. Zootoxins from Insects

Toxicosis is caused by *Hymenoptera* (bees, wasps, and ants). Hymenopteran venoms contain a mixture of proteins, peptides, and small organic molecules. Ant venom has a low protein content and is generally made up of alkaloids, while bee venom has a higher protein content. The toxicity of bee venom is intensified by the presence of hyaluronidases and phospholipases, which account for the allergic responses seen with bee envenomation. Melittin, a membrane-disruptive constituent of bee venom, induces:

- Haemolysis
- Increased cell permeability and capillary blood flow
- Enhanced spread of venom in tissue

Apamin, being a neurotoxin, blocks potassium channels and is accountable for peripheral nerve dysfunction which occurs after bee stings.

c. Zootoxins from Arachnida (Spiders and Scorpions)

Spiders: Various spiders, such as the black widow spider, brown widow spider, red-legged spider, brown recluse spider, etc., produce toxins. Among these, most cases are due to the black widow spider (*Latrodectus mactans*) and brown recluse spider (*Loxosceles reclusa*). Recluse spider venom contains a variety of necrotizing enzymes. Widow spiders produce oily venom composed of a variety of neuro-active proteins, of which α-latrotoxin is the most potent and accountable for most of the signs and symptoms linked with envenomation. Binding of α-latrotoxin to neuronal receptors releases a considerable amount of acetylcholine (ACh) at the neuromuscular junction, resulting in intense muscle contraction. Symptoms include abdominal rigidity, diarrhoea, hyperaesthesia, hypertension, muscular pain and spasms, tachycardia, etc.

Scorpion envenomation is an important public health concern. Scorpions deliver their venom via a barbed appendage in their tail housing two venom glands. It contains a variety of compounds which differ among various scorpion species. Two powerful neurotoxins (α-scorpion toxin and IS-scorpion toxin) that block voltage-sensitive sodium and potassium channels have been identified in the venom of toxic scorpions. These venoms produce allergic responses, cardiac arrhythmia, inflammatory reactions involving the release of cytokines, local pain and oedema at the sting site, myalgia, neurologic effects, respiratory depression, etc.

d. Zootoxins from Toads

Dogs and cats may play with toads and become orally exposed to the toxins of toads secreted by their skin glands. This poison contains adrenaline and nor-adrenaline, budienols, bufotalins, bufotenine, bufotoxins, bufoviridin, catecholamines, and non-cardiac sterols. Bufodienolides are cardio-active steroids synthesized by the parotid gland from cholesterol, with an action similar to digitalis. Its action includes the following:

- Inhibits the activity of ATPase of the Na^+/K^+ pump of the cardiac muscle fibre
- Blocks the activity of Na^+ channels
- Raises the intracellular Ca^{++} concentration
- Increases contractions of the heart
- Reduces the heart rate

Clinical signs of toad poisoning are cardiac irregularities, convulsions, cyanosis, hypersalivation, prostration, pulmonary oedema, etc.

e. Snake Venom

Globally, venomous snakebites result in thousands of deaths annually, with most fatalities occurring in developing nations that have limited access to medical care. Snakebites in animals usually occur while grazing or hunting. Snake venoms are complex mixtures of toxins consisting of:

- Amino acids
- Polypeptides
- Glycopeptides
- Biogenic amines
- Cations such as K^+, Na^+, Ca^{++}, Mg^{++}, Ni^{++}

The most active component of venom is a peptide or polypeptide which may act enzymatically or non-enzymatically. Generally, most snake venoms result in neurotoxicity, cardio-toxicity, or haemo-toxicity. There are more than 3500 species of snakes, of which more than 400 are poisonous and harmful. Snake venoms contain anticoagulant, cardio-toxic, coagulant, haemolytic, necrotizing, and neurotoxic fractions. The venoms of cobra and krait are mostly neurotoxic, while those of vipers and rattlesnakes are haemotoxic in nature. The haemotoxin of pit vipers induces:

- Blood coagulation defects
- Cardiopulmonary dysfunction
- Changes in blood cells
- Hypotensive crisis
- Injury to blood vessels
- Local swelling

Neurotoxins bind to cholinergic receptors at the neuromuscular junction and result in respiratory paralysis, salivation, asphyxia, gasping, convulsions, and death within 2–4 hours. The neurotoxins of cobra and krait induce paralysis of the cranial nerve of eyelids resulting in ptosis and paralysis of muscles of the eyeballs and respiration.

2.2.1 Toxic Effects and Food Poisoning

Natural toxins in food can cause both acute and chronic health effects, with diverse clinical symptoms. Acute symptoms range from mild gastrointestinal upset, neurological symptoms, respiratory paralysis, to fatality. This is more likely among vulnerable groups of the population, e.g., children and the elderly. Acute symptoms are noticed within hours after ingestion of various marine toxins in shellfish and other seafood. Acute poisoning is seen after the consumption of wild mushrooms such as the death cap mushroom or ineffectively treated plants such as ginkgo seeds and bitter apricot seeds. A single death cap mushroom, commonly found during monsoons, can be lethal for an adult. Symptoms of poisoning are mostly GI, but the toxin ultimately kills the victim because of hepatotoxicity. Chronic toxicity is caused more often by plant toxins such as alkaloids. Pyrrolizidine alkaloids present in weeds may cause liver toxicity after prolonged consumption. Health risks from natural toxins in food can be minimized and thus, it is important to:

- avoid storing foods for a long duration
- buy fresh grains and nuts
- avoid damage to grains before and during drying, and in storage, as damaged grains are more likely to be invaded by moulds and consequently result in mycotoxin contamination
- discard food that does not smell or taste fresh, or has an unusual taste
- not to assume that everything that is "natural" it is safe
- make sure that foods are stored suitably – away from insects, dampness, and humidity
- that only mushrooms or other wild edibles identified as non-poisonous are consumed

- spoiled or stained food, particularly mouldy foods, are disposed of
- whole grains (especially corn, wheat, rice), dried figs, and nuts, e.g., peanuts, pistachio, almond, walnut, coconut, and hazelnuts, which are frequently contaminated with aflatoxins due to the presence of mould, must be discarded if they have a mouldy, discoloured, or shrunken appearance

2.3 CLASSIFICATION OF TOXICANTS

Toxic compounds come in all shapes and sizes, both from natural and human-made sources. Depending on the needs of the classifier, toxic compounds are categorized in diverse ways. No classification can properly cover the entire spectrum of toxic substances. Hence, groups of classification systems based on several aspects have been devised. Dioscorides categorized substances using common characteristics, i.e., whether toxic or healing. Orfila classified substances based on the origin, as animal, vegetal, or mineral origin. Classifications may take into consideration both chemical and biological properties of toxic substances, but exposure features are also beneficial in toxicology. Based on following points, different classification schemes for toxic agents are available:

1. **Based on the source of toxicants**:
   - Animal toxicants (zootoxins, venoms)
   - Mineral toxicants (copper, lead, selenium, iron)
   - Plant toxicants (morphine, curare, strychnine)
   - Synthetic toxicants (organophosphates, carbamates, aluminium phosphide)
2. **Based on the physical state of toxicants:**
   - Dust toxicants (asbestos dust, silicon dust, metallic dusts)
   - Gaseous toxicants (HCN, SO_2, CO, phosphine)
   - Liquid toxicants (H_2SO_4, carbon disulphide, nicotine)
   - Solid toxicants (strychnine, opium, atropine)
3. **Based on the physical characteristics**:
   - Inflammable/non-flammable
   - Explosive/non-explosive
4. **Based on the physical effects**:
   - Irritant/non-irritant
   - Corrosive/non-corrosive
5. **Based on the target organ/system:**
   - Dermatotoxicants (coal tar compounds, heavy metals like arsenic, mercury, p-tertiary butyl phenol)
   - Haematotoxins (warfarin, cyanide, phenothiazine, snake venom, pit viper venom)
   - Hepatotoxins (carbon tetrachloride, aflatoxins, phenol)
   - Nephrotoxins (heavy metals and oxalates)
   - Neurotoxins (organophosphorus insecticides, pyrethroids, anaesthetics, nicotine)
   - Pulmonotoxicants (α-naphthylthiourea, hydrogen sulphide, ammonia gas)

6. **Based on the chemical nature/structure of toxicants**:

   Inorganic toxicants: These include metals, metalloids, non-metals and their salts and derivatives, acids, and alkalis.

   - Metals, e.g., lead, copper, mercury, antimony
   - Non-metals, e.g., phosphorus, sulphur, chlorine, nitrate/nitrite
   - Acids and alkalis, e.g., hydrochloric acid, sulphuric acid, potassium hydroxide

   Organic toxicants: These include all carbon compounds other than oxides of carbon, carbonates, metallic carbides, and cyanides.

   - Alcohols and phenols (methyl alcohol, ethyl alcohol, phenol, cresol, pentachlorophenol)
   - Aldehydes and ketones (formaldehyde, paraldehyde)
   - Alkaloids (nicotine, atropine, strychnine)
   - Amides (phenacetin, sulphonamides)
   - Amines (adrenaline, ephedrine, amphetamine)
   - Esters (organophosphorus insecticides, succinylcholine)
   - Ethers (diethyl ether, divinyl ether)
   - Glycosides (digitoxin, cyanogenetic glycosides)
   - Halogen derivatives of hydrocarbons (chloroform, BHC, DOT, carbon tetrachloride)
   - Hydrocarbons (cyclopropane, benzene, paraffin, naphthalene)
   - Organic acids (formic acid, phenoxy acetic acid, salicylic acid)
   - Proteins (ricin, crotin, abrin)

7. **Based on the analytical behaviour of toxicants:** Toxicants can be classified as:
   - Extractive toxicants
   - Metals and metalloids, e.g., copper, mercury, zinc, silver, antimony
   - Toxicants extractable by ether from acid solution, e.g., organic acids, nitro compounds
   - Toxicants extractable by ether from alkaline solution, e.g., alkaloids
   - Volatile toxicants, e.g., hydrocyanic acid, alcohols, acetone, phenol, chloral hydrate
8. **Based on the toxic effects**: It is difficult to classify toxicants on the basis of toxic effects as a single compound may have a number of effects.
   - Carcinogens (thiouracil, vinyl chloride, nickel)
   - Clastogens (UV light, caffeine)
   - Mutagens (ethyl methane sulphonate, UV light, nitrogen mustards, nitroso compounds)
   - Teratogens (phenylmercuric acetate, triazines, thalidomide)

9. Based on the principal uses of toxicants

- Food additives:
  - ✓ Preservatives (ascorbic acid, sodium bisulphite)
  - ✓ Antioxidants (ascorbic acid)
  - ✓ Emulsifying agents (cholic acid, desoxycholic acid)
  - ✓ Colouring agents (amaranth, tartrazine)
  - ✓ Anticaking agents (aluminium-calcium silicate)
- Fungicides (captan, folpet, pentachlorophenol)
- Herbicides (triazine, paraquat, 2,4-d)
- Insecticides (organophosphorus insecticides, carbamates, pyrethroids)
- Rodenticides (warfarin, fluoroacetate, red squill)

10. Based on toxicity/poisoning potential

Group	Lethal dose
Extremely toxic	< 1 mg/kg
Highly toxic	1–50 mg/kg
Moderately toxic	50–500 mg/kg
Slightly toxic	0.5–5 g/kg
Practically non-toxic	5–15 g/kg
Relatively harmless	> 15 g/kg

11. Based on mechanism of action:

- Anticholinesterase agents and cholinesterase inhibitors (organophosphorus, insecticides, carbamates)
- Corrosive toxicants (caustic alkalis, acids, heavy metals, irritant gases)
- Inhibitors of Kerb's cycle (fluoroacetate)
- Inhibitors of mixed-function oxidases (MFO) (pipronyl butoxide)
- Methaemoglobin producers (nitrite)
- Protoplasmic toxicants (heavy metals)
- Sulphydryl (-SH) inhibitors (mercury, arsenic)
- Uncoupler of oxidative phosphorylation (dinitrophenols, chlorophenol fungicides)

12. Based on environmental and human health considerations

Air Pollutants: Air pollutants can be found anywhere in air, i.e., outdoors or indoors. Air pollutants can be divided into three groups:

Criteria Pollutants: The term "*criteria air pollutants*" is used globally to describe air pollutants which have been controlled and used as indicators of air quality, e.g., CO, Pb, NO_2, O_3, particles, and SO_2.

Air Toxins: Referred to as "*hazardous air pollutants*", these include toxic emissions of motor vehicles, solid fuel combustion, industrial emissions, and paints and adhesives in new buildings. The air toxins are benzene, toluene, xylenes, formaldehyde, and polyaromatic hydrocarbons.

Biological Pollutants: This class of pollutants arise from sources such as microbiological contamination, skins of animals and humans, and also remains and droppings of pests such as cockroaches. Biological pollutants have a significant impact on indoor air quality.

Occupational Hazards: Threats experienced in the workplace include many types of hazards, e.g., chemical, biological, psychosocial, and physical hazards. This group includes both long-term and short-term risks. Short-term risks include physical injury, while long-term risks include increased risk of developing cancer or heart disease.

Radiation Hazards: These hazards are caused due to the impact of radiation.

Water Pollutants: This group includes household wastes, insecticides and herbicides, food processing waste, pollutants from livestock processes, volatile organic compounds (VOCs), heavy metals, chemical waste, and others.

Other methods of classification are based on the following:

1. **Analytical Behaviour of Toxicants**
    - Extractive toxicants
    - Metalloids
    - Metals
    - Volatile toxicants
2. **Chemical Nature/Structure of Toxicants**
    - Acids
    - Alkali
    - Metals
    - Non-metals
3. **Organic Toxicants**
    - Carbon compounds other than oxides of carbon
    - Carbonates
    - Cyanides
    - Metallic carbides
4. **Physical State of Toxicants**
    - Dust toxicants
    - Gaseous toxicants
    - Liquid toxicants
    - Solid toxicants
5. **Sources of Toxicants**
    - Animal toxicants
    - Mineral toxicants
    - Plant toxins
    - Synthetic toxicants
6. **Target Organ/System**
    - **Dermato-Toxicants**: This class of toxicants affects the skin and includes antibiotics, hair dye ingredients, metals, local anaesthetics, pesticides, and poisonous plants.
    - **Haemato-Toxicants**: This group of toxicants causes poisoning of the blood and cardiovascular system. These toxicants induce direct (e.g., arsenic, nucleoside reverse transcriptase inhibitors) versus indirect (particulate matter, ozone) oxidative stress mechanisms, and induce cardiovascular disease through both direct and indirect mechanisms (cigarette smoke).

- **Hepato-Toxicants**: These toxicants affect the liver and include inorganic compounds, organic agents, and synthetic drugs, which undergo metabolism in the liver and produce free radicals in liver which damage it. Drugs such as statins, methotrexate, paracetamol, alcohol, etc. are some common hepatotoxic agents.
- **Nephron-Toxicants**: This group includes the toxicants which have a poisonous effect on the functioning of the kidneys and comprises of toxic chemicals and medications.
- **Neuro-Toxicants**: These toxicants attack the nervous system and include heavy metals, such as lead and mercury, as well as pesticides and chemical weapons, which result in slurred speech, loss of muscle control, and even death.
- **Pulmonary-Toxicants**: This class includes the toxicants which affect the respiratory system and mainly the medicinal drugs comprising of gold and nitrofurantoin, as well as the drugs used in cancer chemotherapy.
- **Endocrine Disrupters**: These chemicals disrupt the endocrine system and most often come as drugs or chemicals in plastics. The endocrine system regulates growth, development, sexual maturity, brain function, and even appetite. Endocrine disrupters interrupt the functioning of hormones, which can cause serious problems. Reptiles and amphibians are mainly sensitive to endocrine disrupters.

7. **Based on Toxic Effects**
   - **Allergens**: These chemicals stimulate over-activity in the immune system. When exposed to allergens, the body goes into overdrive, triggering an immune response to get rid of the allergen. This is why pollen and dust cause symptoms similar to being ill.
   - **Carcinogens**: Cancer-causing chemicals, e.g., cigarette smoke, belong to this category for it contains over 4,000 chemicals that can cause cancer.
   - **Clastogens**: These mutagenic agents give rise to or induce disruption or breakages of chromosomes, leading to the deletion, addition, or rearrangement of sections of chromosomes. Acridine yellow, benzene, ethylene oxide, arsenic, phosphine, and mimosine are known clastogens. Exposure to clastogens increases the frequency of abnormal germ cells in paternal males, resulting in developmental defects in the foetus.
   - **Mutagens**: This group includes mutation-causing chemicals. When an organism is exposed to a mutagen, it alters its DNA, causing cancer and other disorders. X-rays are well-known mutagens.
   - **Teratogens**: Chemicals causing harm to unborn babies are placed in this category. *Teras* is a Greek word, meaning *monster*. These chemicals result in birth defects during the development of the foetus, e.g., thalidomide used in a sleeping pill to prevent nausea during pregnancy is a harmful teratogen. Even a single dose is sufficient to produce birth defects in children.
8. **Based on Type of Toxicity**
   - Acute
   - Chronic
   - Sub-acute

9. **Based on Usage Pattern**
   - Food additives
   - Fungicides
   - Herbicides
   - Insecticides
   - Rodenticides

2.4 TYPES OF TOXICITY

Acute Toxicity: This is defined as toxicity caused due to short-term exposure to a toxicant. The occurrence of acute toxicity in the environment is usually linked with an accident (e.g., leakage of a chemical in a river following a train accident) or irrational use of a chemical. Discharge limits placed on industrial and municipal wastes, when obeyed, have been successful in safeguarding against acute toxicity among organisms in waste-receiving areas. The acute toxicity of a chemical is generally quantified as the LC_{50} or LD_{50}. These measures do not offer any perception of ecologically bearable pollutant levels. However, LC_{50} and LD_{50} values do offer analytically sound, replicable procedures for the relative acute chemical toxicity. The acute toxicity of environmental chemicals is governed experimentally by selected species that serve as representatives of certain levels of trophic organization within an ecosystem. It can occur immediately, within seconds or minutes, or hours or days after an exposure. An acute exposure is generally a single dose, or a series of doses received within 24 hours. Sometimes, the acute toxicity may result in death, e.g., in 1989, 5,000 people died and 30,000 were permanently disabled due to exposure to methyl isocyanate from an industrial accident in Bhopal. Likewise, many people die each year after inhaling CO from faulty heaters.

Mechanisms of Acute Toxicity: Environmental chemicals can produce acute toxicity by various mechanisms.

Cholinesterase Inhibition: The inhibition of cholinesterase activity is an attribute of acute toxicity linked with organophosphate and carbamate pesticides. A total of 40–80 per cent inhibition of brain cholinesterase activity is normally reported in lethally poisoned fish. Acute toxicity due to cholinesterase inhibition is relatively common among incidents of acute poisoning of fish and birds owing to high use of organophosphates and carbamates on lawns, farm land, and golf courses. Cholinesterase inhibition in fish may occur after heavy rains in aquatic habitats close to the areas treated with pesticides and subject to runoff from these areas. Acute toxicity generally occurs in birds that feed in areas using excessive pesticides.

Narcosis: This is one of the common methods by which industrial chemicals stimulate acute toxicity, mainly in aquatic organisms. Narcosis occurs when a chemical amasses in cellular membranes hindering the normal functioning of the membranes. The usual responses to narcosis are reduced activity, abridged reaction to external stimuli, and amplified pigmentation. The effects are changeable, and non-moribund organisms generally return to regular activity once the chemical is removed from the organism's environment. Extended narcosis can result in death. Nearly 60 per cent of industrial chemicals that enter the aquatic environment produce acute toxicity through narcosis. These chemicals do not elicit toxicity at specific target sites and are sufficiently lipophilic to accumulate in the lipid phase or the lipid–aqueous border of membranes to satisfactory levels, thus disturbing the membrane function. Alcohols, ketones, benzenes, ethers, and aldehydes induce narcosis.

Chronic Toxicity: Chronic toxicity is caused by long-term exposure to a toxicant. Sub-lethal endpoints, including reproductive, immune, endocrine, and developmental dysfunction are usually linked with chronic toxicity. However, chronic exposure can also cause direct mortality, e.g., exposure to lipophilic chemicals can result in subsequent bioaccumulation of a chemical to a concentration fatal to an organism. Chronic toxicity is measured by endpoints, e.g., the highest level of chemical that does not produce toxicity during a constant, long exposure, the lowest level of chemical that prompts toxicity during non-stop, prolonged exposure (lowest observed effect level, LOEL), or chronic value (CV), which is the geometric mean of NOEL and LOEL. The chronic toxicity of a chemical is often assessed by the acute chronic ratio (ACR), which is calculated by dividing the acute LC_{50} value by the CV. Chemicals that have an ACR of less than 10 generally have low to no chronic toxicity associated with them.

This category of toxicity denotes increasing damage to specific organ systems and takes months or years to manifest as an identifiable disease. Damage due to subclinical individual exposure may go undetected. With repeated exposures or long-term frequent exposure, the damage from this type of exposure slowly builds up until the harm exceeds the threshold for chronic toxicity. Finally, the damage becomes so serious that the organ no longer works normally and may result in chronic toxic effects, e.g., liver cirrhosis after long-term consumption of ethanol for several years or chronic kidney disease after several years of exposure to lead or pulmonary fibrosis or black lung disease in coal miners.

Sub-Chronic Toxicity: This type of toxicity occurs after repeated exposure to toxic chemicals for several weeks or months. This includes exposure to pharmaceuticals and environmental agents, e.g., ingestion of blood thinner tablets for several days as a treatment for venous thrombosis can cause internal bleeding. Workplace exposure to lead over a period of time can result in anaemia.

Carcinogenicity: This is a complex multistage process of abnormal cell growth and differentiation that causes cancer. Chemicals can act as cancer initiators or promoters. The two stages of carcinogenicity are initiation, when a normal cell undergoes irreversible changes, and promotion, when initiated cells are stimulated to develop as cancer.

During the initial transformation, normal cells undergo irreversible changes after mutation of the genes which regulate the normal functioning of cells. Mutation may result in abnormal cell growth and involve a loss of suppressor genes that usually limit abnormal cell growth. Growth factors, immune suppression, and hormones also contribute to carcinogenicity. The cancer manifests differently, e.g., as neoplasm, or a tumour, which is due to uncontrolled growth of cells. A benign tumour grows at the site of origin and does not affect the adjoining tissues, while malignant tumours invade adjacent tissues or migrate to distant sites through metastasis. These are more difficult to treat and often cause death.

Developmental Toxicity: This type of toxicity has an adverse effect on the developing embryo or foetus due to toxicant exposure either of the parents before conception or to the mother and her developing embryo. The three kinds of developmental toxicity are:

- **Embryo lethality**: failure to conceive, spontaneous miscarriage, or stillbirth
- **Embryo toxicity**: growth retardation or delayed growth of specific organ systems
- **Teratogenicity**: permanent conditions that leave long-term birth deformities in live offspring, such as missing limbs

Developmental toxicity is produced in two ways, namely by acting directly on cells of the embryo, causing cell death or cell damage, resulting in irregular organ development and inducing mutation in the parent's germ cell, which is transferred to the fertilized ovum and transmuted fertilized ova, develop into abnormal embryos.

Genetic Toxicity: This category of toxicity is a result of damage to DNA and altered genetic expression, in a process known as mutagenesis. The genetic change is a mutation and the causal agent is called a mutagen. There are three types of genetic changes:

- Gene mutation, i.e., change in DNA sequence within a gene
- Chromosome aberration, i.e., changes in the chromosome structure
- Aneuploidy or polyploidy, i.e., an increase or decrease in the number of chromosomes

If a mutation occurs in a germ cell, the effect is transmissible. This means that there is no effect on the exposed person; rather, the effect is shifted to future generations. If the mutation occurs in a somatic cell, it causes altered cell growth, e.g., cancer or cell death, for example, teratogenesis in an exposed individual.

2.5 FACTORS AFFECTING TOXICITY

The toxicity of a substance is governed by a number of factors, such as the route of entry (dermal, oral, inhalation, intravenous), time of exposure (short or long term), number of exposures (single or multiple doses), physical form of toxicant (solid, liquid, gas), genetic makeup of the individual, the individual's wellbeing, etc. A single exposure to a toxic substance which may result in serious biological harm or death is referred to as acute exposure, and is usually considered as being no longer than a day, while constant exposure to a toxicant over time, often months or years, is referred to as chronic exposure and can cause irreversible side effects. In some instances, individuals can have erratic reactions, or idiosyncratic responses, to a drug or other substance. An idiosyncratic reaction is infrequent, and at times it is difficult to recognize whether it is a consequence of a genetic vulnerability or has some other cause, e.g., immune system status. It could result in an unusually small or short, or large or long, response to a drug or substance, or the reaction could be different to what has been observed in most other individuals. The toxicity of a substance usually depends on the following factors.

2.5.1 Factors Linked with Elements

Physical Characteristic and Chemical Activity: The physical form of a substance affects the toxicity, especially for the metallic elements and heavy metals, e.g., the toxicity of mercury vapour differs significantly from that of methyl mercury, likewise, Cr^{3+} is comparatively non-toxic, while Cr^{6+} can cause skin or nasal corrosion and lung cancer. The distinct chemical activity of a substance also varies significantly. Some substances can rapidly damage cells, resulting in instant cell death, while others slowly affect the functioning of cells, e.g., hydrogen cyanide binds to the enzyme cytochrome oxidase, which causes cellular hypoxia and rapid death, and likewise if nicotine binds to cholinergic receptors in the CNS, it alters nerve conduction and prompts slow onset of paralysis.

Dosage: This is one of the most important and critical factors in defining whether a substance will act as an acute or chronic toxicant. Almost all chemicals can act as an acute toxicant if sufficiently large doses are administered. Often, the toxic mechanism and target organs are diverse for acute and chronic toxicity, e.g., ethanol as an acute toxicant causes CNS depression and as a chronic toxicant it causes liver cirrhosis; similarly, acute toxicity of arsenic results in gastrointestinal damage and chronic toxicity manifests in the form of skin/liver cancer.

Route of Entry: The way an individual encounters a toxic substance or route of entry is important in regulating its toxicity. Some chemicals may be extremely toxic by one route but not by others. Two main reasons for this are variances in absorption and dispersal in the body, e.g., ingested chemicals, when absorbed from the intestine, dispense first to the liver and may be cleared instantly, while inhaled toxicants rapidly enter the blood circulation and circulate all through the body before being cleared by the liver. Different target organs are often affected by diverse routes of exposure.

Capacity to be Absorbed: This is vital for systemic toxicity. Some chemicals are instantly absorbed while others are inadequately absorbed, e.g., almost all alcohols are instantly absorbed when consumed, while most polymers are practically not absorbed. The rate and extent of absorption may differ considerably depending on the route of exposure and the form of the chemical, e.g., ethanol is easily absorbed from the gastrointestinal tract but feebly through the skin; likewise, organic mercury is readily absorbed from the gastrointestinal tract, while inorganic lead sulphate is not.

2.5.2 Factors Linked with Organisms

Species: Toxic responses can vary significantly depending on the species type. While most differences between species are attributable to variances in metabolism, others may be because of anatomical or physiological variations, e.g., rats cannot vomit and expel toxicants before they are absorbed or cause severe irritation, whereas humans and dogs are capable of vomiting. Selective toxicity denotes variances in toxicity between two species when exposed instantly. This is the basis for the efficacy of pesticides and drugs, e.g., an insecticide is lethal to insects but relatively non-toxic to other animals; likewise, antibiotics are selectively toxic to microorganisms and almost harmless to humans.

Life Stage: A person's age or stage of life may be substantially decisive in his or her response to toxicants. Some chemicals are more toxic to children or the aged than to young adults, e.g., parathion is more toxic to young animals; similarly, nitrosamines are more carcinogenic to newborn or young animals.

Gender: Gender plays a significant role in toxicity. Physiologic dissimilarities between men and women, including variances in pharmacokinetics and pharmacodynamics, can affect drug activity. In contrast to men, pharmacokinetics in women can be ascribed to their lower body weight, slower gastrointestinal motility, decreased intestinal enzymatic activity, and slower kidney function. Delayed gastric evacuation in women may result in the need for them to extend the interval between eating and taking medicines that need absorption on an empty stomach. Slower renal clearance in women may result in a need for adjustment of the dose for drugs, e.g., digoxin that is expelled through the kidneys.

Metabolism: Metabolism or biotransformation is the change of a chemical from one form to another by a biological organism. Metabolism is a major factor that defines the level of toxicity. The products of metabolism are known as metabolites. There are two types of metabolism, detoxification and bioactivation. In detoxification, a xenobiotic is changed to a less toxic form. This is a natural defence mechanism of organisms. Usually, detoxification transforms lipid-soluble compounds to polar compounds; meanwhile, in bioactivation, a xenobiotic may be changed to a more reactive or toxic form. Cytochrome P-450 is an example of an enzyme pathway used to metabolize drugs. In the elderly, the metabolism of drugs such as phenytoin and carbamazepine may be decreased. Consequently, the effects of those drugs may be less distinct. Cytochrome P-450 metabolism can also be obstructed by many drugs. Risk of toxicity may be augmented if a cytochrome P-450 enzyme-inhibiting drug is given with one that depends on the metabolism pathway. Even the gut microbiota can impact the toxicity of drugs and other chemicals, e.g., gut microbes can metabolize environmental chemicals and bacteria-dependent metabolism of some chemicals can modify their toxicity. Similarly, environmental chemicals can alter the composition and/or metabolic activity of the gastrointestinal bacteria, thus contributing in a significant way to the shape an individual's microbiome.

Dispersal in the Body: Dispersal of toxicants or toxic metabolites in the body eventually governs the site of toxicity. A key aspect of whether a toxicant will harm cells is its lipid solubility. If a toxicant is lipid-soluble, it readily enters the cell membrane. Many toxicants get accumulated in the body. Adipose tissue, liver, kidneys, and bones are the most common storage sites, while blood and lymph serve as the route for dispersal.

Excretion: The site and rate of excretion are major factors affecting the toxicity of a xenobiotic. The kidney is the main excretory organ, followed by the digestive tract and lungs. Xenobiotics may also be expelled in sweat, tears, and milk. The bulk of blood serum is cleaned through the kidneys. Lipid-soluble toxicants are reabsorbed and accumulated in kidney cells. Reduced kidney function slows the removal of toxicants and increases their toxic potential.

Wellbeing: The health of an individual plays a major role in ascertaining the levels and types of potential toxicity, e.g., an individual may have pre-existing kidney or liver disease. Some situations, e.g., pregnancy, are linked with physiological changes in kidney function that could affect toxicity.

Nutritional Status: Diet can be a main factor in shaping who will or will not develop toxicity, e.g., the consumption of fish containing mercury from polluted water can result in Hg toxicity. A contender for mercury toxicity is nutrient selenium. Some vegetables can store Cd from polluted soil.

Circadian Rhythms: Circadian rhythms play a role in toxicity.

2.5.3 Miscellaneous Factors

The presence of other chemicals, at the same time, earlier, or later may decrease toxicity (antagonism), add to toxicity (additivity), or increase toxicity (synergism or potentiation), e.g., antidotes used to counter the effects of poisons function through antagonism (atropine counteracts poisoning by organophosphate insecticides); likewise, alcohol may increase the effects of many antihistamines and sedatives.

2.6 DOSE–RESPONSE RELATIONSHIP

The dose is the amount of a substance administered at one time. It simply means the total amount of a substance administered to an organism, whereas the term dosage includes a characteristic of the organism, typically body weight or surface area. Therefore, dosage is more precise, and can be related to other organisms, e.g., as mg substance/kg body weight. Response, in this context, generally relates to an observation or effect seen following exposure *in vivo* or *in vitro*. Responses can be either adaptive or adverse. The dose–response relationship is a fundamental concept in toxicology which describes the quantitative relationship between the amount of exposure (dose) to a toxicant and the incidence of adverse effects (response). The normal way to determine and represent the dose–response relation is by determining the percentage of the animals or cells in a particular dosage or concentration group which show the response. The response is then plotted against the dosage or concentration, resulting in a typical sigmoid curve. By using probity analysis the data can be plotted as a straight line. When the response is a graded one the actual values measured are plotted against the dosage or concentration giving a similar type of curve. The effects encountered with acute toxicity commonly include mortality or morbidity. From a quantitative standpoint, these effects are measured as the lethal dose 50 (LD_{50}), effective dose 50 (ED_{50}), lethal concentration 50 (LC_{50}), or effective concentration 50 (EC_{50}). The LD_{50} and ED_{50} represent the dose of the material that causes mortality

(LD_{50}) or some other defined effect (ED_{50}) in 50 per cent of a treated population. The LC_{50} and EC_{50} represent the concentration of the material to which the organisms were exposed that causes mortality (LC_{50}) or some other defined effect (EC_{50}) in 50 per cent of an exposed population. LD_{50} and ED_{50} are typically normalized to the weight of the animal (e.g., milligrams chemical/kilograms body weight); whereas, LC_{50} and EC_{50} are typically normalized to the environment in which the organisms were exposed (e.g., milligrams chemical/litres water for aquatic organisms). These measures of acute toxicity are used to assign a level of toxicity to a chemical.

The dose–response line can also be used to assess the threshold dose, the lowest dose of the chemical that would be expected to elicit a response under the conditions in which the assay was performed. The threshold dose is often empirically assessed as being a dose less than the lowest dose at which an effect was measured but higher than the greatest dose at which no effect was detected.

NOAEL (no observable adverse effect level) is an exposure level at which there is no statistically or biologically significant increase in the frequency or severity of adverse effects between the exposed population and its control. It is the highest tested dose or concentration of a substance at which no adverse effect is found in the exposed test organisms where higher doses or concentrations result in adverse effects.

2.7 ENVIRONMENTAL PERSISTENCE

Many abiotic and biotic processes exist in nature that facilitates the elimination of toxic chemicals. Consequently, many chemicals released into the environment pose minimal hazards simply due to their limited life span in the environment. Chemicals that have factually posed environmental hazards (i.e., DDT, PCBs, etc.) resist degradative processes and thus persist in the environment for a very long time. DDT, e.g., in soil has a half-life of 10 years, likewise, PAH has a half-life of 14 months in soil. Continued disposal of persistent chemicals results in their accumulation in the environment, to a level that may pose toxicity and later a hazard even after their disposal in the environment has been discontinued. Both biotic and abiotic processes add to the degradation of chemicals.

Abiotic Degradation: An excess of environmental forces negotiates the structural integrity of chemicals in the environment. Numerous distinct abiotic degradative processes attributable to the impact of light (photolysis) and water (hydrolysis) occur. Light, mainly in the UV range, has the potential of breaking chemical bonds and can add substantially to the degradation of some chemicals. Photolysis generally occurs in the atmosphere or surface water where light intensity is highest. This depends both on the intensity of the light and the capacity of the pollutant molecules to absorb light. Unsaturated aromatic compounds, e.g. polycyclic aromatic hydrocarbons, are extremely susceptible to photolysis owing to their high capacity to absorb light energy. Light energy can facilitate the oxygenation of environmental pollutants via hydrolytic or oxidative processes. Water, together with light or heat energy, can break chemical bonds. Hydrolytic reactions usually result in the insertion of an oxygen atom in the molecule with an equal loss of some component of the molecules. Ester bonds, like those found in organophosphate pesticides (i.e., parathion), are highly prone to hydrolysis, which radically lowers the environmental half-lives of these chemicals. The hydrolytic rate of chemicals is governed by the temperature and pH of the aqueous media. The rate of hydrolysis increases with increasing temperature and extremes in pH.

Biotic Degradation: While many environmental pollutants are prone to abiotic degradative processes, such processes often occur at extremely slow rates. Environmental degradation of chemical contaminants can occur at considerably faster rates through the action of microbes. Microbes (mainly bacteria and fungi) degrade chemicals to obtain energy from these sources. These biotic degradative processes are enzyme-facilitated and usually occur at rates that far exceed abiotic degradation. Biotic degradative processes can lead to complete mineralization of chemicals to water, carbon dioxide, and basic inorganic constituents. Biotic degradation consists of the processes linked with abiotic degradation (i.e., hydrolysis, oxidation) and processes, namely, dehalogenation (removal of chlorine atoms), ring cleavage (scission of ringed structures), and dealkylation (removal of carbon chains). In addition, bioremediation, the process by which microbes facilitate the removal of environmental contaminants, also occurs.

Non-Degradative Elimination Processes: Many processes are operative in the environment that contribute to the regional removal of a pollutant by altering its distribution. Pollutants with sufficiently high vapour pressure can evaporate from polluted terrestrial or aquatic compartments and travel through the atmosphere to new sites. Such processes of global distillation are largely accountable for the global distribution of relatively volatile organochlorine pesticides such as lindane and hexachlorobenzene. Climbing by wind and upper atmospheric currents of pollutant particles or dust onto which the contaminants are sorbed also redistributes the pollutant. Sorption of pollutants to suspended solids in an aquatic environment corresponding to sedimentation can result in the removal of pollutants from the water column and their relocation into bottom sediments. Sediment sorption of pollutants considerably reduces their bioavailability, since the inclination of a lipophilic chemical to detach from sediments to organisms is substantially less than its tendency to partition from water to an organism. Water-soluble pollutants can be separated and rearranged through runoff and soil percolation. For example, the herbicide atrazine, one of the most widely used pesticides to control broadleaf and weed grasses in both agriculture and landscaping, is universal in surface water owing to its extensive use. Furthermore, atrazine has the tendency to travel into groundwater owing to its comparatively high water solubility and low predilection to reabsorb soil particles.

Bioaccumulation: Environmental persistence alone does not render a chemical in the environment. If the chemical cannot enter the body of organisms, then it would pose no threat of toxicity. Once absorbed, the chemical must accumulate in the body to sufficient levels to produce toxicity. Bioaccumulation is defined as the process by which organisms collect chemicals both directly from the abiotic environment (i.e., water, air, soil) and from dietary sources (trophic transfer). Environmental chemicals are generally taken up by organisms by passive diffusion. Primary sites of uptake include membranes of the lungs, gills, and gastrointestinal tract. While skin and linked structures, e.g., scales, feathers, fur, etc. offer a protective barrier against environmental exploitation, considerable dermal uptake of some chemicals can occur. The reason that the chemicals must cross the lipid bilayer of membranes to enter the body, the bio-accumulation potential of chemicals, is certainly interrelated with lipophilicity or lipid solubility.

The aquatic environment is the main site where lipophilic chemicals traverse the barrier between the abiotic environment and the biota. This is because (1) lakes, rivers, and oceans act as sinks for these chemicals, and (2) aquatic organisms pass a great amount of water across their respiratory membranes (i.e., gills) enabling the removal of chemicals from the water. Aquatic organisms can bio-accumulate lipophilic chemicals and reach a body concentration that is greater than the concentration of the chemical found in the environment. The magnitude to which aquatic organisms accumulate xenobiotics from the environment basically depends on the lipid content of the organism, because the body lipids serve as the primary site of retention of the chemicals.

Chemicals can be transported along food chains from prey to predator organisms (trophic transfer). For highly lipophilic chemicals, this transfer increases the concentrations of the chemical with each progressive link in the food chain (biomagnification). A chemical that bio-accumulates by a factor of two regardless of whether the source of the pollutant is water or food would have the potential to amplify at each trophic level resulting in a higher level in birds of prey relative to that found in the abiotic environment. Bioaccumulation is usually much greater from water than from food, and it is not likely that an organism would gather a chemical to the same degree from both sources. The food-chain transfer of DDT was accountable for the decline in many bird-eating raptor populations that resulted in banning the use of this pesticide. Bioaccumulation can lead to a delayed onset of toxicity since the toxicant may primarily be seized in lipid layers but is mobilized to the target sites of toxicity when these lipid stores are used. For example, lipid stocks are often mobilized in preparation for reproduction. Loss of lipid can result in the release of lipophilic toxicants, rendering them accessible for toxic action and in turn in the death of adult organisms when they approach reproductive maturity. Lipophilic chemicals can also be transmitted to offspring in lipids linked with the yolk of oviparous organisms or the milk of mammals, causing toxicity in offspring that was not marked in the parental organisms. The propensity for an environmental contaminant to bio-accumulate is governed by several factors. The first consideration is environmental persistence. The level to which a chemical bio-accumulates is influenced by the concentration present in the environment. Pollutants that are freely removed from the environment will usually not be available to bio-accumulate. An exception would be cases where the pollutant is continuously introduced into the environment.

Therefore, lipophilicity is a major factor in the bioaccumulation potential of a chemical. However, lipophilic chemicals have a greater tendency to sorb to sediments, thus rendering them less available to bio-accumulate. For example, fish from oligotrophic lakes, with low suspended solid levels, have shown to accumulate more DDT than fish from eutrophic lakes that have high suspended solid contents.

Once absorbed by the organism, the fate of the contaminant will affect its bio-accumulation. Chemicals that are readily bio-transformed are rendered more water-soluble and less lipid-soluble. The bio-transformed chemical is thus less likely to be seized in lipid compartments and more likely to be removed from the body. Chemicals that are liable to biotransform, bio-accumulate are much less than would be expected based on lipophilicity. Differences in chemical removal rates add to species differences in bioaccumulation.

2.7.1 Uptake, Disposition, and Transport Processes

Diffusion and Related Processes: To enter and reach the target site for producing any effect, a foreign substance needs to cross several barriers, including various cells and their membranes. Most toxic substances pass passively through membrane by diffusion. Small water-soluble molecules pass through aqueous channels, or fat-soluble ones, through suspension into and diffusion through the lipid component of the membrane, e.g., ethanol, both water- and fat-soluble, diffuses quickly through cell membranes.

In non-ionized, fat-soluble form, weak acids and bases freely pass membranes while ionized forms, being polar, cannot pass through. The degree of ionization of substances depends on the pH. If the pH gradient occurs across a membrane, they will gather on one side. Urinary excretion of weak acids and bases depends extremely on urinary pH. Foetal or embryonic pH is slightly higher than maternal pH, triggering a slight massing of weak acids in the foetus or embryo.

In the case of facilitated diffusion, passage of a substance is usually facilitated by carriers in the membrane. Facilitated diffusion is identical to enzyme processes as it is protein-mediated and highly selective. Other substances may prevent facilitated transportation of xenobiotics. On the other hand, some substances actively transport across cell membranes, which is mediated by carrier proteins in a procedure comparable to that of enzymes. Active transport and facilitated diffusion are similar, but it occurs against a concentration gradient, needs energy input, and a metabolic inhibitor can block it. Most pollutants are not transported actively, except active tubular secretion and reabsorption of acid metabolites in kidneys. Phagocytosis is a process in which specialized cells, such as macrophages, engulf particles for consequent digestion. This process is important for the removal of particles in the alveoli. During breathing, and movement of blood, lymph, or urine, substances are transferred through the body along with the movement of air in the respiratory system in the process of bulk flow.

Owing to hydrostatic or osmotic pressure, water flows in bulk through pores in the endothelium. Any small solute is filtered together with the water. Filtration occurs to some extent in the capillary bed of tissues and plays an important role in the formation of primary urine in kidney glomeruli.

Absorption: The uptake of a substance from the environment by an organism takes place via absorption. It is furthermore of different types, namely pulmonary absorption, in which the lung is the main route of deposition, and absorption of airborne particles, gases, vapours, and aerosols. A substantial part of uptake occurs in the nose and respiratory tree for highly water-soluble gases and vapours. Deposition and absorption of less soluble substances mainly takes place in the lung alveoli, as it has a very large surface area. Moreover, the diffusion barrier is very small, with only two thin cell layers. As a result, the lungs are efficient not only in the exchange of O_2 and CO_2 but also of other gases and vapours. Generally, diffusion across the alveolar wall is so quick that it does not limit the uptake. Instead, the absorption rate depends on:

- Flow (pulmonary ventilation, cardiac output)
- Solubility (blood:air partition coefficient)
- Metabolic elimination

The importance of these factors in pulmonary absorption differs significantly for different substances. Physical activity results in:

- Increased pulmonary ventilation
- Cardiac output
- Decreased liver blood flow, and hence, biotransformation rate

This increases the pulmonary absorption for many inhaled substances
Another type of absorption is percutaneous absorption. The skin acts as an effective barrier. Besides its thermoregulatory role, it protects organisms from micro-organisms, UV radiation, and other harmful agents, and also against excessive water loss.

Furthermore, the keratin layer is highly resistant to diffusion for most substances. However, significant dermal absorption results in toxicity for highly toxic, fat-soluble substances, e.g. organophosphorus insecticides and organic solvents. Significant absorption occurs after exposure to liquid substances. Percutaneous absorption of vapour is important for solvents with very low vapour pressure and high affinity for water and skin. Gastrointestinal absorption occurs after unintentional or intentional ingestion. Larger particles firstly inhaled and accumulated in the respiratory tract may be absorbed after mucociliary transport to the pharynx. Almost all soluble substances are effectively absorbed in the GI tract, e.g., the low pH of the gut may enable absorption of metals.

In toxicity testing and other experiments, special routes of administration include intravenous (IV), subcutaneous (SC), intraperitoneal (IP), and intramuscular (IM) injections. Generally, substances are totally absorbed at a higher rate and by these routes, particularly after IV injection.

Distribution: Distribution of a substance within an organism is an active process which depends on uptake and removal rates, and blood flow to different tissues and their affinities for the substance. Water-soluble, small, uncharged molecules, univalent cations, and most anions diffuse easily and will ultimately distribute evenly in the body. The distribution capacity is the amount of a substance in the body at a given time, divided by the concentration in blood, plasma, or serum at that time. Accumulation is the gathering of a substance in a tissue or organ to a higher level than in blood or plasma. It may also refer to a steady build-up over the time in the organism. Many xenobiotics are highly fat-soluble and have a tendency to amass in adipose tissue, while others have a special affinity for bone, e.g., calcium in bone may be exchanged for cations of lead, strontium, barium, and radium, and hydroxyl groups in bone may be exchanged for fluoride.

Blood vessels in the brain, testes, and placenta have special anatomical features that prevent the passage of large molecules like proteins. These characteristics often are referred to as blood–brain, blood–testes, and blood–placenta barriers, which gives the incorrect impression that they avert the passing of any substances. These barriers have little or no significance for xenobiotics that can diffuse through cell membrane. Substances may be bound to RBCs or plasma components, or occur open in blood. CO, As, organic Hg ,and hexavalent Cr have a high affinity for RBCs, while inorganic Hg and trivalent Cr show a preference for plasma proteins. Numerous other substances also attach to plasma proteins. Only the unbound fraction is accessible for filtration or diffusion in removing organs.

Elimination: Elimination is the removal of a substance from the body, which may involve excretion from the body or transformation to other substances not caught by a particular method of measurement. The rate of disappearance may be expressed by the removal rate constant, biological half-time, or clearance. A concentration–time curve depicts the concentration of a substance in blood (or plasma) versus time and describes the uptake and disposition of a xenobiotic. The area under the curve (AUC) is the integral of the concentration in blood (plasma) over the time. When metabolic saturation and other non-linear processes are lacking, the AUC is proportional to the absorbed amount of a substance.

The biological half-time (or half-life) is the time required after the end of exposure to decrease the amount in the organism to one-half. Since it is difficult to evaluate the total amount of a substance, measurements such as the concentration in blood (plasma) are used. The half-time is used with caution, as it may change with dose and length of exposure. Moreover, many substances have complex decay curves with several half-times.

Bioavailability is the fraction of an administered dose flowing into the systemic circulation. Without pre-systemic clearance, or first-pass metabolism, the fraction is one. In oral exposure, pre-systemic clearance is due to metabolism within the gastrointestinal content, gut wall, or liver. First-pass metabolism reduces the systemic absorption of a substance and instead increases the absorption of metabolites, which may lead to a different toxicity pattern. Clearance is the plasma per unit time entirely cleared of a substance, while intrinsic clearance is the capacity of endogenous enzymes to transform a substance, and is expressed as volume per unit time. If the intrinsic clearance in an organ is much lower than the blood flow, the metabolism is said to be capacity limited. On the other hand, if the intrinsic clearance is much higher than the blood flow, the metabolism is flow limited.

Excretion: Excretion is the escape of a substance and biotransformation products from an organism. The kidneys are the most important excretory organs. Substances, especially acids with high molecular weights, are expelled with bile. A fraction of substances expelled from the biliary system may be reabsorbed in the intestine. Some substances, such as organic solvents and breakdown products such as acetone, are volatile enough that a substantial fraction may be expelled by exhalation after inhalation. Small water-soluble molecules and fat-soluble ones are freely secreted to the foetus via the placenta, and into milk in mammals. Offspring may be exposed via the mother during pregnancy as well as during lactation. Water-soluble compounds to some extent are expelled in sweat and saliva. These routes are usually of minor importance. However, as a large volume of saliva is produced and swallowed, saliva excretion may add to reabsorption of the compound. Some metals such as mercury are excreted by binding finally to the sulphydryl groups of the keratin in the hair.

Xenobiotic: A xenobiotic is a foreign chemical substance present within an organism that is not naturally produced by or expected to be present in that organism. It includes substances which are present at higher concentrations than the normal concentration.

Drugs such as antibiotics are xenobiotics for humans as these are not produced by the human body, nor they are part of a normal human diet. The term xenobiotic is derived from the Greek words *xenos* = foreigner or stranger and *bios* = life. Xenobiotics are catalysed by enzymes and form metabolite(s) with modified structures. A number of xenobiotics metabolism sites are found in the body, however, the liver is the major site of metabolism. Xenobiotics can be exogenous, i.e. the foreign particles which are not ingested normally or utilized but that gain entry through dietary foodstuffs, or in the form of some medicines/drugs used for therapeutic purposes, or are inhaled through the air, for example, drugs, pollutants, insecticides, etc. These can be endogenous, i.e., they are not foreign substances but have a similar effect as exogenous xenobiotics. These are synthesized in the body or are produced as the metabolites of various body processes, e.g. bilirubin, bile, steroids, and certain fatty acids. Based on their characteristics, xenobiotics can be classified as shown in Figure 2.1.

Usually, xenobiotics are lipid-soluble and they need to be transformed to water-soluble molecules to facilitate their elimination. The process involved in their conversion is biotransformation. Most biotransformation takes place in the liver and involves two phases (see Figure 2.3):

Phase 1: Oxidation of xenobiotics by enzymes in the liver which activates the compound at a higher level than its parent

Phase 2: The oxidized compounds are conjugated with endogenous substances, which make the molecule polar and enable elimination

The main sources of xenobiotic are:

- **Direct sources**: Pharmaceutical industries (phenols), petroleum effluent (hydrocarbons), plastics, paints, dyes, pesticides, insecticides, paper, and pulp effluent
- **Indirect sources**: hospital discharge, pesticide or herbicide residues
- **Products and processes**: Product of the reactions of any processes – on a domestic or industrial scale
- **Intentional and unintentional causes**: Chemicals used in paper and pulp industries accidentally released into the environment

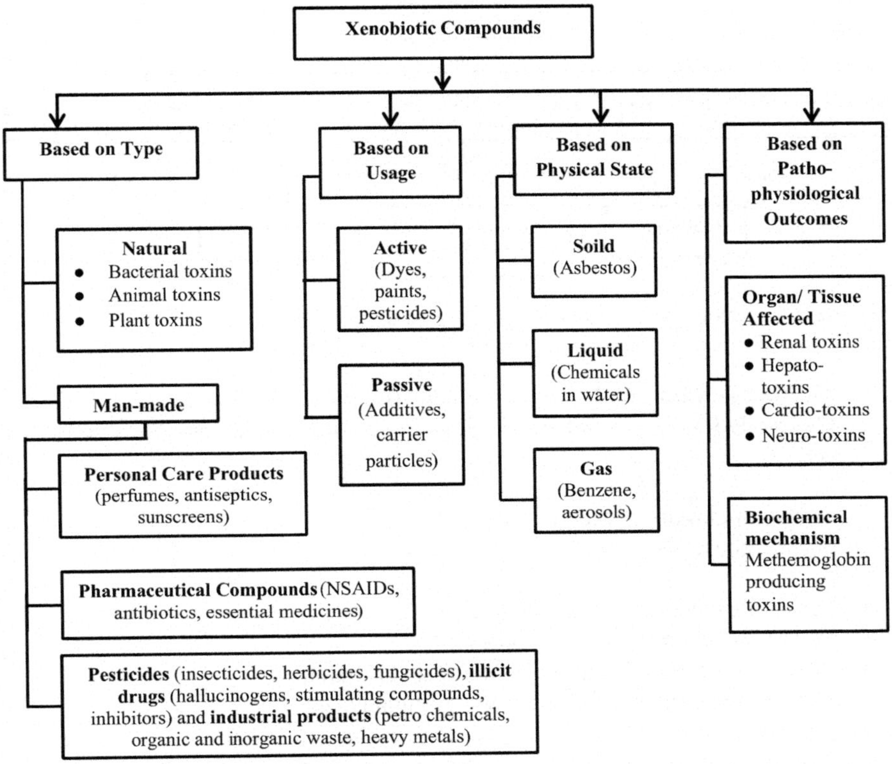

FIGURE 2.1 Classification of xenobiotic compounds.

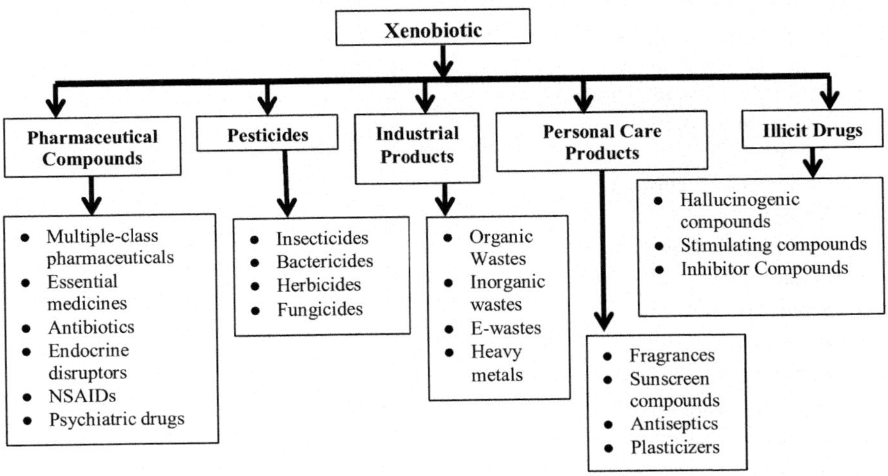

FIGURE 2.2 Sources of xenobiotic compounds.

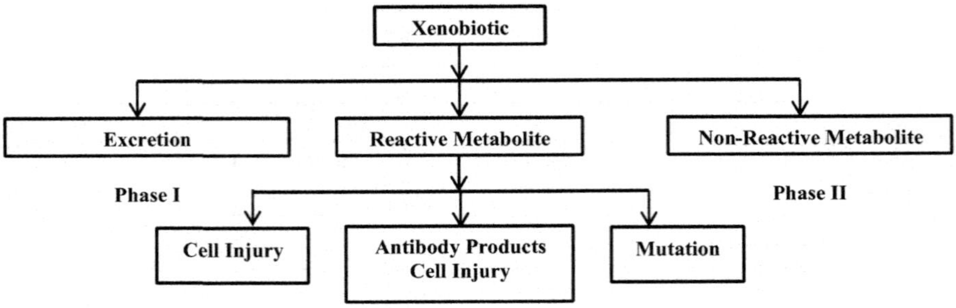

FIGURE 2.3 Toxic effects of xenobiotic compounds.

- **Shifting and immobile**: Cars and industries
- **Controlled and uncontrolled**: large industries and automobiles/household activity

Another way of classifying xenobiotics is depicted in Figure 2.2.
Toxicants are converted to:

- Less toxic chemicals (metabolite), or
- More toxic chemicals (metabolite), or
- Chemicals with different types of effects or toxicities
- They include drugs, food additives, pollutants, etc.

Understanding how xenobiotics are handled at the cellular level is important in learning how to cope with a chemical attack.

Sites of Metabolism: These include wherever appropriate enzymes are present (plasma, kidney, lung, gut wall, and liver), such as:

- Multiple-class pharmaceuticals
- Essential medicines
- Antibiotics
- Endocrine disruptors
- NSAIDs
- Psychiatric drugs
- Insecticides
- Bactericides
- Herbicides
- Fungicides
- Organic wastes
- Inorganic wastes
- E-wastes
- Heavy metals
- Hallucinogenic compounds

- Stimulating compounds
- Inhibitor compounds
- Fragrances
- Sunscreen compounds
- Antiseptics
- Plasticizers

Elimination of toxins accumulated in the body is important. The food humans consume is broken and later taken to the liver which detoxifies it before it enters the circulatory system. The liver is ideally placed to intercept natural ingested toxins (bypassed by injections, etc.) and has a major role in biotransformation.

Scientific developments in the 20th century have led to the discovery of several compounds used to improve daily life (antibiotics, pesticides, dyes, PCPs, additives, etc.) that do not essentially occur naturally in the environment or whose naturally occurring concentration is considerably different from those produced by anthropogenic activity. Xenobiotics are difficult to degrade due to their complex structures, thus they can amass in living organisms. Their incomplete degradation can result in even worse threats (Figure 2.3). To reduce the concentration of xenobiotics in the environment, it is important to know where these compounds are coming from.

Xenobiotic metabolism is the process through which the body responds to a particular toxicant and the factors affecting its metabolism include (see Figure 2.5):

- Age (reduced in the aged and children)
- Gender (women are more sensitive to ethanol)
- Species and genetic makeup
- Clinical or physiological conditions
- Foetuses and neonates have lower drug-metabolizing capacity compared to adults
- Sensitivity to the drug
- Long duration of action (see Figure 2.4)

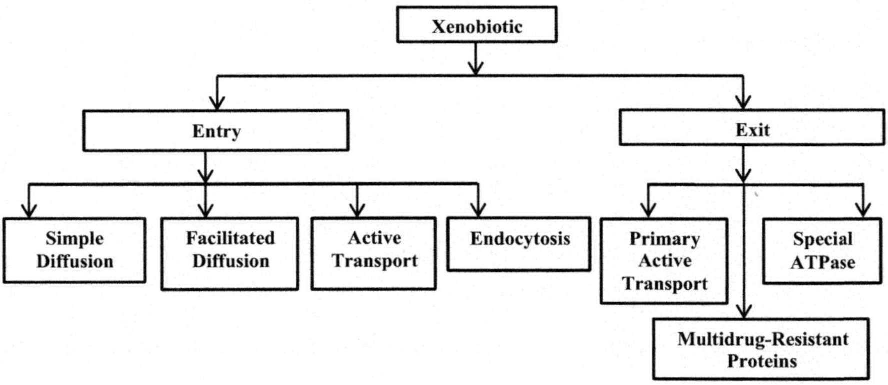

FIGURE 2.4 Entry and exit of xenobiotic compounds.

Genetics

- The activity of xenobiotic metabolizing enzymes can vary between individuals
- The population can be divided into rapid metabolizers and slow metabolizers

Clinical or Physiological Condition

- Liver diseases: Reduced drug metabolism
  - Cirrhosis
  - Acute alcohol exposure, chronic ethanol exposure, alcoholic cirrhosis
- Endocrine disorders: Diabetes
- The body removes xenobiotics by xenobiotic metabolism
- This involves deactivation and excretion of xenobiotics, and happens mostly in the liver
- Excretion routes are urine, faeces, breath, and sweat
- Hepatic enzymes are responsible for the metabolism of xenobiotics by:
  - First activating them (oxidation, reduction, hydrolysis, and/or hydration of the xenobiotic)
  - Secondly, conjugating the active secondary metabolite with glucuronic acid, sulphuric acid, followed by excretion in the bile or urine

Biotransformation: Biotransformation is a route which involves metabolic conversion of foreign compounds (xenobiotics) in the body (Figure 2.5). It is often referred to as metabolism of xenobiotics. It converts lipid-soluble (lipophilic) xenobiotics into large water-soluble metabolites that can be efficiently excreted. The liver is the main biotransformation location. All xenobiotics taken up from the intestine are transferred to the liver by a single blood vessel (vena porta). If taken up in small quantities, a foreign substance may be completely metabolized in the liver before reaching the general circulation and other organs. In most cases it reduces the toxicity of xenobiotics. Inhaled xenobiotics are distributed via the general circulation to the liver. In that case, only a fraction of the dose is metabolized in the liver before reaching other organs. The extrahepatic metabolic sites of biotransformation are:

- Intestinal wall
  - Sulphate conjugation
  - Esterase and lipase
- Lungs, placenta, kidney, brain, skin, and adrenal glands

The basic purposes of biotransformation are that

- It facilitates excretion of lipophilic compounds to hydrophilic compounds
- Detoxification/inactivation or conversion of toxic compounds into less toxic compounds
- Metabolic activation/conversion of chemicals to more active forms

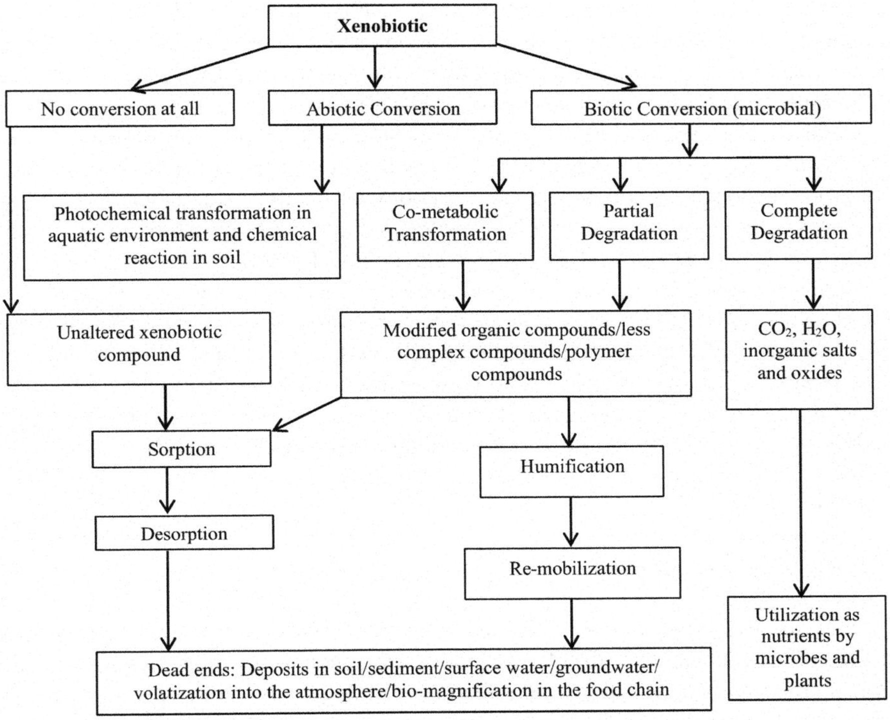

FIGURE 2.5 Mechanism of action of a xenobiotic compound.

Nearly 30 different enzymes catalyse reactions in xenobiotic metabolism. Liver cells contain several enzymes that oxidize xenobiotics. This oxidation generally activates the compound – which turns out to be more reactive than the parent molecule. In many cases, the oxidized metabolite is further metabolized by other enzymes in a second phase. These enzymes conjugate the metabolite with an endogenous substrate, so that the molecule becomes larger and more polar, which enables elimination.

Enzymes involved in xenobiotic metabolism are also present in other organs such as the lungs and kidneys. These enzymes play a precise and qualitatively important role in the metabolism of certain xenobiotics in these organs. Metabolites formed in one organ may be further metabolized in a second organ. Bacteria in the intestine also participate in biotransformation.

Metabolites of xenobiotics can be excreted by the kidneys or via bile. They can also be exhaled via lungs, or attached to endogenous molecules in the body.

The relationship between biotransformation and toxicity is intricate. Biotransformation can be seen as an essential process for survival. It safeguards the organism against toxicity by inhibiting gathering of harmful substances in the body. However, reactive intermediary metabolites may be formed in biotransformation, and these are possibly detrimental. This is called metabolic activation. Therefore, biotransformation may also induce toxicity. Oxidized, intermediary metabolites that are not conjugated can bind to and damage cellular structures. If, for example, a xenobiotic metabolite binds to DNA, a mutation can be induced. If the biotransformation system is burdened, a huge amount of damage to important proteins or lipid membranes may occur, which can cause cell death.

Metabolism is interchangeably used with biotransformation. It connotes chemical breakdown or synthesis reactions catalysed by enzymes in the body. Nutrients from food, endogenous compounds, and xenobiotics are all broken down in the body. Metabolic activation means that a less reactive compound is transformed to a more reactive molecule. This usually occurs during a phase 1 reaction.

Metabolic inactivation means that an active or toxic molecule is converted into a less active metabolite, which generally occurs during phase 2 reactions. In some cases, an inactivated metabolite might be reactivated, e.g., by enzymatic cleavage.

A phase 1 reaction is the first step in xenobiotic metabolism in which the compound is oxidized. Oxidation generally makes the compound more water-soluble and enables further reactions. Phase 1, or functionalization, involves:

- Oxidation
- Reduction
- Hydrolytic reaction

Cytochrome P450 enzymes are a group of enzymes that specially oxidize xenobiotics in Phase 1 reactions. The different enzymes are specialized for managing specific groups of xenobiotics with certain features. Endogenous molecules also act as substrates. Cytochrome P450 enzymes are induced by xenobiotics in a specific way. Obtaining induction data on cytochrome P450 can reveal details about the nature of previous exposures.

A phase 2 reaction is the second step in xenobiotic metabolism and involves conjugation and has the purpose of introducing highly polar conjugates, namely, glucuronic acid and sulphate.

This generally means that the oxidized compound is coupled to an endogenous molecule. This reaction further increases the water solubility. Many conjugated metabolites are vigorously expelled via kidneys. Transferases are a group of enzymes that catalyse phase 2 reactions. They conjugate xenobiotics with endogenous compounds such as amino acids, glucuronic acid, glutathione, and sulphate.

Glutathione is an endogenous molecule, a tripeptide that is conjugated with xenobiotics in phase 2 reactions. Glutathione is present in all cells with high concentrations in liver cells, and usually safeguards from activated xenobiotics. When glutathione is drained, toxic reactions between activated xenobiotic metabolites and proteins, lipids, or DNA may occur.

Induction means involvement of enzymes in biotransformation which increases (activity or amount) as a response to xenobiotic exposure. In some cases, in a few days enzyme activity can be amplified several fold. Induction is often balanced so that both phase 1 and phase 2 reactions are amplified instantly, which may lead to faster biotransformation and can explain tolerance. On the contrary, unbalanced induction may increase toxicity. Inhibition of biotransformation can occur if two xenobiotics are metabolized by the same enzyme. The two substrates have to compete, and usually one of the substrates is chosen. In that case, the second substrate is not metabolized, or slowly metabolized. As with induction, inhibition may increase and decrease toxicity.

Oxygen activation can be triggered by metabolites of certain xenobiotics. They may auto-oxidize under the creation of stimulated oxygen species. The oxygen-derived species, which include superoxide, hydrogen peroxide, and the hydroxyl radical, may damage DNA, lipids, and proteins in cells. Oxygen activation is also involved in inflammatory processes. Genetic adaptability between individuals is seen in many genes coding for phase 1 and phase 2 enzymes. Genetic variability may explain why certain individuals are more prone to the toxic effects of xenobiotics than others. While comparing the two phases, it can be seen that a phase 1 reaction involves hydrolysis, oxidation, and reduction where the increase in hydrophilicity is small, exposes functional groups, and as a consequence may result in metabolic activation; whereas phase 2 reactions involve conjugation, where the increase in hydrophilicity is large and polar compounds are added to the functional groups and eventually facilitate excretion.

2.8 TOXICITY TESTING

Acute Toxicity Test: Acute toxicity tests are those designed to determine the effects which occur within a short period after dosing. These tests can determine a dose–response relationship and the LD_{50} value if required. The exact conduct of toxicity studies varies depending on the compound, its eventual use, and the particular regulations to be satisfied. Usually, at least four dosages are used which may be in logarithmic progression, especially if no range-finding studies have been done. However, the traditional LD_{50} determinations are now no longer required by most regulatory authorities. The information to be gained from an acute toxicity test is the nature of the dose–response relationship and observations on the toxic effects and time to death, if any of the animals die. At the end of the toxicity test the surviving animals are killed and undergo a post-mortem with a pathological examination of tissues. Animals dying during the study should also undergo a post-mortem.

Sub-Chronic Toxicity Test: Following acute toxicity tests, sub-chronic toxicity tests are usually carried out. These involve exposing the animals to the substance under test for a prolonged period, usually 28 or 90 days. The exposure is frequent and usually daily. These tests provide information on the target organs affected by the compound and the major toxic effects. Toxic effects which have a slow onset can be detected, and reversible and adaptive responses may become apparent during the test. Measurements of levels of the compound in blood and tissues can be made and this information correlated with any toxic effects seen. At the end of the study a pathological examination is carried out and during the study clinical chemical measurements should indicate the development of any pathological lesions. Data derived from sub-chronic toxicity studies also help in designing chronic toxicity studies.

Chronic Toxicity Test: This test involves exposure of animals to the compounds of interest for at least 12 months to 2 years in rodents (about 50 per cent lifespan) and

6–12 months in non-rodents. Chronic toxicity tests may be combined with *in vivo* carcinogenicity tests, in which case the exposure of rodents will be for the lifetime, and satellite groups may be used for interim chronic toxicity information. As with sub-chronic toxicity tests, the chronic toxicity test will terminate with a pathological examination and there may also be clinical chemical measurements made at intervals throughout. These clinical chemical measurements may indicate the development of pathological changes which can then be detected at post-mortem. Chronic toxicity studies are important for drugs administered over long periods of time, for food additives to which humans may be exposed for their entire lifetimes, and for environmental and industrial chemicals where exposure to low levels for long periods may take place.

Heavy Metals Toxicity Test: Heavy metals toxicity is tested simply by a blood test in which a small blood sample is collected to check the concentration of heavy metals present in the body. Additional tests include kidney function tests, liver function tests, urine and hair analyses, fingernail analyses, electrocardiograms, and X-rays

BIBLIOGRAPHY

Calow, P. (1998) *Textbook of Ecotoxicology*. Oxford, UK: Blackwell Science.

Cockerham, L. G. and Shane, B. S. (1994) *Basic Environmental Toxicology*. Boca Raton: CRC Press.

Hayes, A. W. (1989) *Principles and Methods of Toxicology*. New York, NY: Raven Press.

Hodgson, E. (2010) *A Textbook of Modern Toxicology*. North Carolina: John Wiley and Sons.

Hughes, W. W. (2005) *Essentials of Environmental Toxicology*. USA: Taylor & Francis.

Johnson, B. L. (1983) Occupational toxicology: NIOSH perspective. *Journal of the American College of Toxicology* 2(1): 43–50.

Mathew, B. B., Singh, H., Biju, V. G., and Krishnamurthy, N. B. (2017) Classification, source, and effect of environmental pollutants and their biodegradation. *Journal of Environmental Pathology, Toxicology and Oncology* 36: 55–71.

Timbrell, J. (1999) *Principles of Biochemical Toxicology*. London: Taylor & Francis.

Williams, P. L. and Burson, J. L. (1985) *Industrial Toxicology: Safety and Health Applications in the Workplace*. New York: Lifetime Learning Publications.

Wright, D. A. and Welbourn, P. (2002) *Environmental Toxicology*. Cambridge: Cambridge University Press.

Zakrewski, S. F. (2002) *Environmental Toxicology*. New York: Oxford University Press.

3 Heavy Metal Toxicity
An Overview

Heavy metal is always going to be there. At its core, it's all about a primitive connection we all need to keep in our lives.

Rob Halford

After going through this chapter, the reader will learn about: sources of heavy metals, the basic mechanism of heavy metal contamination, thermodynamics of heavy metals, the transmission of heavy metals from the soil to plants, heavy metals in biota and food chains, and heavy metals toxicity.

FACT FILE

- The industrial and anthropogenic activities of the last century have considerably increased human exposure to heavy metals (HMs)
- Contamination of water and air by toxic metals is an environmental concern for hundreds of millions of people who are affected by this around the world
- Food contaminated with HMs is a matter of concern for human and animal health
- HMs such as mercury, lead, and cadmium are of the highest concern due to their ability to travel long distances in the atmosphere
- Hg exposure can harm the brain, heart, kidneys, lungs, and immune system of humans of all ages
- Hg accumulation in fish may harm fish and other animals that consume them
- Birds and mammals that consume fish are more exposed to Hg than other animals living in aquatic ecosystems
- According to the United Nations Environment Programme (UNEP), Pb causes neurodevelopmental defects in children, even at low levels of exposure
- Pb bio-accumulates in most organisms and is toxic to plants, animals, and microbes
- Young fish are more susceptible to Pb poisoning than mature fish or eggs
- Pb toxicity causes spinal deformity and blackening of the caudal region in fish
- Cd is toxic to humans and exposure can cause pulmonary irritation, kidney disease, bone weakness, and possibly lung, prostate, and kidney cancers

DOI: 10.1201/9781003343721-3

- Food and cigarette smoke are potential sources of cadmium exposure for humans
- Cd is toxic to plants, animals, and micro-organisms and, according to UNEP, Cd amasses mainly in the kidneys and liver of vertebrates, aquatic invertebrates, and algae
- The acute toxic effects of Cd on fish, birds, and other animals may include death or foetal malformations
- Cd affects plants, resulting in decreased growth rates and even death
- Cr often accumulates in aquatic life, with risks caused by the eating of fish exposed to high levels of Cr
- In 1932, sewage containing Hg was released by Chisso's chemicals works into Minimata Bay in Japan, which accumulated in sea creatures, leading ultimately to Hg poisoning in the population
- In 1952, the first incident of Hg poisoning appeared in the population of Minimata Bay in Japan, caused by the consumption of fish polluted with Hg, with over 500 fatalities

3.1 INTRODUCTION

Due to their toxic and bio-accumulative nature and permanence in the environment, heavy metals (HMs) are identified as environmental pollutants. Their natural sources include the weathering of rocks and volcanic eruptions, while mining, industrial, and farming activities are anthropogenic causes. Mining and industrial activities linked with the extraction of minerals and their subsequent uses for economic development have increased the mobilization of HMs in the environment in addition to disrupting the respective biogeochemical cycles. Chemically, metals are the "elements, which conduct electricity, have a metallic shine, are malleable and ductile, form cations, and have basic oxides".

Elements having comparatively high density as compared to water are known as HMs. Csuros and Csuros defined heavy metal as "a metal with specific gravity greater than 5". According to Duffus the term "HMs" is often used for the metals and metalloids that are linked with pollution and probable toxicity or eco-toxicity. With the concept that weight and toxicity are interrelated, HMs also include metalloids. Though HMs are naturally occurring elements present throughout the Earth's crust, most environmental pollution and human exposure are the consequence of man-made activities. In recent years, there has been growing environmental and public health concern linked with environmental pollution caused by HMs. Human exposure to HMs has increased intensely in response to an exponential increase in their use in numerous industrial, farming, household, and scientific applications. Environmental pollution is quite perceptible in point source areas, e.g., quarrying areas, foundries and smelters, and other metal-based techniques.

Some metals and metalloids are essential for life and play significant physiological and biochemical roles in the body as they may be the part of biomolecules, namely enzymes, which catalyse biochemical reactions in the body. The deficiency or excess of these micro-nutrients can disturb the metabolism, resulting in various deficiency diseases. Metals, e.g., cobalt (Co), copper (Cu), chromium (Cr), iron (Fe), magnesium (Mg), manganese (Mn), molybdenum (Mo), nickel (Ni), selenium (Se), and zinc

(Zn), are vital nutrients that are important for various biochemical and physiological processes. Owing to their presence in trace concentrations (ppb range to less than 10 ppm) in various ecological matrices, HMs are also regarded as trace elements. Their bioavailability is governed by physical factors, e.g., temperature, phase relationship, adsorption, and sequestration. Chemical factors, namely thermodynamic equilibrium, kinetics, lipid solubility, and water partition coefficients, also affect their bioavailability. Biological factors, e.g., features of species, trophic interactions, and biochemical/physiological adaptation, also play a significant role in their bioavailability.

Essential HMs have biochemical and physiological roles both in plants and animals. Besides acting as the key ingredients of numerous enzymes, HMs play an important role in many oxidation-reduction reactions. For example, Cu serves as an important co-factor for several oxidative stress-related enzymes including catalase, peroxidase, etc. Thus, it is a vital nutrient that is integrated into a number of metallo-enzymes involved in the formation of haemoglobin, carbohydrate metabolism, catecholamine biosynthesis, and cross-linking of collagen, elastin, and hair keratin. The ability of Cu to cycle between an oxidized and reduced state is used by cupro-enzymes involved in redox reactions. This property of Cu makes it possibly toxic because changes between Cu (II) and Cu (I) generate superoxide and hydroxyl radicals. Excessive exposure to Cu is linked to cellular damage leading to Wilson disease. Like Cu, numerous other vital elements are required for biologic functioning. For metals like Cu and Cr, there is a very thin range of concentrations between useful and noxious effects. Metals, such as, aluminium (Al), antinomy (Sb), arsenic (As), barium (Ba), beryllium (Be), bismuth (Bi), cadmium (Cd), gallium (Ga), germanium (Ge), gold (Au), indium (In), lead (Pb), lithium (Li), mercury (Hg), nickel (Ni), platinum (Pt), silver (Ag), strontium (Sr), tellurium (Te), thallium (Tl), tin (Sn), titanium (Ti), vanadium (V), and uranium (U) have no conventional biological role and are believed to be the non-essential metals.

Due to their role in causing toxicity, these elements are among the main metals having immense public health significance. However, all of them are known to be systemic toxicants, which can induce multiple organ damage, even at lower exposures. In biological systems, HMs are reported to affect the cellular organelles and components, e.g., cell membrane, mitochondrial, lysosome, endoplasmic reticulum, nuclei, and some enzymes involved in metabolism, decontamination, and repair of damage. Metal ions interact with cell components, namely DNA and nuclear proteins, causing DNA damage and conformational changes that may alter cell cycle modulation, causing carcinogenesis or apoptosis. Studies have revealed that reactive oxygen species (ROS) production and oxidative stress play an important role in the toxicity and carcinogenicity of metals like As, Cd, Cr, Pb, and Hg. Depending on the dose and duration of exposure, some HMs are also harmful for living organisms.

HM-induced toxicity and carcinogenicity have many mechanistic aspects, some of which have not been clearly explained or understood. Yet, each metal is known to have distinct features and physico-chemical properties that confer to its definite toxicological mechanisms of action. HM toxicity is basically the deposition of HMs, in lethal amounts, in body soft tissues. Symptoms and physical manifestations linked with heavy metal toxicity differ according to the quantity of metal deposited. Many of the HMs, such as zinc, copper, chromium, iron, and manganese, are essential for functioning of the body but only in trace amounts. However, if these metals are stored in the body in concentrations high enough to cause toxicity, then considerable harm may occur.

HMs are considered lethal for the environment. Though non-essential HMs are lethal to plants, animals, and humans even at very low concentrations, the essential HMs may also result in adverse health conditions at high concentrations. Noxious metals which are both permanent and bio-accumulative are more harmful. The pollution of aquatic and terrestrial ecosystems with toxic metals is not only an environmental problem but also a matter of public health concern. Being permanent contaminants, HMs amass in the environment and subsequently pollute food chains. An increase of possibly toxic HMs in biota poses threats to human health, and aquatic and terrestrial food chains. It is very important to evaluate and observe the concentration of potentially toxic HMs and metalloids in the environment.

Concentrations of HMs above threshold levels affect the microbial balance of soil and reduce its fertility. Bioaccumulation of toxic metals in the biota of riverine ecosystems may have adverse outcomes on animals and humans. Higher levels of HMs can negatively affect the ecological health of aquatic animals and reduce their populations. HMs have a neurotoxic effect on fish. The interface of HMs with chemical stimuli may disturb the interactions of fish with their environment. HMs are found to be linked with fish malformations.

3.2 SOURCES OF HMS

Sources of HMs in the environment can be either natural and geological, or anthropogenic. Natural sources include the weathering of rocks containing metals and volcanic eruptions. Industrialization and urbanization have increased the amounts of HMs in the environment. The release of wastewaters, such as through industrial and domestic runoffs, increases HMs in the environment. The application of chemical fertilizers and burning of fossil fuels in industries, homes, and transportation are manmade sources of HMs in the environment, e.g., during burning of coal, Cd, Pb, and As are produced which are incompletely volatile, while Hg is fully volatile. Thus, vehicular traffic is among the main sources of Cr, Zn, Cd, and Pb. A higher quantity of ecologically important HMs has been reported in soil and plants adjacent to roads in urban and suburban areas. For example, globally, about 50,000 tons/year of Cr are emitted due to the combustion of coal, wood, and refuse incineration. Likewise, fertilizers usually contain significant Cr contents. Similarly, approximately 60,000 tons/year of Ni are generated globally as a result of the combustion of coal. Natural sources of Cd in the environment include volcanic action and weathering of rocks, whereas an anthropogenic source is non-ferrous metal mining, mainly processing of Pb–Zn ores. Worldwide, about 7,000 tons/year of Cd may be emitted through coal combustion, in addition to sewage sludge incineration, which is a source of Cd. The anthropogenic increase in Cd has been caused by excessive use of chemical fertilizers. Phosphorus-containing fertilizers contain Cd as a pollutant at concentrations ranging from trace quantities to 300 ppm on a dry weight basis and therefore these are a main source of input of this metal to farming systems. Similarly, Pb is released into the environment from diverse sources including acid batteries, old plumbing systems, and lead shot used for shooting. Burning of Pb-containing gasoline is a source of Pb in the environment. Pb poisoning is caused by exposure, such as chewing or ingestion of lead paint in older houses. Lead poisoning may also occur through water contamination from Pb pipes, in addition to Ca products, progressive hair dyes, kohl, and external digestive remedies.

HM toxicity is a consequence of lethal deposition of some metals, which contend with and substitute certain vital minerals over the course of time and affect the body's organ systems. Arsenic toxicity may be caused by medicines, such as Fowler's solution (potassium arsenite) and some topical creams used for treating skin conditions. The ingestion of herbicides, insecticides, pesticides, fungicides, or rodenticides containing arsenic may result in arsenic poisoning. Occupational contact with arsenic during the manufacture of paints, enamels, glass, and metals can cause arsenic poisoning.

Occupational toxicity of Hg is due to exposure to large amounts of Hg in manufacturing thermometers, mirrors, X-ray machines, vacuum pumps, etc. An additional cause of mercury poisoning is polluted water and fish. Children come into contact with Hg through paint, teething powder, thermometers, batteries, etc. Phosphorus poisoning may be caused by insecticides. Thallium toxicity is caused by the ingestion of rodenticides containing thallium. Thallium in pesticides, insecticides, metal alloys, and fireworks can be absorbed through the dermal, oral, and respiratory routes. Similarly, antimony is used for strengthening lead, and in the production of batteries and cables. Cu is used in the manufacture of electrical wires. Likewise, Li is used for making glasses and pharmaceuticals. Cd poisoning may be due to ingestion of food including food grains, cereals, and leafy vegetables and also cigarette smoke. Mn poisoning is caused by chronic inhalation and consumption of manganese particles, while industrial exposure to manganese can occur in the mining and separation of manganese ore. Broadly, the sources and exposures to HMs and environmental contamination include:

- Agricultural activities
- Atmospheric sources
- Battery and plastics industries
- Brass or copper foundries
- Burning of coal and petroleum in thermal power stations
- Combustion of fossil fuels
- Construction of bridges, tunnels, and elevated highways
- Electroplating industries
- Firearms instruction
- Galvanizing and soldering processes
- Gasoline manufacturing
- Geologic activities
- Glass manufacturing
- High-tension lines
- Household discharges
- Incandescent lights manufacturing
- Industrial emissions

- Insecticides, fungicides, rodenticides, and pesticides
- Lead plating and lead shot
- Leather tanneries
- Manufacturing units, including metal handling in factories
- Metal evaporation from water resources, soil, and groundwater
- Metal-plating units
- Micro-electronics and mirror manufacturing units
- Mining
- Natural processes (weathering and volcanic eruptions)
- Nuclear power stations
- Old plumbing systems
- Paper processing industries
- Percolation of HMs
- Pharmacological units
- Plastics manufacturing units
- Polluted water, sea food, fish, and algae
- Printing units
- Rusting of metal
- Smelting works
- Soil erosion of metal ions
- Steel and textile industries
- Weaving units
- Wood preservation
- X-ray machines

3.3 BASIC MECHANISM OF HEAVY METAL CONTAMINATION

Toxic HMs threaten both aquatic and terrestrial ecosystems. Once released from natural and anthropogenic sources, HMs reach and pollute natural water bodies, sediments, and soils. HMs released into the atmosphere due to volcanic eruptions or industrial processes eventually return to the land and pollute water and soil. Being long-lasting in the environment, HMs are either collected in biota or percolate down into the groundwater resulting in contamination of these, which in turn affects human health. Several physicochemical and climatic aspects affect dynamics and biogeochemical cycling of HMs in the environment.

The pollution of water bodies with HMs because of ecological permanence, bio-accumulation, and bio-magnification in food chains and heavy metal toxicity is a matter of concern. It is believed that water is the elixir of life and, being a universal solvent, it dissolves different organic and inorganic chemicals and environmental pollutants. Aquatic ecosystems, both freshwater and marine, are open to contamination. Aquatic ecosystems are polluted by HMs from different sources including discharge from mining operations. Industrial emissions without treatment into aquatic bodies are the main source of contamination of surface water and groundwater. Other sources of water pollution with HMs include domestic sewage and farm run-off. Pollution of water sources by HMs is a key environmental issue which adversely affects plant, animal, and human health. HMs are toxic to aquatic organisms even at very low concentrations and can cause substantial histopathological changes in the tissues of aquatic organisms.

Likewise, contamination of sediments with HMs is an ecologically significant concern with effects on aquatic organisms and human health. Sediments act as the pool of metals in the aquatic environment and their quality illustrates the status of water contamination. Sediments act both as a sink and source of HMs, releasing them into water. Permanent deposition of HMs in sediments can pollute groundwater. Adsorption, desorption, and successive concentrations of HMs in sediments are governed by several physicochemical factors, for example:

- Content of organic matter and microbes
- Hydrodynamic environments
- Redox state
- Salinity
- Size of particles
- Temperature

On the other hand, dissemination of HMs in sediments is governed by:

- Chemical composition of sediments
- Total organic matter content
- Grain size

The bioavailability of metals in sediments is also determined by pH. A decrease in pH increases the competition between metal ions and H^+ for binding sites in sediments and may affect dissolution of metal complexes, thus liberating free metal ions into the water column. Higher concentrations of toxic HMs in riverine sediments threaten the benthos, the bottom-living creatures. HMs and metalloids are discharged in soil from lithogenic and manmade sources. Factors governing occurrence and dispersal of HMs in soils include:

- Structure of rock
- Extent of weathering
- Physico-chemical and biotic features of soil
- Climatic conditions

A substantial quantity of HMs has been reported in soils where fertilizers and Cu fungicides are used as compared to soil receiving low inputs. In metropolitan areas, soils may be polluted with HMs due to vehicular traffic on roads, e.g., in urban areas concentrations of bio-accessible lead is high in soil. The bioavailability of HMs in soil is a key factor accountable for their destiny in the environment and uptake by plants. The bioavailability of HMs in soils differs and depends on metal speciation and the physicochemical properties of the soil.

HMs enter the body of an organism from an abiotic environment, i.e., water, sediments, and soil, or from its food/prey, e.g., HMs may enter the body of a fish directly from water or sediments through the fish gills or skin, or from its food through its digestive tract. Withholding of HMs in the body of an organism depends on the speciation of the metal concerned and the physiological mechanisms evolved by the organism for the control, homeostasis, and decontamination of toxins. Methylated HMs, e.g. Hg, are collected in biota to a greater extent and thus are biomagnified in food chains because of their lipophilicity. Some plants have the capacity to flourish on metal-rich habitations, these are called metallophytes. Metallophytes have evolved distinct mechanisms for handling higher metal concentrations in soil and are divided into three groups: excluders, indicators, and hyper-accumulators. Since HMs remain permanently in the environment, they are stored in living organisms and are transported from one trophic level to another in food chains. The level of amassing of HMs in the biota depends on the rate of accumulation and removal from bodies. Consequently, different metals have diverse half-lives in different species.

3.4 THERMODYNAMICS OF HMS

3.4.1 Bio-Dilution

Sometimes known as "bloom dilution", bio-dilution is a reduced concentration of an element or pollutant with an increasing trophic level. This phenomenon is seen mainly during algal blooms, whereby an increase in algal biomass decreases the concentration of contaminants in organisms higher up the food chain. Primary elements and pollutants of concern are HMs such as Hg, Cd, and Pb. These toxins are known to bio-accumulate in the food web. In some cases, metals, such as Hg, can bio-magnify.

3.4.2 Biomagnification

Biomagnification, also known as bioamplification or biological magnification, is the concentration of a toxin, such as pesticides, in the tissues of tolerant organisms at a constantly higher level in a food chain. This increase occurs because of:

- Persistence – where the substance is not broken down by environmental procedures
- Food chain energetics – where the substance's concentration increases gradually as it goes up the food chain
- Low or non-existent rate of internal degradation or excretion of substance – mostly due to water-insolubility

Biological magnification is the process whereby particular substances, e.g., pesticides or HMs, work their way into lakes, rivers, and ocean, and then move up the food chain in gradually higher concentrations after combining with the diet of aquatic organisms, e.g., zooplankton, which are sequentially eaten by fish, for example, and later consumed by larger fish, followed by large birds, animals, or humans. Substances become progressively concentrated in the tissues or internal organs as they go up the food chain.

3.4.3 Bioaccumulation

This is the gradual amassing of substances, such as pesticides or other chemicals, in an organism. This happens when an organism takes up a substance at a rate higher than the substance is lost by catabolism and excretion. Thus, the longer the biological half-life of a toxic constituent, the greater is the risk of chronic poisoning, even though the environmental concentration of toxin may not be very high.

3.4.4 Bioconcentration

Bioconcentration is the uptake and increase of a substance from water alone. The term is used in the field of aquatic toxicology. It is the process by which a chemical concentration in an aquatic organism goes beyond water after contact with a water-borne chemical. Though used interchangeably with "bioaccumulation", a significant difference is drawn between the two, and with bioconcentration. Bioaccumulation occurs within a trophic level and is an increase in the concentration of a substance in certain tissue of organisms' bodies owing to absorption from food and the environment, while bioconcentration occurs when uptake from water is greater than excretion. Therefore, bioconcentration and bioaccumulation occur within an organism, and biomagnification occurs across trophic levels. Lipid- or fat-soluble substances cannot be diluted, broken down, or excreted in urine, a water-based medium, and hence collect in the fatty tissues of an organism if the organism lacks enzymes to reduce them. When consumed by another organism, fats are absorbed in the gut, transferring the substance, which later accumulates in the fats of the predator. Since at each level of the food chain there is a lot of energy loss, a predator must consume many prey, including all of their lipophilic substances, e.g., although Hg is only present in small quantities in seawater, it is absorbed by algae as methyl mercury, which is the most injurious variant of Hg. It is easily absorbed, but slowly excreted by organisms. Bioaccumulation and bioconcentration result in its build-up in adipose tissue of subsequent trophic levels: zooplankton, small nekton, fish, etc. Anyone who consumes fish also consumes the higher level of Hg accumulated in fish. This process describes why predacious fish or birds have higher concentrations of Hg in their tissue than could be accounted for by direct exposure. DDT, believed to bio-magnify, is one of the most significant reasons it was thought to be detrimental to the environment. DDT is stored in the fat of animals and takes years to break down, and if the fat is consumed by predators, the quantity of DDT bio-magnifies.

There are two main groups of substances that bio-magnify. Both are lipophilic and they are not broken-down easily. Novel organic substances are not easily destroyed or reduced because organisms lack previous exposure and have thus not developed definite decontamination and excretion methods. These substances are therefore recognized as "persistent organic pollutants", or POPs. Metals are not degradable, as they are elements. Organisms, generally those subject to naturally high levels of exposure to metals, have systems to separate and excrete metals. Difficulties arise when organisms come into contact with higher concentrations than usual, which they cannot remove fast enough to avoid harm. Some HMs are hazardous to the organism's reproductive system.

3.4.5 Transference of HMs from Soil to Plants

Transference of HMs from soil to plants is an important step in the trophic transfer of metals in food chains. Metals are taken up by the plants from contaminated soil and afterwards transported to herbivores in the food chain. Concerning the adulteration of the human food chain, pollution of crops such as cereals and vegetables is a critical subject. Consumption of cereals polluted with toxic HMs poses a threat to human health. A higher concentration of HMs has been reported in vegetables cultivated with wastewaters, as compared to those cultivated using groundwater. Moreover, a higher concentration of HMs is found in leafy vegetables as compared to bulbs and tubers. Some terms have been used to measure the degree or level of accumulation of HMs in biota, e.g., bioconcentration factor (BCF), bioaccumulation factor (BAF), and bioaccumulation coefficient (BAC).

BCF is indicative of the amount of heavy metal in an organism with regard to its habitation. It is expressed as "the ratio of the concentration of a heavy metal in an organism tissue to its concentration in the abiotic medium, viz. water and sediments".

Alternative terms like transfer factor (TF), metal transfer factor (MTF), accumulation factor (AF), bioaccumulation factor (BAF), and biota-sediments accumulation factor (BSAF) express the amount of accumulation of a heavy metal in the organism subsequent to the habitat in which it grows or lives.

BAC is defined as the ratio of metal concentration in plant biomass to that in the soil.

(TF is defined as the ratio of metal concentration in the shoots to that in the roots.

It is clear that the values of BCF, BAF, BAC, etc. depend on the concentration of heavy metal in the organism in the related medium. Values of these indices are inversely linked with the metal concentration in water, sediments, and soil, and they are used with care for the calculation of heavy metal pollution in biota.

Molecular and physiological bases of the interface of plants with the environment have caught the attention of the scientific world. Since plants remain fixed in one place, they are continuously exposed to adverse environmental conditions during their entire lifecycle that adversely affect their growth, development, and yield. Toxic compounds, e.g. HMs, can damage plants by altering their physiological and metabolic processes. The elements having a specific gravity above 5 are included in HMs and a vast range of metals and metalloids, such as copper (Cu), iron (Fe), manganese (Mn), zinc (Zn), nickel (Ni), cobalt (Co), cadmium (Cd), arsenic (As), etc., are toxic to plants. Essentially, a few HMs and transition metals like sodium (Na), potassium (K), calcium (Ca), magnesium (Mg), Fe, Cu, Zn, Co, and Ni are, at certain concentrations, essential micronutrients and play a significant role in the functional activities of large numbers of proteins meant for sustaining growth and development. Although HMs are normal constituents of soil and occur naturally in the environment, currently, soil polluted by toxic metals and metalloids is a matter of global concern. The problem of HM pollution is constantly worsening because of anthropogenic activities; hence research into phytotoxicity is gaining importance. Since industrialization is an expanding activity, these toxic metal contaminations reach farm soil and water bodies. The sources of different HMs are discussed in Table 3.1.

TABLE 3.1
Different Sources of Heavy Metals in Agriculture and the Environment

#	Sources	HMs
1.	Aerosols	As, Zn, Cd, Pb
2.	Anthropogenic and industrialization activities	Pb, Zn, Cu, Co, Ni, As, Zn
3.	Atmospheric aerosols, windblown dust	Pb, Hg, Ni
4.	Chemical industries and geological minerals	Cd, Pb, Zn, As
5.	Coal mining and steel processing industries	Zn
6.	Earth's crust	Cd, Hg, Pb, A
7.	Fossil fuels combustion, chemical industries, electroplating	Ni, As, Cr
8.	Geological and anthropogenic activities	As, Al, Fe, Mn, Si, As Cr, Hg, Ni Co, Pb, Zn, Cd
9.	Industrial pollution/industrial waste pipes/additives and geological minerals	As, Cu
10.	Mining activities/smelting	Cd, Cr, Pb, Zn
11.	Ores/mineral dissolution	Fe, Mn, Cr, Cd, Hg
12.	Pesticides/phosphate fertilizers	Hg, As, Cu, Zn, Cd
13.	Poultry waste, feed additives	As
14.	Sewage irrigation/manure, sludge, limes	Cd, Pb, Cr, Hg
15.	Smelting/agricultural/industrialization/urbanization activities	Ni, Cu, Cr, Mn, Zn, Fe
16.	Sulfide mineral deposits	As
17.	Surface runoff from rain/snow	Hg
18.	Volcanic emissions, forest fires, atmospheric deposition	Pb, Hg, Ni
19.	Waste dumping	Cd, Pb, Cu
20.	Weathering of sedimentary rocks like sandstone and dolomite	Mn, Cr, As, Fe, Zn

Source: Khanna et al. (2018).

Based on their oxidation states, HMs can be extremely reactive, causing toxicity of plant cells. At the cellular and molecular levels, HM toxicity changes the diverse plant physiological processes, including:

- Inactivation and denaturation of enzymes and proteins
- Obstructing the functional groups of metabolically important molecules
- Displacement/substitution of essential metal ions from biomolecules and functional cellular units
- Conformational modifications and disruption of membrane integrity

Alterations in the physiological processes further:

- Affect the plant metabolism
- Inhibit the physiological processes like photosynthesis, respiration, etc.
- Alter the activities of several key enzymes

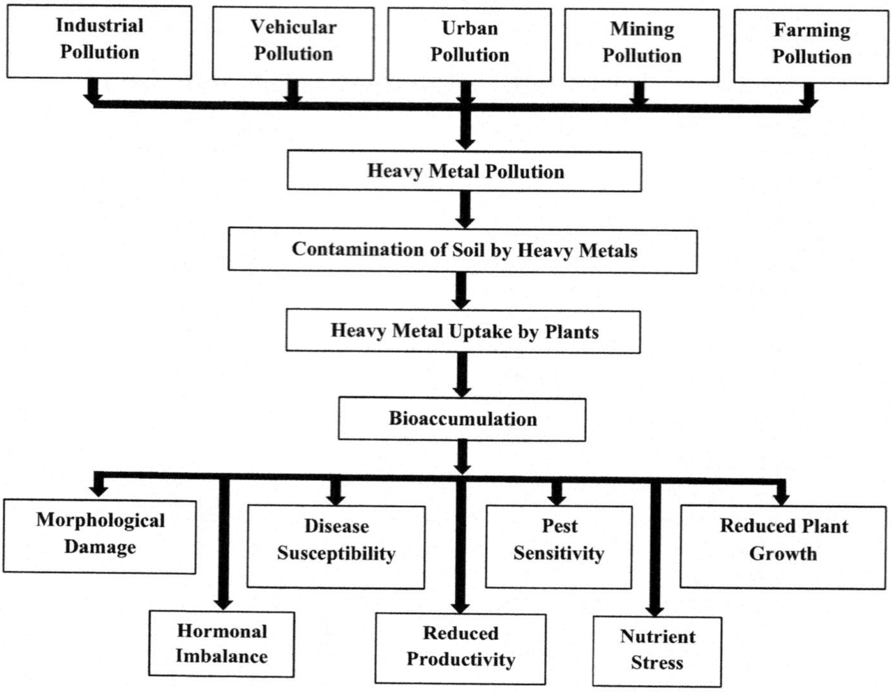

FIGURE 3.1 General impact of heavy metals on plants.

Furthermore, HMs are known to interrupt the redox homeostasis by encouraging the formation of free radicals and reactive oxygen species (ROS), such as singlet oxygen ($^1O_2$), superoxide radicals (O^{2-}), hydrogen peroxide (H_2O_2), and hydroxyl radicals (-OH). The general impact of HMs on plants is summarized in Figure 3.1. An increase in methylglyoxal (MG), a cytotoxic compound, in plant cells in response to HM stresses further increases the production of ROS by inhibiting different plant physiological and metabolic processes, including:

- Inactivation of the antioxidant defence system
- Interference in vital plant physiological processes, e.g., photosynthesis
- Exposes plant cells to oxidative stress, leading to:
  - ✓ Lipid peroxidation
  - ✓ Biological macromolecule deterioration
  - ✓ Membrane dismantling
  - ✓ Ion leakage
  - ✓ DNA-strand cleavage
  - ✓ Eventual death of plants

Coordination of complex physiological and biochemical processes, including changes in global gene expression, is required by the plant to develop tolerance mechanisms.

Plants use various strategies to deal with the toxic effects of metals or metalloids. Resistance to HM stress can be achieved by "avoidance", where the plants restrict the uptake of metal, or by "tolerance", where plants are able to endure the presence of high metal concentrations. Avoidance involves:

- Reduction in metal concentration entering the cell by extracellular precipitation
- Biosorption to cell walls
- Reduced uptake, or increased efflux

Tolerance involves:

- Intracellular chelation of HMs through the synthesis of amino acids, organic acids, glutathione (GSH), etc.
- Binding of ligands to HMs, such as metallothioneins (MTs), phytochelatins compartmentation (PCs)
- Up-regulation of antioxidant defence and glyoxalase systems to counter the deleterious effects caused by ROS and MG

Thus, GSH by itself and its linked metabolizing enzymes, proteins, and peptides play a vital role in HM tolerance by regulating diverse plant physiological processes, such as:

- ROS and MG detoxification
- HM uptake
- Translocation
- Chelation
- Detoxification

Mechanism of Action of Toxic HMs in Plant Cells

The toxicity of HMs is demonstrated in numerous ways when plant cells store them at high levels. HMs can be divided into two groups:

Redox Active HMs (Fe, Cu, Cr, Co): Involved in redox reaction in cells and result in the formation of O^{2-} and subsequently in H_2O_2 and -OH production

Redox Inactive HMs (Cd, Zn, Ni, Al, etc.): Results in oxidative stress through indirect mechanisms such as:

- Interface with an antioxidant defence system
- Disturbance in the electron transport chain
- Initiation of lipid peroxidation

Another important mechanism of HM toxicity is the capacity of HMs to intensely attach to oxygen, nitrogen, and sulphur atoms which is related to the free enthalpy of formation of the product of the HM and ligand with low solubility of these products. Due to these qualities, HMs can incapacitate enzymes by fastening to cysteine residues, e.g., Cd binding to sulfhydryl groups of structural proteins and enzymes leads to misfolding and inhibition of activity and/or interference with redox-enzymatic regulation. Many enzymes demand cofactors to work accurately for both HM ions (such as Fe^{2+}, Mg^{2+}, Cu^{2+}, Ca^{2+}) and organic molecules (such as haem, biotin, flavin adenine dinucleotide [FAD], nicotinamide adenine dinucleotide [NAD], or coenzyme A). Displacement of one HM ion by another inhibits enzyme activities. Divalent cations such as Co^{2+}, Ni^{2+}, and Zn^{2+} displace Mg^{2+} and result in a loss of activity of related enzymes. Moreover, HMs result in damaged membrane through various mechanisms including:

- Oxidation of and cross-linking with protein thiols
- Inhibition of key membrane protein, namely H^+-ATPase
- Changes in the composition and fluidity of membrane lipids

The accumulation of MG, a cytotoxic compound, was found to increase in response to HM stress in plants owing to impairment of the glyoxalase system that stimulates oxidative stress by reducing the GSH content. Thus, HM toxicity is attributed to three mean causes:

a. Stimulation of ROS and MG production by auto-oxidation or by alteration of the antioxidant defence system and the glyoxalase system
b. Direct interface with proteins attributable to their empathies for thioyl-, histidyl-, and carboxyl-groups, causing the HMs to target structural, catalytic, and transport sites of the cell
c. Dislocation of important metal ions from specific binding sites, triggering function to collapse

Consecutive steps of ROS-induced damage development in plants in response to HM stress are outlined in Figure 3.3.

Heavy Metal Stress and Response of Plants

Toxic HMs pose a threat to living organisms and plants as they are highly persistent and remain present in the soil for a long time (Table 3.2). In terrestrial plants, the roots are the main sites for contact and exposure to toxic HMs. For aquatic plants, the entire plant body is exposed to growing medium where crucial and non-essential metals that may be toxic in excess, inhibit its growth and development, and may even cause plant death (Figure 3.3). To survive and deal with HM stresses, plants have evolved some adaptive mechanisms, such as the following:

- Chelation and sequestration by specific ligands
- Elimination of plasma membranes
- Initiation of stress proteins
- Restrained absorption and transport
- Synthesis of specific heavy metal transporters

TABLE 3.2
Heavy Metal Toxicity Symptoms in Plants

Metal	Symptoms/Impacts
Aluminium (Al)	• Affects photosynthetic activity • Decreases cellular respiration in roots and formation of DNA • Inhibits cell division in roots; hence, roots remain undersized and fragile • Reduces formation and transfer of cytokines • Restricts water and nutrient uptake and transport across cell membrane • Poor development of root hair and deformed, puffy, and brittle root apex
Arsenic (As)	• Affects plant cell, chlorophyll content, plant growth • Inhibits root extension and propagation • Interferes with critical metabolic processes, which can cause plant death • Reduces root development and reproductive capacity • Translocation to shoot inhibits plant growth
Cadmium (Cd)	• Enhances tendency to produce ROS, causing oxidative stress • Growth retardation and reduced root development • Inhibits carbon fixation, formation of lateral root, and stomatal density • Reduces synthesis of chlorophyll and photosynthetic activity • Stunted growth, chlorosis, necrosis, and wilting
Chromium (Cr)	• Affects transpiration and nitrogen assimilation • Altered enzymatic activities • Reduced pigment content and chlorosis in leaves • Decreased cell division and shoot growth due to ultrastructural modifications of cell membrane and chloroplast • Root cells damage
Cobalt (Co)	• Affects photosynthetic activity, plant cells, and growth of seedling • Iron deficiency in plants • Pale-coloured leaves, discoloured veins, and loss of leaves
Iron (Fe)	• Leaf discoloration and chlorosis • Stunted root system, poor growth, and reduced yield
Lead (Pb)	• Affects chlorophyll production • Delays production of ATP, seed germination, and root elongation • Lipid peroxidation and DNA damage due to ROS • Lowers transpiration and chlorophyll synthesis and prevents plant growth
Mercury (Hg)	• Affects growth and metabolism of plants • Increased lipid peroxidation • Low exchange of gases • Reduced photosynthesis, transpiration, water uptake, chlorophyll synthesis, root development, and stunted growth
Nickel (Ni)	• Chlorosis and necrosis • Enzymatic imbalance and impaired cell membrane • Inhibits physiological processes • Oxidative damage • Reduces seed germination, cell division, and root/shoot growth
Zinc (Zn)	• Affects biomass yield, plant growth, photosynthetic activity • Chlorosis due to decreased chlorophyll synthesis or chloroplast degradation • Reduced distribution of photo-assimilates from leaves to roots, reduced yield, and stunted growth

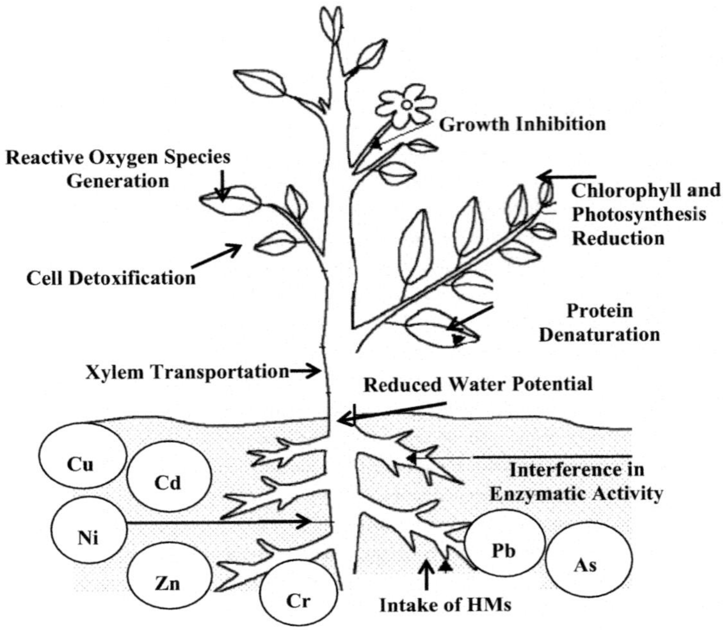

FIGURE 3.2 Effects of heavy metal toxicity on plants.

There are three ways of keeping concentration of metal ions low in cytoplasm:

- Preventing toxic metals from being transported across the plasma membrane
- A cellular mechanism for heavy metal tolerance
- Detoxifying toxic metal ions by inactivating them through chelation or altering toxic metal ions to a lower toxic concentration

This can be achieved by:

- Increasing the metal ion binding to the cell wall (Figure 3.2)
- Pumping the metal out of the cell through active efflux pumps

Plants are capable of solubilizing and absorbing different soil nutrients which may help in generating chelating agents and changing pH and redox reactions (oxidation and reduction reactions). There exist precise mechanisms for different nutrients or metals being translocated and accumulated, which assist in the uptake of heavy metals and translocation. The accumulation of HMs in plants is significantly affected by a number of factors such as:

- Carbohydrate level
- Depth of the root system
- Interface with nutrients

- Lifecycle of the plant
- Partial oxygen pressure
- pH of the soil
- Plant vigour
- Presence of microbes
- Respiration rate
- Structure of the plant
- Temperature

Plants can precipitate HMs intensely by changing the rhizosphere's pH or by expelling anions. During the adsorption process, the root surface can bind many HMs. These HMs (Cd, Ni, Pb, and Sr) quickly assemble in the root tissue of the plant. Based on the mechanism of action evolved for survival under stress, plants can be generally classified as:

- Accumulators
- Excluders
- Indicators

The accumulation of HMs in excess of 0.1–1 per cent of the dry weight of plants is referred to as hyperaccumulation. As an ecological adaptation, some plants have evolved the ability to grow in soil contaminated with HMs and accumulate high quantities of metals. The mechanism involved in the hyperaccumulation of toxic HMs in plants includes bio-activation of HMs in the rhizosphere through:

- A root–microbe crossing point
- Enhanced activity of metal conveyor proteins in the cell membranes
- Detoxification of metals by restraining them to apoplasts
- Chelation of HMs in the cytoplasm by some ligands
- Sequestration of metals in the vacuole via several ligands

Exposure of plants to toxic concentrations of HMs produces a variety of physiological and metabolic alterations. However, different HMs have different sites of action within plants; and the general visual toxic response varies between HMs. The most common evidence of heavy metal toxicity is reduced plant growth, including leaf chlorosis, necrosis, turgor loss, decreased rate of seed germination, and crippled photosynthetic apparatus, frequently linked with systematic senescence routes or plant death (see Figure 3.3). All these effects are related to ultrastructural, biochemical, and molecular changes in plant tissues and cells caused by the presence of HMs.

Additionally, HMs impact homeostatic events, including water uptake, transport, transpiration, and nutrient metabolism, and also interfere with the uptake of Ca, Mg, K, and P. Increased levels of HMs generally decrease photosynthesis because of their direct effect on the photosynthetic apparatus, including thylakoids. Studies have shown that HMs regulate the release of proteins, lipids, and element components of thylakoid membranes, causing damage to light-harvesting complexes

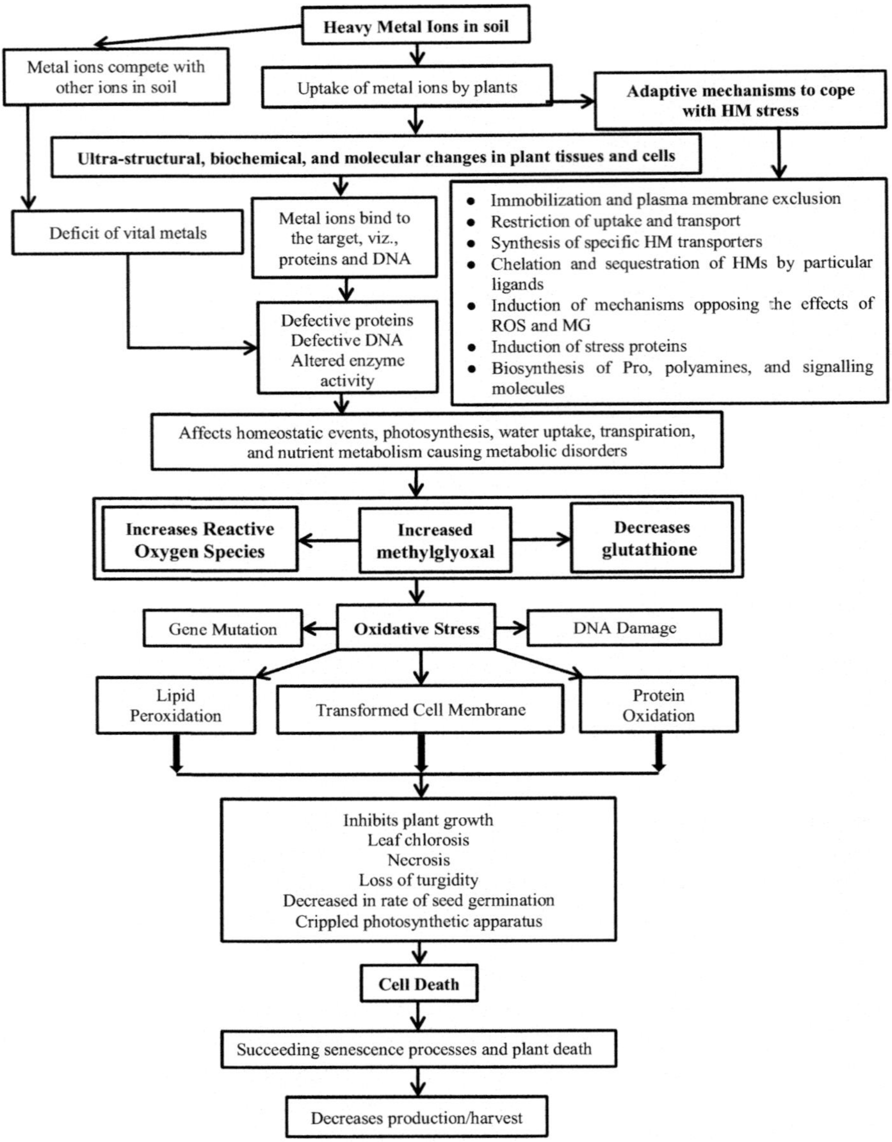

FIGURE 3.3 Mechanism of action of heavy metal toxicity in plants.

and photosystem II (PS II). Additionally, some HMs can replace Mg in the chlorophyll. Reduced synthesis of chlorophyll after HM stress inhibits the enzyme involved in its synthetic pathway. Additionally, HM reduces carbon absorption by hindering enzymes involved in CO_2 fixation. HMs also disturb carbohydrate metabolism and their separation in growing plants. HMs that bind to the cell nucleus cause promutagenic damage including:

- DNA base modification
- Inter- and intra-molecular cross-linking of DNA and proteins
- DNA strand breaks
- Rearrangement and depurination

Other impacts of HMs include:

- HMs can affect microtubule assembly/disassembly, thus modifying cell cycle and cell division. High rates of mutation and malformed embryos have been observed in *Arabidopsis* plants exposed to Cd
- HM toxicity results in the accumulation of excess ROS inside the cell, which depends on the particular metal
- HM toxicity results in the accumulation of excessive ROS in cells, e.g., Cu can directly generate ROS, whereas Cd is a redox-inactive HM and can only generate ROS indirectly by enzyme inactivation and therefore cause oxidation of polyunsaturated fatty acids
- ROS action may cause cell disruptions which increase senescence processes in association with ethylene and jasmonic acid, although the activation of antioxidant machinery can help plants to overcome HM stress
- A plant's molecular response to heavy metal stress is illustrated by the synthesis of stress-related amino acids, proteins, genes, and signalling molecules. Recent studies have shown that HMs increase the synthesis of heat shock proteins (HSPs)

Bioaccumulation of HMs in Living Systems

The presence of essential or non-essential trace elements beyond safe levels or in excess may produce physiological or morphological anomalies or genetic mutations, thereby slowing or preventing growth or resulting in mutations. Crops and vegetables that form the most important component of the human diet may contain a variety of both useful and harmful metals, depending on the characteristics of the growth medium employed. Humans are exposed to HMs when they consume vegetables, which account for approximately 90 per cent of the overall intake, while the remaining 10 per cent comes from dermal contact and inhaling contaminated dust. Owing to increasing demand for food, food safety has become a major public health concern, especially in regard to human health. The food chain is continuously replenished with essential and non-essential materials due to the excessive use of agrochemicals, municipal wastewater, industrial effluents, and sewage for irrigation. Mineral resources and elements, such as copper, chromium, iron, manganese, and zinc, among others, are essential both for animals and humans as they are involved in various metabolic functions, enzyme activities, receptor sites, hormonal functions, and protein transport at specific concentrations. Another group of elements, including, arsenic, cadmium, lead, and mercury, are non-essential and have no role in plants, animals, or humans. They have no nutritive use since they are very toxic.

To set quality standards and ascertain the hazards to human safety and health, it is essential to define the sources and quantities of HMs in soil. Environmental pollution caused by HMs is constant, and lasting. As metals are non-biodegradable and have a long half-life, biological species are incapable of decomposing themselves, and thus these remain in their body parts and environment, posing health risks. The bioaccumulation of HMs in vegetables poses a health hazard due to the potential of transmission from polluted land and water into the food chain.

Soil properties are vital in food production, and HM pollution, as well as their resultant absorption and bioaccumulation in crops, poses considerable environmental and health concerns, particularly in developing nations. The concentration of HMs is governed by soil type, plant genotype, and their interface with nature. Mineral fertilizers have high concentrations of HMs as compared to organic manure; consequently, the use of mineral fertilizers causes heavy metal pollution.

Health risks linked with toxic metals depend on the concentrations of metals in certain media and the duration of exposure. Even at low concentrations, long-term and chronic exposure of hazardous metals causes health problems.

3.4.6 HMs in Biota and Food Chains

Since HMs are permanent in the environment, they enter from the surroundings to the organisms and are stored therein. As mentioned earlier, the uptake and bioaccumulation of HMs in biota depend on several factors. For example, the uptake of HMs in plants depends on the bioavailability of the metal in soil, which in turn depends on several factors such as metal speciation, pH, and organic matter contents of the soil. Metals which are more bioavailable in soil may be accumulated in plants more easily and therefore will have greater bioaccumulation potential. A measurement of the bioaccumulation of HMs in plants is used for the assessment of the bioavailability of the metals in soil. Such an evaluation may also be used for analysing the contamination status of the environment. It has been described that plants are more susceptible to ecological changes than soils. Different plant and animal species are regarded as bioindicators of heavy metal pollution in the environment.

Bioaccumulation of HMs in biota is significant from ecological and health perspectives and has significant repercussions for wildlife and human health. Contamination of aquatic and terrestrial food chains with possibly toxic HMs threatens the health of consumer organisms, including humans. In aquatic ecosystems, organisms instantly come into contact with different metals, which may have stabilizer, synergistic, or incompatible interfaces.

Aquatic biota come into contact with HMs through different routes, e.g., water, sediments, and food. Freshwater fish come into contact with different toxic HMs discharged to freshwater bodies from diverse natural and manmade sources. Intoxication of fish by HMs has become a global issue because it not only threatens fish but poses health risks to fish consumers. Assessment of the bioaccumulation of heavy metals in fish from different aquatic habitats is important because the estimation of heavy metal concentration in fish tissues is vital for management of aquatic ecosystems and human intake of fish. Fish possess a greater amount of unsaturated fatty acids and low levels of cholesterol and are major source of proteins. The use of edible fish in the human diet is useful and helps provide a balanced diet. However, contamination of fish by toxic HMs is believed to be a threat to human health and has raised concerns about their consumption, particularly by women, children, and those with comorbid conditions.

Bioaccumulation of HMs in freshwater fish depends on various aspects, both in regard to the features of fish and ecological factors. Features related to fish include the age, size in terms of weight and length, feeding practices, and physiology, but other ecological factors include the concentration and bioavailability of metals in the water column, physico-chemical properties of water, and other climatic factors. The extent of the bioaccumulation of HMs in different fish tissues usually depends on the structure and function of the tissue. Usually, metabolically functional tissues, e.g., gills, liver, and kidneys have greater HM deposits than other tissues, such as the skin and muscles. Comparatively higher heavy metal deposits in metabolically active tissues of fish are mostly explained by the induction/occurrence of metal-binding proteins called metallothioneins (MTs) in these tissues on exposure to HMs. Fish gills are the main target tissue for the deposition and exclusion of HMs such as nickel. Fish muscles have reduced deposition of heavy metals and are vital from the perspective of intake by humans. Bioaccumulation of trace metals in the muscles of fish is usually species-specific.

The bioaccumulation of toxic HMs in freshwater fish raises significant ecological and social concerns. Aquatic HMs are assimilated into fish and enter the human body via the food chain, and have an impact on human health. Moreover, toxic HMs also govern fish health. It has been reported that river contamination with HMs induces stress in freshwater fish, making them weak and more susceptible to disease. Heavy metal contamination is considered to be one of the probable causes of the population decline of freshwater fish and other aquatic organisms in freshwater ecosystems. Humans are omnivorous and come into contact with toxic metals through diverse foodstuffs, e.g., fish, cereals, and vegetables. Heavy metal pollution in freshwater bodies results in the bioaccumulation of these metals in freshwater fish, while such toxic metals in farmlands lead to bioaccumulation in crops. The adulteration of human food chains with toxic metals threatens human health. Minamata and itai-itai disease are known to be caused by the intake of Hg-contaminated fish and Cd-contaminated rice, respectively. In the biomagnification of metals in food chains, organisms at higher trophic levels are at greater risk. To safeguard human health from the detrimental effects of HMs, human food chains must be continuously examined for the bioaccumulation and biomagnification of HMs. HM toxicity later results in the production of reactive oxygenic species and consequently induces oxidative stress, resulting in the development of various diseases and abnormal conditions. HMs at times act as metabolic poisons. Heavy metal toxicity is mainly due to their reaction with sulfhydryl enzyme systems and their consequent inhibition. Toxicity is the attribute of a chemical to alter and modify the existence, development, and reproduction of an organism. Some HMs have carcinogenic, mutagenic, and/or teratogenic effects on different species, but this depends on the dose and extent of exposure. HMs affect both animal and human health, but some species are more susceptible to HMs than others. The manner in which a heavy metal affects different organs, tissues, and system in different organisms is very intricate, and some of these are not yet fully studied. Chronic exposure to HMs is a real threat to living organisms.

Humans come into contact with toxic HMs in the natural environment and workplace through different routes, including consumption, breathing, and dermal absorption. Workers in mining and industrial operations are exposed to metals where they may inhale dust and particulate matter containing metal particles. Those involved in the extraction of gold through an amalgamation process are exposed to Hg vapour. Welders with occupational exposure to welding fumes have considerably high concentration of Cr, Ni, Cd, and Pb in their blood and show increased oxidative stress. Cigarette smoke is a major source of human exposure to Cd and other toxins present in tobacco. HMs reduce the major antioxidants of cells, mainly the antioxidants and enzymes having a thiol group (-SH), and increase the generation of reactive oxygen species (ROS) like hydroxyl radical ($HO^{\cdot}$), superoxide radical (O_2^{-}), and hydrogen peroxide (H_2O_2). The increased generation of ROS can upset the intrinsic antioxidant protection of cells and lead to oxidative stress.

3.4.7 Health Effects of Heavy Metals

Neurotoxicity: As an essential element, manganese is involved in several physiological functions of the body. Its acute exposure exerts a potential neuro-protective action by decreasing apoptotic cellular death but exposure to a large quantity may induce damaging neurological problems, such as Alzheimer and Parkinson diseases, which not only result in apoptotic cell death but also alter homeostasis. Homeostasis of cellular Mn relies on adequate intake, storage, as well as excretion via different cell receptors and ion channels. Receptors linked with metal uptake are down-regulated by homeostatic pathways in terms of excessive Mn exposure, while those involved in its discharge from the cell become up-regulated. The accumulation of Mn increases ROS formation, adding to mitochondrial dysfunction, which results in the discharge of cytochrome C, subsequently stimulating the apoptosis precursor caspase-9, by which caspase-3 splits. The sliced caspase-3 fragment fixes to a pro-apoptotic protein, PKCδ (protein kinase C delta). Proteolytic cleavage of PKCδ induced by caspase-3 adds to DNA fragmentation, as well as apoptosis (Figure 3.4).

The CNS suffers from cognitive impairment after the consumption of arsenic. It is also linked with a number of neurological illnesses, including neuro-developmental changes, and leads to a number of neuro-degenerative diseases. Ar toxicity alters the synaptic transmission and affects neurotransmitter balance.

Moreover, the neurotoxic effects of arsenic are ascribed to induce multiple apoptotic mechanisms. Primarily, arsenic and its methylated metabolites enable caspase-induced apoptosis in neural cells via mitogen-activated protein kinase (MAPK) signalling pathways and follow the intrinsic mitochondrial–apoptotic mechanisms. Moreover, arsenic starts an upsurge in intracellular calcium that determines apoptosis. In contrast, cellular apoptosis can be facilitated by activating autophagy. Neurodegenerative defects and neurotoxicity induced by Cd include:

- Amyotrophic lateral sclerosis
- Parkinson's disease
- Alzheimer's disease
- Multiple sclerosis

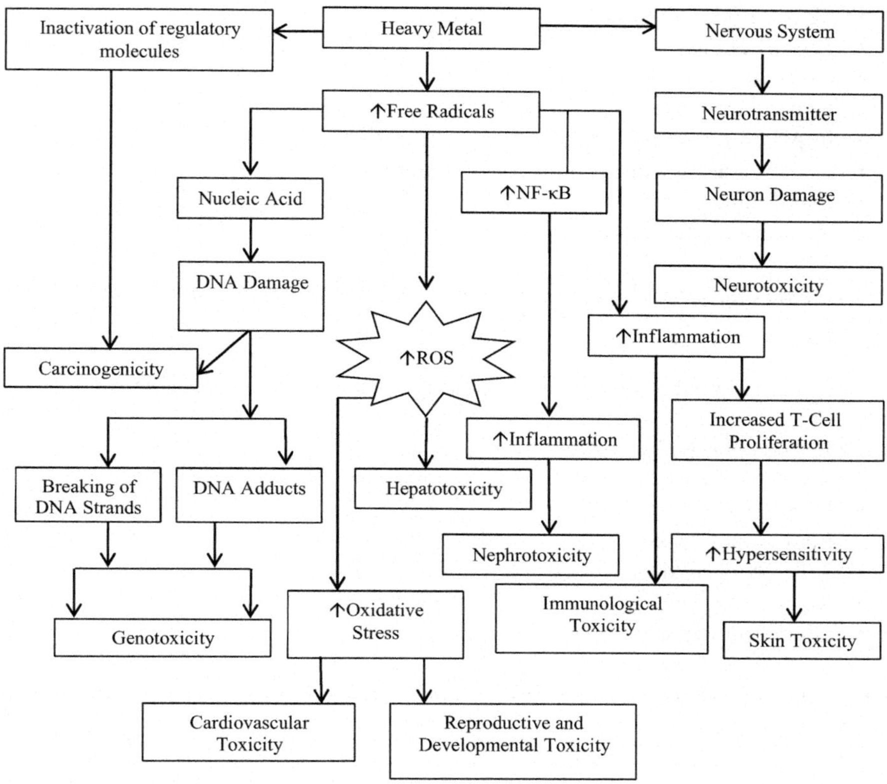

FIGURE 3.4 Mechanism of heavy metal toxicity in humans.

Cd seriously affects the functionalities of PNS and CNS, with many clinical manifestations:

- Peripheral neuropathy
- Olfactory dysfunctions
- Neurological disturbances
- Impairment of motor function and behavioural changes

Furthermore, it affects cellular activity, such as, cell differentiation, proliferation, and cell death. In addition to Mn, As, and Cd, many other HMs result in neurotoxic consequences. Cu and Zn, like iron, obstruct neurodevelopment. Excess Cu retention causes Wilson disease, a hereditary condition which causes neurobehavioural abnormalities similar to schizophrenia. Zn deficiency has an adverse influence on neurodevelopment.

Nephrotoxicity: Cd-induced nephrotoxicity clinically manifests as glucosuria, Fanconi-like syndrome, phosphaturia, and aminoaciduria. Proximal tubular epithelium is affected by direct/excessive exposure to the kidneys, causing:

- β2-Microglobulinuria
- Aminoaciduria
- Glucosuria
- Hypercalciuria
- Impaired renal tubular phosphate reabsorption
- A significantly increased level of cadmium in urine
- Renal failure
- Renal tubular acidosis

Pb has toxic effect on all organs, but more so on the kidneys. Acute Pb nephropathy causes:

- Proximal tubular dysfunction, resulting in Fanconi-like syndrome

Chronic Pb nephropathy results in:

- Hyperplasia
- Interstitial fibrosis
- Atrophy of tubules
- Renal failure
- Glomerulonephritis

Also, acute exposure of kidneys to Hg causes acute tubular necrosis with clinical symptoms, such as:

- Abdominal pain
- Acute dyspnoea
- Altered mental status
- Hypotension
- Profuse salivation
- Tremors

In contrast, chronic exposure to Hg causes:

- Higher urine excretion of albumin and retinol-binding protein
- Injury of epithelium and necrosis in the pars recta of the proximal tubule
- Mercury-induced chronic kidney injury
- Nephritic state with a characteristic of membranous nephropathy
- Tubular failure

Carcinogenicity: Poisoning increases the risk of cancer by fastening to DNA-binding proteins and reducing the DNA-repair process. Toxicity causes:

- Changes in the p53 protein's expression
- DNA damage and methylation

- Epigenetic alterations and histone modifications
- Reduced p21 expression

Likewise, Pb as a carcinogenic substance:

- Affects the DNA repair mechanism
- Affects cellular tumour-regulating genes, and chromosomal structure and sequence by releasing ROS
- Disrupts transcription by shifting zinc from certain regulatory proteins

Mercury's peroxidative activity produces ROS, which aids in pro-tumourigenic signalling and cancerous cell growth. ROS adds to carcinogenesis by damaging cellular proteins, lipids, and DNA, causing cell damage. Ni as a carcinogen regulates a variety of carcinogenic mechanisms, including:

- Gene regulation
- Transcription factor management
- Free radical generation

It also controls the expression of particular long non-coding RNAs, mRNAs, and microRNAs. It takes part in the methylation of promoter and down-regulation of gene 3 (MEG3) to increase modulation of hypoxia-inducible factor-1, both of which result in carcinogenesis.

Hepatotoxicity*:* Pb toxicity in liver cells is a well-established phenomenon. Exposure to Pb increases oxidative stress, which damages the liver. Organic solvents, together with Pb, also cause liver injury. Chronic exposure to Pb can be toxic to liver cells, resulting in depletion of glycogen and cellular infiltration, which manifests as chronic cirrhosis.

Cd on other hand has two target tissues, namely, the renal cortex and the liver. During acute exposure, Cd accumulates in the liver and causes various hepatic dysfunctions. It alters the cellular redox balance, causing oxidative stress and hepatocellular damage. Acute and chronic hepatotoxicity are induced by Cd, eventually leading to liver failure and increasing the risk of cancer. Cu is well-known to accumulate in the liver. Increased levels of Cu may cause oxidative stress; leading to deposition of hepatic Cu which is both pathognomonic and pathogenic. An elevated level of hepatic Cu is also observed in cholestatic liver diseases. Cr (VI) can harm the liver, causing hepatotoxicity and histopathological changes, such as the following:

- Apoptosis
- Arresting of cell growth
- Decreased activity of antioxidant enzyme
- DNA damage
- Elevated ROS levels
- Lipid peroxidation

- Mitochondrial dysfunction including impaired mitochondrial bioenergetics
- Necrosis
- Suppression of DNA, RNA, and protein synthesis

Immunological Toxicity: Acute and chronic exposure of Pb leads to several toxic effects on the immune system and causes immune responses, such as:

- Autoimmunity
- Cancer
- Immunosuppressive effects
- Increased allergies and susceptibility to the development of autoimmunity and hypersensitivity
- Infectious diseases
- Influences cellular and humoural responses
- MHC activity
- Modified role of T-cells

Occupational and environmental exposure to cadmium may induce:

- Phagocytosis, natural killer cell activity, and host resistance
- Production of B- and T-cells

Cr is known to have adverse effects on the human immune system, causing:

- Allergic contact dermatitis
- Humoural immune response
- Hypersensitivity reactions: type I (anaphylactic type) and type IV (delayed type)
- Obstruction of lymphocyte proliferation
- Phagocytic action of alveolar macrophages

Cardiovascular Toxicity: Chronic exposure to Pb may cause:

- Adjusted vascular response to vasoactive agonists
- Alteration of the renin–angiotensin system
- Arteriosclerosis
- Atherosclerosis
- Cardiac disease
- Disruption of vascular smooth muscle Ca^{2+} signalling
- Hypertension
- Long-term exposure increases arterial pressure
- Lowered vasodilator prostaglandins
- Thrombosis

Cd is a toxicant and carcinogenic metal that induces diseases related to bones, the cardiovascular system, and kidneys. Low to moderate Cd exposure results in:

- Hypertension
- Diabetes

- Carotid atherosclerosis
- Chronic kidney disease
- Stroke
- Heart failure

Mercury has been shown to cause neurotoxicity, nephrotoxicity, and hepatotoxicity in humans. Paraoxonase, an extracellular anti-oxidative enzyme linked to HDL dysfunction, is inactivated by mercury; which is directly linked to:

- Acute myocardial infarction
- Cardiovascular disease
- Coronary heart disease
- Increased risk of coronary heart disease
- Progression of atherosclerosis

Dermal toxicity: Chronic exposure to arsenic results in numerous skin diseases, including:

- Hyperkeratosis (affecting the soles and palms, but it can also affect the legs, toes, fingers, arms, etc.)
- Hyperpigmentation
- Several types of skin cancer (Bowen's disease, a type of early skin cancer)

Exposure to Cr causes various acute and chronic severe dermatological conditions, including:

- Contact dermatitis
- Skin cancer

Several skin infections are caused by Hg or Hg-containing compounds, such as:

- Acrodynia (Pink disease)
- Irritation
- Moderate swelling
- Scaling
- Vesiculation

Reproductive and Developmental Toxicity: HMs are known to have various impacts on reproductive and developmental toxicity. For example, arsenic causes abnormalities in experimental animals, particularly neural tube anomalies. Inorganic arsenic impairs male reproduction by reducing the weight of testes, the accessory sex organs, and the number of sperm in the epididymis. The incidence of endometrial cancer in females is linked to arsenic exposure. Ovulation disturbance in females is another impact of heavy metal toxicity which is characterized by irregular or absent menstrual cycles. The risk of infertility in women is elevated because of increasing levels of toxin exposure, which cause hormonal disruption, delayed ovulation, and chromosomal abnormality in oocytes. Female infertility caused by hormonal imbalance is aggravated by endocrine disruption due to heavy metal poisoning.

Genotoxicity: Several studies have illustrated inter-individual variability in receptivity to arsenic poisoning, with genetic factors as the fundamental source of this variability. Genotoxicity of arsenic results in:

- Chromosomal abnormalities
- Deletion and sister chromatid exchange
- Deoxyribonucleic acid alteration
- Induction of oxidative stress and disruption of DNA repair
- Micronuclei production
- Mutation

Arsenic is regarded as a weak mutagen because, despite its low mutagenicity, it affects the mutagenicity of other carcinogens, especially when exposed to UV light. Genotoxic effects from chemical compounds bring about changes in genetic material mainly via two processes: carcinogenesis and teratogenesis.

3.5 HM TOXICITY

Antimony, aluminium, barium, bismuth, copper, gold, iron, lithium, platinum, silver, tin, and zinc cause metal toxicity. Common symptoms of metal toxicity may be gastrointestinal, renal, and/or neurological. Metal toxicity of some common metals is as follows:

Aluminium: Aluminium occurs naturally in air, water, and soil. Mining and processing of aluminium increase its presence in the environment. Aluminium is a major threat to humans, animals, and plants, and causes many diseases. Water pH and organic matter content significantly affect aluminium toxicity. With decreasing pH, the toxicity is increased. The mobilization of toxic aluminium ions, due to a change in the pH of soil and water caused by acid rains and increasing acidification of the surrounding atmosphere, adversely affects the environment, which results in drying of forests, plant poisoning, crop decline or failure, death of aquatic animals, and various imbalances in the function of human and animal systems. The usual manifestations of aluminium toxicity are root growth inhibition, cellular modification in leaves, small and dark green leaves, yellowing and death of leaves, chlorosis, purpling, and foliar necrosis. Aluminium in high concentrations is toxic for aquatic animals, especially for fish, causing osmoregulatory failure by damaging the plasma and haemolymph ions. Living organisms in water, such as seaweeds and crawfish, are also affected by aluminium toxicity. Enzymes such as hexokinase, phosphodiesterase, alkalic phosphatase, and phosphoxidase are stopped by aluminium as it has a greater affinity to DNA and RNA.

Mechanism of Aluminium Toxicity: Aluminium inhibits most physical and cellular processes. The exact mechanism of absorption of aluminium by the gastrointestinal tract is not understood entirely. Aluminium toxicity possibly results from an interaction between aluminium and the plasma membrane. In humans, Mg^{2+} and Fe^{3+} are replaced by Al^{3+}, which causes many disturbances linked with intercellular communication, cellular growth, and secretory functions.

Iron: Iron is one of the vital components of organisms like algae and of enzymes, such as cytochromes and catalase, as well as of oxygen-transporting proteins, including haemoglobin and myoglobin. The source of iron in surface water is anthropogenic and is related to mining activities. In Lithuania, many people were exposed to higher levels of iron through drinking water, as the collected groundwater exceeded the acceptable limit set by the European Union Directive on the quality of drinking water. An abundance of species, such as periphyton, benthic invertebrates, and fish have their diversity seriously affected by the direct and indirect effects of iron contamination.

Mechanism of Iron Toxicity: A number of harmful free radicals are formed when absorbed iron fails to bind to the protein, which in turn affects the concentration of iron in mammalian cells and biological fluids. This circulating unbound iron has a corrosive effect on the gastrointestinal tract and biological fluids. A higher level of iron enters into the body and it becomes saturated. These free ions enter the cells of the heart, liver, and brain. Due to disruption of oxidative phosphorylation by free iron, the ferrous iron gets changed to ferric iron that releases hydrogen ions, thus increasing the metabolic acidity. Free iron can also lead to lipid peroxidation, resulting in severe damage to mitochondria, microsomes, and other cellular organelles. Iron toxicity in cells causes iron-mediated tissue damage, cellular oxidizing and reducing mechanisms, and their poisonousness in intracellular organelles. The iron-produced hydrogen free radicals attack DNA, resulting in cellular damage, mutation, and malignant changes which cause an increased incidence of diseases. Some common ailments linked with heavy metal toxicity are:

- Aluminium containers used in manufacturing and processing of some foods, cosmetics, and medicines, and water purification cause brain damage
- Antimony causes lung disease and skin cancer, particularly among smokers
- Copper causes a flu-like reaction called metal fume disease and blood disorders
- Lithium causes ailments of the stomach, intestinal tract, central nervous system, and kidneys
- Overexposure to zinc causes flu-like symptoms of metal fume fever; stomach and intestinal disorders; and/or liver dysfunction
- Overexposure to bismuth causes life-threatening drowsiness and neurologic disorders, e.g., misperception, reduced attentiveness, illusions, misconceptions, myoclonic jerks, shivers, seizures, and decreased ability to coordinate voluntary movements, and/or the incapacity to stand or walk
- Overexposure to gold used in the treatment of rheumatoid arthritis may cause skin flushes, depression in bone marrow, stomach and abdominal bleeding, headaches, vomiting, and yellowing of the skin and mucous membranes, and jaundice

- Overexposure to nickel is linked with an increased risk of lung cancer, while overexposure to selenium may cause irritation of the respiratory system, digestive tract, and eyes, inflammation of the liver, hair loss, skin depigmentation, and peripheral nerve damage
- Overexposure to silver manifests as discoloration of the skin and hair, together with nausea, vomiting, and diarrhoea
- Overexposure to tin may harm the nervous system and result in psychomotor disorders including tremor, convulsions, illusions, and psychotic behaviour

Metal toxicity of arsenic, lead, chromium, cadmium, and mercury are discussed in the following chapters.

BIBLIOGRAPHY

Baatrup, E. (1991) Structural and functional effects of HMs on nervous system, including sense organs, of fish. *Comparative Biochemistry and Physiology Part C: Comparative Pharmacology* 100(1–2): 253–257.

Barabasz, W., Albinska, D., Jaskowska, M., and Lipiec J. (2002) Ecotoxicology of aluminium. *Polish Journal of Environmental Studies* 113: 199–203.

Bernard, A. (2008) Cadmium & its adverse effects on human health. *Indian Journal of Medical Research* 1284: 557–564.

Beyersmann, D. and Hartwig, A. (2008) Carcinogenic metal compounds: Recent insight into molecular and cellular mechanisms. *Archives of Toxicology* 828: 493–512.

Bradl, H. (2002) *HMs in Environment: Origin, Interaction and Remediation.* London: Academic Press.

Chandra, P. and Kulshreshtha, K. (2004) Chromium accumulation and toxicity in aquatic vascular plants. *Botanical Review* 703: 313–327.

Chang, L. W., Magos, L., and Suzuki, T. (1996) *Toxicology of Metals.* Boca Raton, USA: CRC Press.

Csuros, M. and Csuros, C. (2002) *Environmental Sampling and Analysis for Metals.* Boca Raton, USA: Lewis Publishers.

Dalcorso, G., Farinati, S., Maistri, S., and Furini, A. (2008) How plants cope with cadmium: Staking all on metabolism and gene expression. *Journal of Integrative Plant Biology* 5010: 1268–1280.

Dissanayake, C. B. and Chandrajith, R. (2009) Phosphate mineral fertilizers, trace metals and human health. *Journal of the National Science Foundation of Sri Lanka* 373: 153–165.

Duffus, J. H. (2002) HMs – A meaningless term? *Pure Applied Chemistry* 745: 793–807.

Ernst, W. H. O., Krauss, G. J., Verkleij, J. A. C., and Wesenberg, D. (2008) Interaction of HMs with sulphur metabolism in angiosperms from an ecological point of view. *Plant, Cell and Environment* 311: 123–143.

Fergusson, J. E. (1990) *Heavy Elements: Chemistry, Environmental Impact and Health Effects.* Oxford: Pergamon Press.

Fine, J. S. (2000) Iron poisoning. *Current Problems in Pediatrics* 303: 71–90.

Gamalero, E., Lingua, G., Berta, G., and Glick, B. R. (2009) Beneficial role of plant growth promoting bacteria and arbuscular mycorrhizal fungi on plant responses to heavy metal stress. *Canadian Journal of Microbiology* 555: 501–514.

Gupta, N., Gaurav, S. S., and Kumar, A. (2013) Molecular basis of aluminium toxicity in plants: A review. *American Journal of Plant Science* 4: 21–37.

Hamelink, J. L., Landrum, P. F., Harold, B. L., and William, B. H. (1994) *Bioavailability: Physical, Chemical, and Biological Interactions*. Boca Raton, FL: CRC Press Inc.
Han, J. X, Shang, Q., and Du. Y. (2009) Effect of environmental cadmium pollution on human health. *Health* 13: 159–166.
He, Z. L., Yang, X. E., and Stoffella, P. J. (2005) Trace elements in agro-ecosystems and impacts on the environment. *Journal of Trace Element in Medicine and Biology* 192(3): 125–140.
Hossain, M. A., Piyatida, P., da Silva, J. A. T., and Fujita, M. (2012) Molecular mechanism of heavy metal toxicity and tolerance in plants: Central role of glutathione in detoxification of reactive oxygen species and methylglyoxal and in heavy metal chelation. *Journal of Botany* 2012: 1–31.
Järup, L. (2003) Hazards of heavy metal contamination. *British Medical Bulletin* 681: 167–182.
Johnston, R. (1986) Aquatic chemistry and the human environment. *Chemistry and Ecology* 22: 125–169.
Khanna, K., et al. (2018) Role of micro-organisms in modulating antioxidant defence in plants exposed to metal toxicity. In: *Plants Under Metal and Metalloid Stress*, Mirza H., Kamrun N., Masayuki F. (eds.). Singapore: Springer, pp. 303–335.
Kumar, B., et al. (2010) Distribution, partitioning, bioaccumulation of trace elements in water, sediment and fish from sewage fed fish ponds in eastern Kolkata, India. *Toxicological & Environmental Chemistry* 922: 243–260.
López Carnelo, L. G. de Miguez, S. R. de, and Marbán, L. (1997) HMs input with phosphate fertilizers used in Argentina. *Science of the Total Environment* 2043: 245–250.
Malik, D. S. and Maurya, P. K. (2014) Heavy metal concentration in water, sediment, and tissues of fish species *Heteropneustis fossilis* and *Puntius ticto* from Kali River, India. *Toxicological & Environmental Chemistry* 96(8): 1195–1206.
Martin, S. and Griswold, W. (2009) Human health effects of HMs. *Environmental Science and Technology Briefs for Citizens* 15: 1–6.
Mortvedt, J. J. (1996) Heavy metal contaminants in inorganic and organic fertilizers. *Fertilizer Research* 431(3): 55–61.
Mudgal, V., et al. (2010) Effect of toxic metals on human health. *Open Nutraceuticals Journal* 3: 94–99.
Nagajyoti, P. C., Lee, K. D., and Sreekanth, T. V. M. (2010) HMs, occurrence and toxicity for plants: A review. *Environmental Chemistry Letters* 83: 199–216.
Nriagu, J. O. (1989) A global assessment of natural sources of atmospheric trace metals. *Nature* 338: 47–49.
Pacyna, J. M. (1996) Monitoring and assessment of metal contaminants in air. In: *Toxicology of Metals*, Chang L. W., Magos L., Suzuli T. (eds.). Boca Raton: CRC Press.
Patrick, L. (2003) Toxic metals and antioxidants: Part II. Role of antioxidants in arsenic and cadmium toxicity. *Alternative Medicine Review* 82: 106–128.
Perfus-Barbeoch, L., Leonhardt, N., Vavasseur, A., and Forestier, C. (2002) Heavy metal toxicity: Cadmium permeates through calcium channels and disturbs the plant water status. *Plant Journal* 324: 539–548.
Rascio, N. and Navari-Izzo, F. (2011) Heavy metal hyperaccumulating plants: How and why do they do it? And what makes them so interesting? *Plant Science* 1802: 169–181.
Sutton, D, Tchounwou, P. B., Ninashvili, N., and Shen E. (2002) Mercury induces cytotoxicity, and transcriptionally activates stress genes in human liver carcinoma cells. *International Journal of Molecular Science* 39: 965–984.
Tchounwou, P. B., Centeno, J. A., and Patlolla A. K. (2004) Arsenic toxicity, mutagenesis and carcinogenesis – a health risk assessment and management approach. *Molecular and Cellular Biochemistry* 255: 47–55.

Villiers, F., Ducruix, C. Hugouvieux, V., et al. (2011) Investigating plant response to cadmium exposure by proteomic and metabolomic approaches. *Proteomics* 119: 1650–1663.

Wang, D., Dang, Z., Feng, H., and Wang, R. (2015) Distribution of anthropogenic cadmium and arsenic in arable land soils of Hainan, China. *Toxicological & Environmental Chemistry* 973(4): 402–408.

Wang, S. and Shi, X. (2001) Molecular mechanisms of metal toxicity and carcinogenesis. *Molecular and Cellular Biochemistry* 222: 3–9.

WHO. (1996) *Trace Elements in Human Nutrition and Health.* Geneva, Switzerland: World Health Organization.

4 Heavy Metal Toxicity
Lead

Alchemists associated the metal with the planet Saturn and quested for a way to transmute lead into gold.

Anonymous

After going through this chapter, the reader will learn about the sources and route of exposure of lead, aetiology, pathophysiology, and clinical manifestations, mechanism of lead toxicity, and toxicological effects of lead.

FACT FILE

- The name of the element "lead" comes from the Anglo-Saxon word used for metal. Its chemical symbol is Pb, which is based on the word "*plumbum*", the Latin name for lead. The Latin word for the metal, *plumbum*, is also the origin of the word "*plumbing*"
- Pb, with an atomic mass of 207 and atomic number of 82, is a dense, soft, and malleable heavy metal
- The abundance of lead in the Earth's crust is 14 parts per million by weight, while its abundance in the solar system is 10 parts per billion by weight
- Lead is a basic, or post-transition, metal
- The most common oxidation state of lead is 2^+, followed by 4^+. The 3^+, 1^+, 1^-, 2^-, and 4^- states also occur
- Lead is obtained from the mineral galena by a roasting process
- Lead has a relatively low melting point
- Freshly cut lead is bluish-white but it turns dull grey when exposed to air
- Lead has three isotopes, i.e., lead-206, lead-207, and lead-208
- Lead is a relatively unreactive post-transition metal
- Its weak metallic character is illustrated by its amphoteric nature
- Lead and its oxides react with acids and bases, and it tends to form covalent bonds
- It tends to form a bond with itself; it can form chains, rings, and polyhedral structures
- Lead is easily extracted from its ores

DOI: 10.1201/9781003343721-4

- Lead's toxicity was recognized in the late 19th century
- Metallic lead beads from 7000 to 6500 BCE were discovered in Asia Minor and represent the first example of metal smelting
- Ancient Egyptians for the first time used lead minerals in cosmetics. They also used lead for sinkers in fishing nets, glazes, glasses, enamels, and ornaments
- People of the Indus valley civilization and Mesoamericans used lead for making amulets
- Lead was used for making water pipes by the Roman Empire
- Its resistance to corrosion ensured its widespread use in pharmaceuticals, insecticides, hair dyes, and as an anti-knocking additive for petrol, and also in roofing, currency, and warfare
- Lead is widely used in manufacturing car batteries, pigments, bullets, cable sheathing, weights, and weight belts for diving, lead crystal glass, radiation protection, and repairs
- Lead is often used to store corrosive liquids

According to the World Health Organization:

- The daily intake of lead from all sources is about 0.1 milligrams
- An average human body stores about 120 milligrams of lead in the bones
- Lead as a toxicant affects multiple organs and is particularly injurious to children
- Lead is circulated in the brain, liver, kidney, and bones and is stored in the teeth and bones, where it accumulates over the years. Human exposure to lead can be measured in the blood
- Lead is toxic, teratogenic, and carcinogenic
- During pregnancy, if lead in bone is released into the blood, it can affect the developing foetus
- Lead is a neurotoxin that is stored in the soft tissues and bones, damages the nervous system, and causes blood disorders
- Exposure to lead is generally avoidable

4.1 INTRODUCTION

Lead is a naturally occurring toxic metal found in the Earth's crust; however, its extensive use has resulted in environmental pollution, with human exposure causing health problems. Children are more susceptible to its lethal effects, mainly affecting the development of brain. It may result in long-term effects in adults, including increased risk of high blood pressure and renal failure. Exposure of pregnant women to lead can lead to miscarriage, stillbirth, premature birth, and low birth weight, in addition to minor deformities.

4.2 SOURCES AND ROUTE OF EXPOSURE

People are exposed to lead through occupational and environmental sources. Important sources of environmental contamination include:

- Mining
- Smelting (zinc and copper smelting, processing of ore)
- Use of leaded paint, leaded gasoline, and leaded aviation fuel
- Use of lead-acid batteries in motor vehicles

4.3 AETIOLOGY

The most significant lead exposure among adults usually occurs in the workplace, whereas for children, other modes of environmental exposure are more important. Although lead toxicity can occur after a single event, it is usually a result of chronic exposure. Lead is a constituent of dyes, paints, soldering material, tainted glass, lead crystal glassware, grenades, ceramic coatings, jewellery, and toys, besides being a constituent of cosmetics and traditional medicines. Drinking water supplied through lead pipes may contain lead. Inhaling dust containing lead particles produced by burning, smelting, recycling, and peeling-off of materials containing lead and consumption of lead-contaminated dust, water, and food, can affect human health. Exposure to lead-contaminated soil and dust released during the recycling of batteries and mining has resulted in mass lead poisoning and multiple deaths in children in Nigeria, Senegal, and other developing nations.

In adults, occupational exposure is the main cause of lead poisoning. Those working with lead-containing products, lead miners and smelters, plumbers and fitters, auto mechanics, glass manufacturers, construction workers, battery producers, firing range instructors, and plastic manufacturers are at risk for exposure to lead. Lead exposure risks include welding, the rubber and printing industries, and combustion of solid waste.

4.4 PATHOPHYSIOLOGY AND CLINICAL MANIFESTATIONS

Exposure occurs through inhalation, ingestion, or rarely, dermal contact. Lead may be taken in through direct contact with the mouth, nose, and eyes, or through cracked skin. Tetraethyl lead, a gasoline additive used in aviation fuel, permeates through the skin; however, inorganic lead found in paint, food, and most lead-containing products is only absorbed through the skin. The main source of inorganic lead is ingestion and inhalation. In the case of adults, about 35–40 per cent of lead-containing dust gets deposited in the lungs, and about 95 per cent reaches the bloodstream. Lead is stored in the blood, soft tissue, and bone. Lead in bones, teeth, hair, and nails is bound tightly and not available to other tissues, and therefore is considered harmless. The half-life of lead in bone is 20–30 years, and bone can introduce lead into the bloodstream once the initial exposure goes away. Other tissues storing lead include the brain, spleen, kidneys, liver, and lungs. Lead is slowly removed from the body in urine and smaller amounts of lead are removed through faeces and very small amounts though the hair, nails, and sweat. It has no physiological role in the body, and its damaging effects are innumerable. The primary causes of lead's toxicity are due to:

- Its interference with enzymes as it attaches to sulfhydryl groups found on many enzymes

- Its ability to mimic other metals that take part in biological processes
- Distortion in coordination of ligands because of the presence of a lone pair in the electronic structure of lead ions
- Inhibition of the activity of enzymes, namely delta-amino levulinic acid dehydratase, or ALAD (aminolevulinate dehydratase), and ferrochelatase required for the biosynthesis of haem, the cofactor found in haemoglobin
- Lead can easily pass through endothelial cells at the blood–brain barrier by substituting calcium ions by calcium-ATPase pumps. Therefore, it affects normal development of the child's brain and nervous system
- Loss of the myelin sheath of neurons, reduced numbers of neurons, interference with neurotransmission, and decreased neuronal growth

4.5 MECHANISM OF LEAD TOXICITY

- Lead toxicity is due to the increased generation of reactive oxygen species (ROS) which interferes with the production of antioxidants (Figure 4.1). Later, ROS are stabilized by glutathione in the body. A total of 90 per cent of glutathione in cells is in reduced form and 10 per cent in oxidative form, which acts as an antioxidant defence mechanism
- Glutathione stabilizes ROS, but after oxidizing, it is converted to glutathione disulfide, and reduced back to GSH by glutathione reductase
- Lead inactivates glutathione by binding to GSH's sulfhydryl group, which hampers the replacement of GSH and increases the oxidative stress
- Increased oxidative stress due to lipid peroxidation damages the cell membrane
- Lead blocks the activity of 5-aminolevulinic acid dehydratase leading to haemoglobin oxidation, which together with lipid peroxidation results in red cell haemolysis
- Lead entering intravascular space binds quickly to RBCs, from where it diffuses into the soft tissues, including the kidneys, brain, liver, and bone marrow
- Lead diffused into bone is stored for several decades. Increased bone turnover with pregnancy, menopause, lactation, or immobilization can increase the level of lead in the blood

4.6 TOXICOLOGICAL EFFECTS OF LEAD

Haematological Toxicity: This manifests itself in the form of:

- Impaired synthesis of haem
- Haemolysis of red cells
- Shortened red cell survival
- Mild anaemia in adults

Gastrointestinal Toxicity: This manifests itself in the form of:

- Lead colic
- Nausea and vomiting

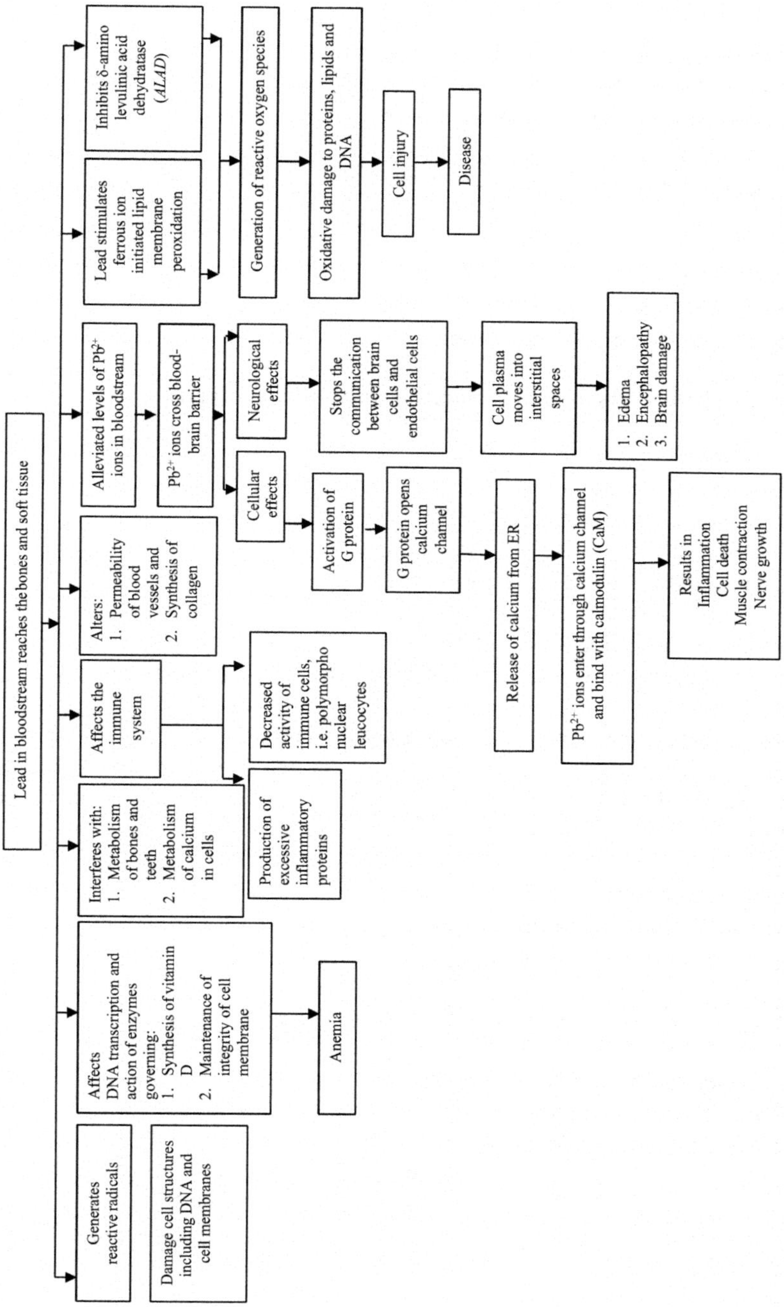

FIGURE 4.1 Effects of lead toxicity on the human body.

- Anorexia or eating disorders resulting in very low weight
- Occasionally, severe diarrhoea and dehydration

Renal Toxicity: This manifests itself in the form of:

- Acute nephropathy with tubular Fanconi's syndrome

Cardiovascular Impacts: This manifests itself in the form of:

- Hypertension
- Oxidative stress
- Chronic nephropathy
- Cardiovascular disease and stroke

Reproductive Effects: In men, this manifests itself in the form of:

- Reduction in libido
- Abnormal spermatogenesis
- Chromosomal damage and infertility

In women, it manifests itself in the form of:

- Increase in the incidence of stillbirth
- Miscarriage and pregnancy-induced hypertension and prematurity

Paediatric Lead Poisoning: Lead poisoning in children is a somewhat newly recognized problem, which is believed to be transmitted from a mother to her child through breast milk. Though there are no pathognomonic symptoms, but if they appear, they are usually nonspecific and include the following:

- Erratic behaviour
- Hyperactivity or decreased activity
- Loss of or delayed developmental milestones

Long-term exposure to lead may cause symptoms including:

- Abdominal pain and loss of appetite
- Vomiting and constipation
- Headache, fatigue, and seizures

Excretion of Lead: Lead is mainly expelled in the urine and bile, although the rate of removal differs. The kidneys excrete lead through glomerular filtration and tubular secretion, with the clearance of lead ranging from 1–3 ml/min.

4.7 DIAGNOSIS

The diagnosis of lead toxicity involves establishing the clinical signs and recording the medical history. The diagnosis of lead poisoning can be done by laboratory analysis of the blood lead level. Examination of the blood film may relate the basophilic stippling of RBCs, in addition to the changes usually associated with anaemia, known as sideroblastic anaemia.

- Pb exposure can also be assessed by measuring erythrocyte protoporphyrin (EP) in blood samples. EP is a part of RBCs which increases when the amount of lead in blood is high. Thus, EP levels together with blood lead levels can suggest the time period of exposure; if blood Pb levels are high but EP remains normal, the finding suggests a recent exposure
- Lead in bones can be assessed by X-ray fluorescence. The radiographic sign of raised Pb levels is the presence of radio-dense lines called "*lead lines*" at the metaphysis in the long bones of growing children, mainly around the knees
- Lead lines, caused by increased calcification due to interrupted metabolism in developing bones, become broader if the exposure time increases. X-rays may reveal Pb-containing foreign materials such as paint chips in the gastrointestinal tract. Faecal content of Pb may be a precise method to evaluate the overall amount of lead intake, especially when taken orally.

4.8 TREATMENT AND MANAGEMENT OF LEAD POISONING

The first step in treating Pb poisoning is to remove the source of contamination. In addition, chelation therapy is beneficial in which a medication given orally binds with the lead and helps in its elimination in urine. This might be suggested for children with high blood levels of Pb or symptoms of Pb poisoning. Another option is ethylene diamine tetra acetic acid (EDTA) chelation therapy, in which the doctors treat adults and children who cannot tolerate the drug used in conventional chelation therapy. In this therapy, a chemical called calcium disodium EDTA is given through an injection. Pb poisoning is avoidable, and there are some safeguards that can be taken to ensure that children are safe from Pb poisoning.

4.9 IMPACT OF LEAD ON THE ENVIRONMENT

Industrialization, urbanization, mining, and many other anthropogenic activities have resulted in redistribution of lead from the Earth's crust to soil and the environment. Lead forms various complexes with soil components, and only a small fraction of lead present as these complexes in soil solutions are phyto available. Despite it lacking a role in plants, lead is absorbed by them mainly through the roots from soil solution and thus enters the food chain. The absorption of lead by roots occurs via an apoplastic pathway or via Ca^{2+}-permeable channels. The behaviour of lead in soil and uptake by plants are controlled by its speciation and by soil pH, soil particle size, cation-exchange capacity, root surface area, root exudation, and degree of mycorrhizal transpiration. After uptake, lead mainly accumulates in root cells, because of the blockage by Casparian strips within the endodermis. Lead is also entrapped due to the negative charges that exist in roots' cell walls. Excessive accumulation of lead in plant tissue impairs various morphological, physiological, and biochemical functions in plants, either directly or indirectly, and induces a range of deleterious effects. It causes phyto-toxicity by changing cell membrane permeability, by reacting with active groups of different enzymes involved in plant metabolism, and by reacting with phosphate groups of ADP or ATP, and by replacing essential ions. Lead toxicity causes inhibition of ATP production, lipid peroxidation, and DNA damage by overproduction of ROS. Additionally, lead strongly inhibits seed germination, root elongation, seedling development, plant growth, transpiration, chlorophyll production, and water and protein contents. Negative effects that lead has on plant vegetative growth mainly result from the following factors:

- Distortion of chloroplast ultrastructure
- Obstructed electron transport
- Inhibition of Calvin cycle enzymes
- Impaired uptake of essential elements, such as Mg and Fe, and induced deficiency of CO_2 resulting from stomatal closure

Under lead-induced stress, plants possess several defence strategies to cope with lead toxicity. These strategies include:

- Reduced uptake into the cell
- Sequestration of lead into vacuoles by the formation of complexes
- Binding of lead by phytochelatins, glutathione, and amino acids
- Synthesis of osmolytes

In addition, the activation of various antioxidants to combat increased production of lead-induced ROS constitutes a secondary defence system. Not only leaded gasoline causes increased lead concentrations in the environment but other human activities, such as fuel combustion, industrial processes, and solid waste combustion, also contribute.

Lead can end up in water and soil through the corrosion of leaded pipelines in water-transporting systems and through the corrosion of leaded paints. It cannot be broken down, but is converted to other forms. Lead accumulates in the bodies of aquatic and soil organisms. These organisms also experience the adverse health effects of lead poisoning. Health effects on shellfish can take place even when only a small concentration of lead is present. The body functioning of phytoplankton is disturbed when lead interferes. Phytoplankton are an important source of oxygen production in the oceans and many larger sea-animals consume them. Soil functions are disturbed by lead intervention, mainly near highways and farmlands, where extreme concentrations are present. Soil organisms also suffer from lead poisoning. Lead is a particularly dangerous chemical, as it can accumulate in individual organisms and the entire food chain.

BIBLIOGRAPHY

Assi, M. A., et al. (2016) Detrimental effects of lead on human and animal health. *Veterinary World* 9(6): 660–671.

ATSDR. (2007) *Toxicological Profile for Lead.* Division of Toxicology and Environmental Medicine, Georgia: Agency for Toxic Substances and Disease Registry.

Baird, C. and Cann, N. (2012) *Environmental Chemistry.* New York: W. H. Freeman and Company.

Ekong, E. B., Jaar, B. G., and Weaver, V. M. (2006) Lead-related nephrotoxicity: A review of the epidemiologic evidence. *Kidney International* 70: 2074–2084.

Flora, G., Gupta, D., and Tiwari, A. (2012) Toxicity of lead: A review with recent updates. *Interdisciplinary Toxicology* 5: 47–58.

Hodge, T. A. (1981) Vitruvius, lead pipes and lead poisoning. *American Journal of Archaeology* 85(4): 486–491.

Jacobs, D. E., et al. (2002) Prevalence of lead-based paint hazards in U.S. housing. *Environmental Health Perspectives* 110: 599–606.

Mahaffey, K. R. (1990) Environmental lead toxicity: Nutrition as a component of intervention. *Environmental Health Perspectives* 89: 75–78.

Mason, L. H., Harp, J. P., and Han, D.Y. (2014) Pb neurotoxicity: Neuropsychological effects of lead toxicity. *BioMed Research International* 8.

Moore, M. R. (1977) Lead in drinking water in soft water areas—health hazards. *Science of Total Environment* 7(2): 109–115.

Navas-Acien, A., Guallar, E., Silbergeld, E. K., and Rothenberg, S. J. (2007) Lead exposure and cardiovascular disease—a systematic review. *Environmental Health Perspectives* 115: 472–482.

Patrick, L. (2006) Lead toxicity: A review of literature. Part 1: exposure, evaluation, and treatment. *Alternative Medicine Review* 11: 2–22.

Sanborn, M. D., Abelsohn, A., Campbell, M., and Weir, E. (2002) Identifying and managing adverse environmental health effects: 3. Lead exposure. *Canadian Medical Association Journal* 166: 1287–1292.

Sinha, S. P., et al. (1993) Neurotoxic effects of lead exposure among printing press workers. *Bulletin of Environmental Contamination and Toxicology* 51(4): 490–493.

Staudinger, K. C. and Roth, V. S. (1998) Occupational lead poisoning. *American Family Physician* 57: 719–726.

Waldron, H. A. (1985) Lead and lead poisoning in antiquity. *Medical History* 29(1): 107–108.

WHO. (2018) *Lead Poisoning and Health.* Geneva, Switzerland: World Health Organization.

Woolf, A. D., Goldman, R., and Bellinger, D. C. (2007) Update on the clinical management of childhood lead poisoning. *Pediatric Clinic of North America* 54: 271–294.

5 Heavy Metal Toxicity
Arsenic

Cyanide is natural. So is arsenic.

Chuck Palahniuk

After going through this chapter, the reader will learn about the sources and routes of exposure to arsenic, the aetiology, pathophysiology, and clinical manifestations, the mechanism of arsenic toxicity, and the toxicological effects of arsenic.

FACT FILE

- Arsenic, with the atomic symbol As and atomic number 33, has an atomic weight of 74.92 and density of 5.776 grams per cubic centimetre
- Arsenic is solid at room temperature, with a melting point of 817°C and boiling point of 603°C
- Of the 33 radioactive isotopes, the most common isotope of arsenic is As-75, which is stable
- Arsenic has been in use since ancient times, mainly by ancient Greeks, Egyptian, and Chinese civilizations
- During the Bronze Age arsenic was used to toughen alloys and it is still used to strengthen alloys of lead and copper
- Albertus Magnus in 1250 BCE isolated arsenic as an element
- Though unconfirmed, there is a theory that Napoleon Bonaparte was gradually poisoned with arsenic, which eventually led to his death in the year 1821
- During the Victorian period, white arsenic, or arsenic trioxide (As_2O_3), was commonly available and sold in grocery stores. Women used to consume or rub arsenic mixed with vinegar or chalk to improve their complexion
- Arsenic has three common allotropes, grey, yellow, and black arsenic; grey arsenic is most stable, while yellow arsenic is most unstable and highly toxic
- Fumes of heated arsenic smell like garlic
- Arsenic is difficult to find in the natural environment; however, arsenic as a natural metalloid is present in groundwater and its inorganic form is very toxic

DOI: 10.1201/9781003343721-5

- Contaminated water used for drinking, preparation of food, and irrigation poses a threat to public health
- Arsenic is readily absorbed by plants from groundwater and soil, which results in food poisoning
- For environmental reasons, arsenic mining has been banned
- Arsenic is produced as a by-product of copper purification
- Arsenic toxicity is useful for tackling fungi, bacteria, and insects in wood preservation
- It is used in pesticides and added to livestock feed to increase the weight of animals and combat infections
- Arsenic was used in chemical weapons during the First World War
- Despite its toxic quality, arsenic has been used for therapeutic purposes, e.g., as a skin bleaching agent, in cancer treatment, and for curing syphilis and leukaemia. It is used as an indicator in cancer-detecting PET scans
- Long-term exposure to arsenic can cause cancer, skin lesions, cardiac diseases, and diabetes. In utero and early childhood exposure it affects the cognitive development and result in increased deaths

5.1 INTRODUCTION

Arsenic, a naturally occurring metalloid with universal distribution in the Earth's crust and groundwater, is also found at lower levels in the air and food products, with crustaceans and seafood posing the main threats of poisoning. Arsenic toxicity is a worldwide health problem affecting billions of people. Arsenic poisoning, or arsenicosis, occurs if a person is exposed to a high level of arsenic. Arsenic manifests its toxicity by deactivating enzymes, mainly those involved in cellular energy pathways, and DNA synthesis and repair. Primarily, acute arsenic poisoning is linked with nausea, vomiting, abdominal pain, severe diarrhoea, etc. Encephalopathy and peripheral neuropathy have also been reported due to arsenic toxicity. Chronic arsenic toxicity causes multisystem disease. Arsenic is a well-documented carcinogen affecting numerous organs. Evidence-based treatment is not available to treat chronic arsenic poisoning but the use of antioxidants has been encouraged. The focus is on decreasing arsenic consumption from drinking water and there is a growing emphasis on using alternate water supplies.

5.2 SOURCES AND ROUTES OF EXPOSURE

Arsenic is a natural semi-metallic element found across the world in groundwater. Arsenic in the environment does not affect humans directly, and lethal quantities of arsenic are only found sporadically in nature. Arsenic from natural sources escapes from mining and other industrial processes into aquifers, polluting drinking water. Arsenic has been used in some traditional medicines as one of the ingredients of medicinal formulations. Greek physicians such as Hippocrates and Galen used it in the healing of various diseases. Some Chinese medicines contain arsenic sulphide. In India, arsenic is used in homoeopathic formulations. Arsenic trioxide is also used to treat acute promyelocytic leukaemia. Arsenic is absorbed from the small intestine, dermal contact, and breathing.

Long-term chronic exposure to Fowler's solution containing arsenic cause haemangiosarcoma, angiosarcoma of the liver, and nasopharyngeal carcinoma. Arsphenamine containing arsenic was used intravenously to cure syphilis and protozoan infections till the Second World War. Arsenic in food occurs as comparatively non-toxic organic compounds. Seafood, fish, and algae can increase arsenic levels in blood but are quickly excreted unaltered in urine. Organic and inorganic arsenic compounds may enter the plant food chain from agriculture goods or from soil watered with arsenic-polluted water.

5.3 AETIOLOGY

Consumption of toxic arsenic beyond the permissible limit is the main cause of arsenic poisoning. If consumed in large quantities it can instantly kill a person. In cases where it is consumed in smaller quantities over a number of years, it can result in serious illness or fatality. The main cause of arsenic poisoning across the world is the drinking of contaminated groundwater containing high levels of arsenic released by rocks. Moreover, groundwater contains runoff from industrial effluents. Other possible causes of arsenic poisoning include:

- Exposure to landfill or waste disposal sites
- Breathing contaminated air from industries or mines using/releasing arsenic
- Inhalation of smoke or dust from wood or waste formerly treated with arsenic
- Consuming arsenic-contaminated food, e.g., seafood and animal products
- Smoking tobacco-containing products
- Staying near industrialized zones
- Working with arsenic at the workplace (glass production, smelting, wood treatment, pesticide plants); if appropriate safety measures are not taken into consideration, workers in some industries face a higher risk of toxicity

Humans exposure to arsenic depends on the manner in which it is used in the industries, e.g., arsenic may be inhaled in the smelting industry as there is inorganic arsenic in the emissions. Moreover, there may be traces of arsenic in some foods, for example meat, fish, and poultry (due to antibiotics in chicken feed). Rice has been found to contain higher levels of arsenic than water.

5.4 PATHOPHYSIOLOGY AND CLINICAL MANIFESTATIONS

The main site of absorption of arsenic is the small intestine. Absorption occurs by an electrogenic process involving a proton (H^+) gradient. The ideal pH for arsenic absorption is 5, although the pH of the small intestine is approximately 7 because of pancreatic bicarbonate secretion. Absorbed arsenic undergoes hepatic biomethylation to form monomethylarsonic acid and dimethylarsinic acid that are less toxic but not totally harmless. Nearly half of the dose is excreted in the urine in 3–5 days.

Dimethylarsinic acid is the leading urinary metabolite (60–70 per cent) when compared with monomethylarsonic acid. Some inorganic arsenic is also excreted unchanged. The highest concentration of arsenic is in the kidneys and liver. There are instances which highlight the multiple organ system effects due to arsenic. In 1900, beer contaminated with arsenic resulted in 6,000 poisonings and approximately 71 deaths in northern England. Sulphuric acid polluted with arsenic used to make sugar from sugarcane for brewing beer was the source of contamination, which affected nearly 100 breweries. Common clinical presentations were anorexia, brown pigmentation, muscular weakness, pain and paresthesias in the extremities, hepatic lesions, localized oedema, and fatty degeneration of the heart.

Gastrointestinal Effects: Arsenic affects different parts of body, e.g., the gastrointestinal (GI) effects of arsenic are usually the result of ingestion; but GI effects may also occur after heavy exposure through other routes. GI implications after arsenic ingestion are intense, and less often after inhalation or dermal absorption. The fundamental GI lesion appears to be increased penetrability of the small blood vessels, leading to fluid loss and hypotension. Extensive inflammation and necrosis of the mucosa and sub-mucosa of the stomach and intestine may occur, which may lead to perforation of the gut wall. A haemorrhagic gastroenteritis may develop, with bloody diarrhoea. Acute arsenic toxicity is linked with hepatic necrosis and elevated levels of liver enzymes. Arsenic intoxication may result in toxic hepatitis with elevated liver enzyme levels.

Renal Effects: Arsenic is capable of producing renal effects. Systemic toxicity may include acute tubular necrosis with acute renal failure. Also, glomerular damage can result in proteinuria; however, the kidney is not the most sensitive target organ for chronic arsenic toxicity.

Cardiovascular Effects: Both acute and chronic exposures to high levels of arsenic may produce a range of adverse cardiovascular effects. Acute poisoning may result in capillary leakage and cardiomyopathy and shock.

Hypertension is reported with long-term exposure to arsenic. Myocardial damage results in various electrocardiographic findings, including broadening of the QRS complex, and atypical, multifocal ventricular tachycardia. Gangrene of the extremities, known as "blackfoot disease", was reported after drinking arsenic-contaminated well water in Taiwan. Individuals with blackfoot disease had a higher incidence of arsenic-induced skin cancers.

Neurologic Effects: Arsenic may cause encephalopathy at acute high doses. Arsenic poisoning can cause peripheral neuropathy. In high-level arsenic exposures, the onset of neuropathy may occur after 7–14 days, with increased sweating, muscle cramps, muscle tenderness, numbness, and spontaneous pain.

Dermal Effects: Pigment changes are characteristic of chronic arsenic exposure. Delayed effects of acute or chronic exposure may be seen as Mee's lines in the nails.

The skin lesions occurring most frequently in arsenic-exposed humans are hyperkeratosis, hyperpigmentation, and skin cancer.

Respiratory Effects: Inhalation of concentrated arsenic compounds results in irritation of the respiratory mucosa. Lung cancer has been linked with chronic arsenic exposure in smelter workers and pesticide workers.

Haematopoietic and Haematologic Effects: Bone marrow depression may result from acute or chronic arsenic intoxication which may manifest as pancytopenia. Both acute and chronic arsenic poisoning may affect the hematopoietic system.

Reproductive Effects: Increased frequency of spontaneous abortions and congenital malformations are linked to arsenic exposure. Arsenic is a reproductive toxicant and a teratogen, which readily gets transferred across the placenta.

Lung Cancer: There is a link between lung cancer and occupational exposure to inorganic arsenic, and it is regarded as a cause of lung as well as skin cancer. There is a risk of lung cancer among workers exposed to arsenic trioxide in smelters and to pentavalent arsenical pesticides. In addition, chronic arsenic ingestion may cause cancers of the bladder, kidney, liver, lung, prostate, and skin in humans.

5.5 MECHANISM OF ARSENIC TOXICITY

Occurrence of Arsenic: Several chemical forms and oxidation states
Impacts: Acute and chronic health effects, including cancer

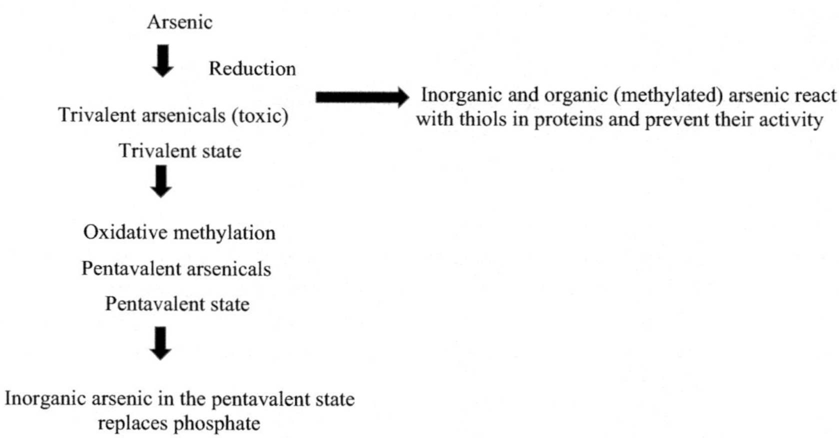

Regarding cancer, possible mechanisms include:

- Altered cell proliferation
- Altered DNA methylation
- Co-carcinogenesis

- Geno-toxicity
- Oxidative stress
- Tumour promotion

The basic mechanism involves:

1. Inhibition of formation of acetyl-CoA and enzyme succinic dehydrogenase by arsenite
2. Replacement of phosphate in many reactions by arsenate
3. Formation of Glc-6-arsenate
4. Muscle weakness in chronic arsenic poisoning
5. It attacks enzyme-bound thioester in the glyceraldehyde 3-phosphate dehydrogenase reaction
6. Formation of 1-arseno-3-phosphoglycerate, which is unstable and suddenly hydrolyses
7. Consequently, inhibition of ATP formation and escaping of phosphoglycerate kinase reaction
8. Formation of 2,3-bisphosphoglycerate in erythrocytes is affected, followed by higher oxygen affinity of haemoglobin and consequently greater cyanosis
9. Synthesis of sub-mitochondrial particles adenosine-5'-diphosphate-arsenate from ADP and arsenate in the presence of succinate
10. Loss of cell respiration and reduced ATP formation due to arsenate
11. Mitochondrial and cell swelling due to ATP depletion
12. Fatty changes in the liver, heart, and kidney due to glycogen depletion in liver cells

Arsenic suppresses the replication of DNA with transformed repair of enzymes. It prevents enzymatic activity in mitochondria due to inhibition of dehydrogenase and stimulation of mitochondrial adenosine triphosphatase activity by separating oxidative phosphorylation and tissue damage, which triggers cytotoxicity. Arsenic alters mitochondrial membrane potential and ATP formation during glycolysis and induces apoptosis in various cells. It has a geno-toxic effect on mammalian cells. It results in oxidative DNA damage based on iron release from ferritin with the generation of reactive oxygen species (ROS). It has a clastogenic effect in several cell types. Another mechanism of arsenic toxicity is methylation of inorganic arsenic. Documentary evidence suggests that ROS are key players in the inducting oxidative stress in cells and exposure to arsenic generates nitric oxide (NO) and superoxide anions (O^{2-}), which are consequently converted to more damaging species such as hydroxyl radicals (OH). Biological outcomes of arsenic toxicity are illustrated in Figure 5.1.

5.6 TOXICOLOGICAL EFFECTS OF ARSENIC

The symptoms of arsenic poisoning include:

- Changes in fingernail pigmentation, called leukonychia striata
- Confusion

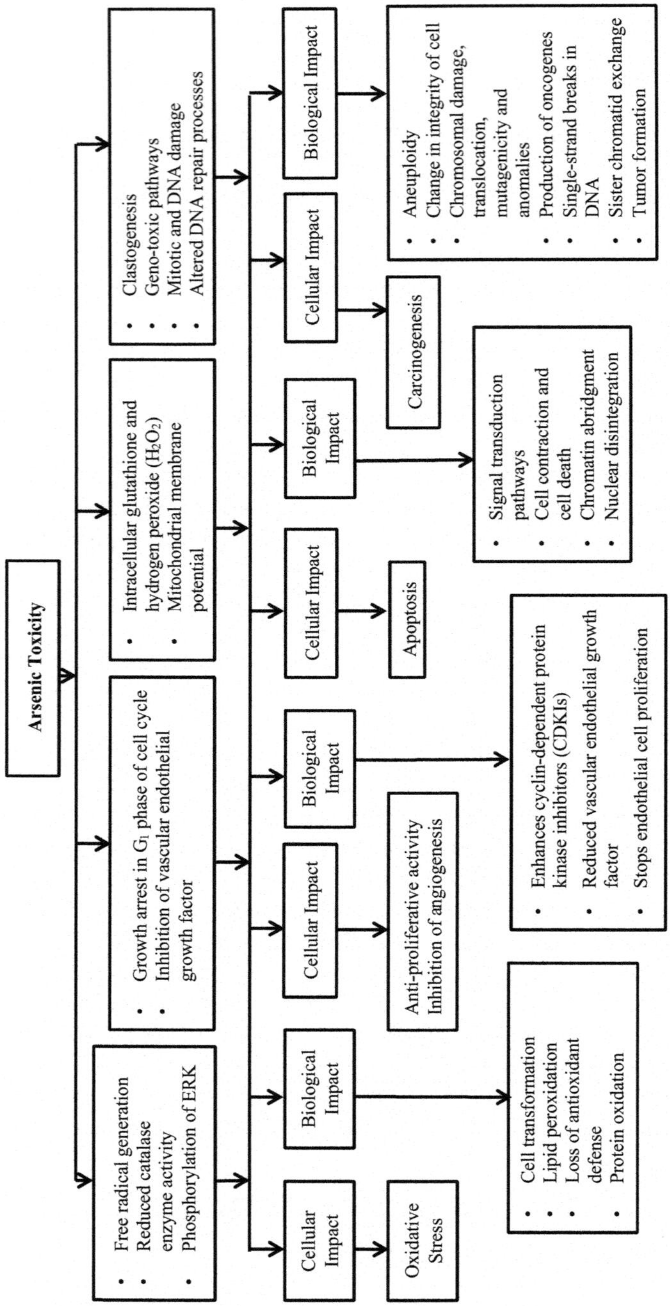

FIGURE 5.1 Mechanism of arsenic toxicity and its cellular and biological outcomes.

- Convulsions
- Drowsiness
- Headaches
- Severe diarrhoea

After chronic exposure, arsenic accumulates in the liver, kidneys, heart, and lungs, with smaller amounts in muscles, the nervous system, gastrointestinal tract, and spleen. Though most arsenic is removed from these sites, a quantity remains in the keratin-rich tissues, nails, hair, and skin. After about two weeks of ingestion, arsenic deposition starts in the hair and nails. The final outcome of arsenic poisoning is coma and death. Arsenic also causes:

- Cancer
- Cerebrovascular diseases
- Chronic lower respiratory diseases
- Diabetes
- Heart disease/cardiovascular disease
- Stroke

Skin effects include skin cancer after long-term exposure. Other dermal effects include darkening and thickening of the skin. The comutagenic effect of arsenic can cause interference with base and nucleotide excision repair, and finally interaction with zinc finger structures.

5.7 DIAGNOSIS

Arsenic may be assessed in the blood or urine to screen for environmental or occupational exposure and to confirm toxicity in the victim or to help in forensic investigation in the case of a fatal overdose. Analytical techniques can distinguish between organic and inorganic forms of element. Organic arsenic compounds are eliminated in the urine in unaffected form, while inorganic forms mostly are converted to organic arsenic compounds in the body before excretion. Tests diagnose poisoning by determining arsenic in the blood, urine, hair, and fingernails. The urine test is the most reliable test for arsenic exposure. In acute exposure, urine testing needs to be done within 24–48 hours. Testing of hair and fingernails can assess exposure to high levels of arsenic over the past 6–12 months. These tests can determine whether an individual has been exposed to above-average levels of arsenic and whether these arsenic levels will affect health. Chronic arsenic can stay in the body for a longer duration than a short-term or more isolated exposure and can be detected in a long-term arsenic exposure.

Owing to its ability to store trace elements from blood, hair is a potential bioindicator of arsenic exposure. Assimilated elements retain their position during hair growth. Thus, for a temporal assessment of exposure, an assay of hair composition can be carried out with a single hair. This type of bio-monitoring has been accomplished with micro-analytical techniques such as synchrotron radiation-based X-ray fluorescence (SXRF) spectroscopy and micro-particle induced X-ray emission (PIXE).

5.7.1 Treatment and Management of Arsenic Poisoning

Arsenic toxicity is a life-threatening condition and requires aggressive treatment with specific attention to fluid resuscitation, blood pressure support, and cardiac monitoring. Patients requiring hospitalization must be in an intensive care unit (ICU) for advanced life support monitoring. Chelation therapy has been shown to improve the outcomes if commenced within minutes to hours after exposure. Dimercaprol and dimercaptosuccinic acid are chelating agents that isolate arsenic away from blood proteins and are used in treating acute arsenic poisoning.

5.8 IMPACT OF ARSENIC ON THE ENVIRONMENT

Arsenic is found in the natural environment, in soil, groundwater, and plants. Epidemiological evidence has showed that it is carcinogenic both for humans and animals. Asia and other non-European locations have the most significant amounts of inorganic arsenic in the environment, with the exception of areas with industrial processes. Different areas around the world are affected by arsenic pollution of groundwater; the largest population at risk is in Bangladesh, followed by West Bengal in India. Arsenic concentrations in drinking water have severe health implications worldwide. Arsenic contamination in drinking water has been reported in Nepal, Myanmar, China, Inner Mongolia, Thailand, Vietnam, and Japan, Chile, Bolivia, central northern Mexico, Peru, Argentina, and Cambodia.

Aquatic and terrestrial biota indicates diverse sensitivities to arsenic, which is controlled by biological and abiotic factors. Inorganic and organic compounds of arsenic have varying bioavailabilities. Inorganic arsenicals are more toxic than organo-arsenicals, and arsenite is more toxic than arsenate. The uptake and toxicity mechanism of arsenate differ significantly among living organisms. The basic mechanism of arsenite toxicity is believed to result from its binding to protein sulfhydryl groups. Arsenate is known to affect oxidative phosphorylation by its opposition to phosphate. Arsenate toxicity to biota is usually abridged in environments where the phosphate concentration is high. Since arsenate is a phosphate equivalent, organisms living in elevated arsenate environments obtain the nutrient phosphorus to evade arsenic toxicity. Groundwater contamination of arsenic has become a global problem in the last few decades. Soil is being contaminated with arsenic through irrigated water, and rice, vegetables, and other plants are thus contaminated with arsenic through its uptake to a toxic level. Arsenic enters the food chain though drinking water, food, meat, milk, and eggs. The intake of bovine milk is one of the main exposure paths for chemicals and stock of lasting organic chemicals in tissues in the agrarian food chain.

Arsenic compounds have acute and chronic effects in individuals, populations, and communities, at concentrations ranging from a few micrograms to milligrams per litre, depending on the species, time of exposure, and end-points measured. These effects include lethality, inhibition of growth, photosynthesis, and reproduction, and behavioural effects. Arsenic-contaminated environments are characterized by limited species abundance and diversity. If levels of arsenate are high enough, only species which exhibit resistance may be present.

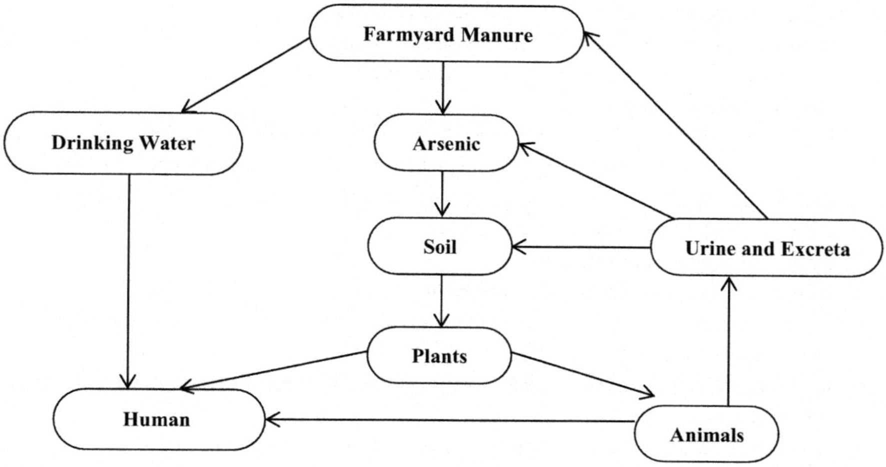

FIGURE 5.2 Diagrammatic representation of arsenic contamination in humans and animals.

The arsenic cycle (Figure 5.2) has broadened as a consequence of human interference and, because of this, large amounts of arsenic end up in the environment and living organisms. Arsenic is mainly emitted by the copper-, lead-, and zinc-producing industries, and also agriculture. It cannot be destroyed once it has entered the environment; hence, the amount that is added can spread and cause health effects in humans and animals in diverse locations.

Plants absorb arsenic fairly easily; hence, the concentration of the same in food can be quite high. Arsenic is non-essential and usually toxic to plants. Roots are generally the first tissue to be exposed to arsenic, where the metalloid obstructs root extension and proliferation. After translocation to the shoot, As can rigorously inhibit plant growth by reducing or stopping the expansion and biomass accumulation, and compromising plant reproductive capacity through losses in fertility, yield, and fruit production. At sufficiently high concentrations, arsenic inhibits the metabolic processes, which leads to cell death. Most plants have mechanism to hold much of their arsenic burden in the root. However, a genotype-dependent proportion of arsenic is translocated to the shoot and other tissues of the plant. Various physiological processes are vulnerable to arsenic toxicity, which are mainly ROS induced, such as:

- Accumulation of soluble sugars
- Breaking of DNA strands
- Decline in the number, area, and water potential of leaves
- Decrease in nitrate and nitrate reductase activity
- Delayed antioxidant defence mechanisms
- Disorganized function of integral photosynthetic processes
- Disturbed respiration
- Disturbed uptake and water uptake
- Imbalance in nutrient intake

- Impaired membrane veracity and nutrient acquisition
- Increased accumulation of anthocyanin in leaves, acid invertase, peroxidation, membrane leakiness, damaged cellular membranes in plants exposed to As, causing electrolyte leakage
- Increased synthesis of ascorbate, the γ-Glu-Cys-Gly tripeptide glutathione (GSH), and the GSH oligomer ([γ-Glu-Cys]n-Gly) phytochelatin (PC) in the plant
- Inhibition of DNA repair mechanism
- Lowered pigment synthesis
- Mutation and chromosomal aberrations
- Overexpression of sucrose phosphate synthase and sucrose synthase genes
- Oxidation of free amino acids
- Reduced plant growth, productivity, plant transpiration intensity, stomatal conductance
- Reduction in alpha and beta amylase and stress-responsive genes
- Repressed ATPase synthesis
- Increase in malondialdehyde, a product of lipid peroxidation, causing oxidative stress in As toxicity
- Suppression of biomass and starch hydrolysing enzymes
- Wilting, curling, and necrosis of leaf blades

Arsenic remains in the environment for a long time, and eventually accumulates to levels that could harm the physiochemical properties of soils, resulting in loss of soil fertility and reduced crop yield. Arsenic, if not detoxified, may prompt a sequence of reactions leading to growth inhibition, disruption of photosynthetic and respiratory systems, and stimulation of secondary metabolism. Plants respond to arsenic toxicity through various mechanisms including hyper-accumulation, antioxidant defence systems, and phyto-chelation. Only a few plant species are arsenic-tolerant; among these are a group of plants including *Pteris vittata* and other members of Pteridaceae that hyperaccumulate arsenic.

The concentration of hazardous inorganic arsenic present in surface water increases alterations in the genetic material of fish. This is mainly caused by the accumulation of arsenic in the bodies of plant-eating freshwater organisms. Birds can eat fish that contain high amounts of arsenic and die through arsenic poisoning. As arsenic compounds are toxic to microbes and lower aquatic organisms, they have been used in pesticides, wood preservatives, and paints. Arsenates are used in power-plant cooling towers for controlling fungi attacks, which can cause corrosion of structural wood. However, arsenates are used rarely for this purpose, owing to their comparatively high toxicity; instead, a preservative containing arsenate is used, which may enter the aquatic environment. Arsenic has been found to be toxic to invertebrates and is consequently used for controlling shipworms and other wood-borers. It tends to accumulate in mollusks and may have chronic effects on them.

The effects of longer exposure to low concentrations of arsenic separately or in combination with other metals are generally unknown. Pollutants, however, are rarely found in the environment in isolation. Arsenic renders selenium less toxic and has been experimentally added to cattle and poultry feeds in areas high in selenium. Although substantial information has been published on the effects of arsenic on aquatic organisms, most research is focussed on freshwater organisms, and very little is known about its effects on marine organisms.

The clinical features of acute arsenic poisoning among animals include intestinal distress, nausea, and diarrhoea, followed by rapid circulatory collapse. Death may occur within a few days and at times in less than one hour. Arsenic causes gastritis in monogastric animals, with dysentery and nervous signs. Chronic arsenic toxicity in cattle results in fibrosis, producing toughness and irregular enlargement of the hocks or other limb joints. Acute and sub-acute arsenic toxicity in experimentally induced goats were categorized by increased heart rate and respiration, diarrhoea, dribbling saliva, inflexible gait, lameness, tremor, convulsion, coffee-coloured urine, congested mucous membranes, etc., whereas chronic arsenic exposure caused reduced body weight, coffee-coloured urine, and clogged mucous membranes. Inorganic arsenic in edible tissues of mammals and birds is usually low, and thus, foods obtained from these tissues generally cause slight intoxication in humans. After intake, the inorganic and organic arsenic compounds accumulate in various tissues (higher to lower concentration: kidneys > lungs > urinary bladder > skin > blood > liver).

Arsine, a non-irritating, colourless, and nearly tasteless gas, is a by-product of the mining process, and is used in the production of semiconductors. It is lethal at low doses, and affected individuals may not observe any toxicity until hours after exposure. The classic symptoms of arsine gas toxicity include headache, nausea, vomiting, diarrhoea, "*port wine*"-hued urine, and jaundice. The prognosis of arsenic toxicity is extremely variable and depends on the arsenic state, amount ingested, and duration of exposure. Exposure can lead to a variety of different systemic illnesses. Peripheral neuropathy may settle once the exposure is terminated; however arsenic exposure as a carcinogen can cause cancer of the skin, lungs, and bladder, and hepatocellular carcinoma. Arsenic can cross the placenta and is teratogenic. Arsenic shows poor excretion through lactation, and research has shown that exclusive breastfeeding decreases arsenic exposure risk to infants in endemic regions. Handling of arsenic and related toxicities needs an inter-professional team approach, including inputs from physicians, toxicologists, specialty-trained nursing, and pharmacists, working jointly to improve the patient outcome. Thus, a combination of physiological and biochemical studies can give an insight into the mode of action of As in humans, plants, and the environment.

BIBLIOGRAPHY

Aposhian, H. V. (1989) *Biochemical Toxicology of Arsenic*. New York, NY: Elsevier/North Holland.

Armstrong, C. W., et al. (1984) Outbreak of fatal arsenic poisoning caused by contaminated drinking water. *Achieves of Environmental Health* 39: 276–279.

Beckett, W. S., Moore, J. L., Keogh, J. P., and Bleecker, M. L. (1986) Acute encephalopathy due to occupational exposure to arsenic. *British Journal of Industrial Medicine* 43: 66–67.

Centeno, J. A., et al. (2002) Pathology related to chronic arsenic exposure. *Environmental Health Perspectives* 110: 883–886.

Chakraborti, D., Hussam, A., and Alauddin, M. (2003) Arsenic: environmental health aspects with special reference to ground water in South Asia. *Journal of Environmental Science and Health* 38(1): xi–xv.

Chakraborti, D., et al. (2003) Arsenic toxicity from homeopathic treatment. *Journal of Clinical Toxicology* 417: 963–967.

Engel, R. R., Hopenhayn, Rich C., Receveur, O., and Smith, A. H. (1994) Vascular effects of chronic arsenic exposure: A review. *Epidemiology Reviews* 16: 184–209.

Flora, S. J. S., et al. (2008) Combined administration of taurine and monoisoamyl DMSA protects arsenic induced oxidative injury in rats. *Oxidative Medicine and Cellular Longevity* 11: 39–45.

Garvey, G. J., Hahn, G., Lee, R. V., and Harbison, R. D. (2001) Heavy metal hazards of Asian traditional remedies. *International Journal of Environmental Health Research* 111: 63–71.

Gorby, M. S. (1988) Arsenic poisoning. *Western Journal of Medicine* 149: 308–315.

Guha, M. D. N. (2003) Chronic arsenic toxicity: clinical features, epidemiology, and treatment experience in West Bengal. *Journal of Environmental Science and Health Part A Toxic/Hazardous Substance and Environmental Engineering* 38: 141–163.

Hertz-Picciotto, I., Arrighi, H. M., and Hu, S. W. (2000) Does arsenic exposure increase the risk for circulatory disease? *American Journal of Epidemiology* 151: 174–181.

Hu, J., Fang, J., Dong, Y., Chen, S. J., and Chen, Z. (2005) Arsenic in cancer therapy. *Anti-Cancer Drugs* 16: 119–127.

Kyle, R. A. and Pease, G. L. 1965. Hematologic aspects of arsenic intoxication. *New England Journal of Medicine* 273: 18–23.

Miller Jr., W. H., et al. (2002) Mechanisms of action of arsenic trioxide. *Cancer Research* 6214: 3893–3903.

Mitra, S. R., et al. (2004) Nutritional factors and susceptibility to arsenic-caused skin lesions in West Bengal, India. *Environmental Health Perspectives* 112: 1104–1109.

National Academy of Sciences. (1977) *Medical and Biologic Effects of Environmental Pollutants, Arsenic*. Washington, DC: National Research Council, National Academy of Sciences.

National Research Council. (2000) *Arsenic in Drinking water*. Washington, DC: National Academy Press.

Obinaju, B.E. (2009) Mechanisms of arsenic toxicity and carcinogenesis. *African Journal of Biochemistry Research* 3(5): 232–237.

Paul, M. K., Kumar, R., and Mukhopadhyay, A. K. (2008) Dithiothreitol abrogates the effect of arsenic trioxide on normal rat liver mitochondria and human hepatocellular carcinoma cells. *Toxicology and Applied Pharmacology* 226(2): 140–152.

Peraza, M. A., et al. (2006) Morphological and functional alterations in human proximal tubular cell line induced by low level inorganic arsenic: Evidence for targeting of mitochondria and initiated apoptosis. *Journal of Applied Toxicology* 26(4): 356–367.

Ratnaike, R. N. (2003) Acute and chronic arsenic toxicity. *Postgraduate Medical Journal* 79: 391–396.

Rosenman, K. (2007) Occupational heart disease. In: *Environmental and Occupational Medicine,* 4th edition, Rom, W., Markowitz, S. (eds.), p. 688. Hagerstown, MD: Lippincott Williams & Wilkins.

Rossman, T. (2007) Arsenic. In: *Environmental and Occupational Medicine* 4th edition, Rom, W., Markowitz, S. (eds.), pp. 1006–1007. Hagerstown, MD: Lippincott Williams & Wilkins.

Schoen, A., Beck, B., Sharma, R., and Dube, E. (2004) Arsenic toxicity at low doses: Epidemiological and mode of action considerations. *Toxicology and Applied Pharmacology* 198: 253–267.

Simeonova, P. P. and Luster, M. I. (2004) Arsenic and atherosclerosis. *Toxicology and Applied Pharmacology* 198: 444–449.

Tseng, C. H., et al. (2002) Epidemiologic evidence of diabetogenic effect of arsenic. *Toxicology Letters* 133: 69–76.

Vahter, M. (2002) Mechanisms of arsenic biotransformation. *Toxicology* 181: 211–217.

Welch, K., Higgins, I., Oh, M., and Burchfiel, C. (1982) Arsenic exposure, smoking and respiratory cancer in copper smelter workers. *Archives of Environmental Health* 37: 325–335.

Yoshida, T., Yamauchi, H., and Sun, G. F. (2004) Chronic health effects in people exposed to arsenic via the drinking water: dose–response relationships in review. *Toxicology and Applied Pharmacology* 198: 243–252.

Yu, H. S., Liao, W. T., and Chai, C. Y. (2006) Arsenic carcinogenesis in skin. *Journal of Biomedical Science* 135: 657–666.

6 Heavy Metal Toxicity
Chromium

I guess chromium can really extend the life of what we see around us...

Janet Kuypers

After going through this chapter, the reader will learn about the sources and routes of exposure of chromium, its aetiology and clinical manifestations, the mechanism of chromium toxicity, the diagnosis, treatment, and management of chromium poisoning, and the environmental impacts of chromium.

FACT FILE

- Chromium (Cr) is the 17th most abundant element in the Earth's mantle, with an average concentration of 125 mg/kg
- The word "*chromium*" is derived from the Greek word "*chroma*," which means colour, which connotes the colourful attribute of chromium compounds and the popularity of its pigments
- Chromium is a hard, lustrous, and steel-grey metal that is mostly obtained by mining the mineral chromite
- Chromium with the atomic number 24 has an atomic weight of 51.996 and a density of 7.19 grams per cubic cm
- Like other transition metals, Cr also has a high melting point (1,907°C or 3,465°F) and a high boiling point (2,671°C or 4,840°F)
- Chromium has multiple oxidation states, ranging from Cr (0), i.e. the elemental chromium, to Cr (VI), the hexavalent chromium
- The trivalent form, Cr (III), is the most common as well as most stable form of chromium
- Chromate (CrO_4^{2-}) and dichromate ($Cr_2O_7^{2-}$) ions are the most important species of chromium
- Chromium occurs naturally as chromite ($FeCr_2O_4$) in ultramafic and serpentine rocks, or with other metals such as crocoite, bentorite, tarapacaite, vauquelinite, etc.

DOI: 10.1201/9781003343721-6

- Chromite ore is used for manufacturing chromium metal and mono- or di-chromates
- Trivalent chromium is generally found in food supplements and is non-toxic, whereas hexavalent chromium is found in industrial pollutants and is toxic and carcinogenic when inhaled
- Chromium is the only element that shows antiferromagnetic property in its solid state below room temperature. Chromium becomes paramagnetic above 38°C
- Chromium occurs as a mixture of three stable isotopes: Cr-52, Cr-53, and Cr-54
- Cr-52 is the most abundant isotope, accounting for 83.789 per cent of its natural availability
- Chromium is used for making yellow, red, and green pigments, as a decorative and protective metal coating, and as a catalyst
- The green colour of emerald, serpentine, chrome mica, and the red colour of ruby are due to traces of chromium metal
- Cr is extensively used in industry for plating, alloying, tanning of animal hides, prevention of water corrosion, textile dyes and mordents, pigments, ceramic glazes, refractory bricks, and pressure-treated lumber
- Chromium is added to iron and nickel in the form of ferrochromium used for producing alloys that are mainly characterized by their high resistance to corrosion and oxidation
- Approximately 75–85 per cent of commercially produced chromium is used for making alloys, such as stainless steel, and by adding chromium, stainless steel becomes harder and more resistant to corrosion
- Chromium (VI) and chromium (III) are used for chrome plating, dyes and pigments, leather tanning, and wood preservation
- Chromium is released in the air mainly by combustion processes and metallurgical industries
- Non-occupational sources of chromium include contaminated soil, air, and water
- Differences in the bioavailability and bioactivity between Cr (III) and Cr (VI) account for their differences in toxicity
- Cr (VI) contamination of drinking water was first widely shown in 2000 in the movie Erin Brockovich featuring the southern California town of Hinkley
- When compared to Cr (III), Cr (VI) is better absorbed from lungs, gut, and skin and, after absorption, Cr (VI) is reduced to Cr (III)
- Cr (III) is an essential dietary nutrient, whereas Cr (VI) compounds when inhaled act as respiratory tract irritants, causing pulmonary sensitization and increasing the risk of lung, nasal, and sinus cancers
- Dermal contact with Cr (VI) compounds causes severe dermatitis and painless skin ulcers
- Cr (VI) compounds, such as potassium dichromate and chromium trioxide, are corrosive and irritating to the gastrointestinal mucosal lining and can cause mild to severe liver abnormalities
- Ingestion of a lethal dose of chromate results in cardiovascular collapse

- Oral exposure to Cr (VI) compounds may result in haematological toxicity
- Cr (VI) compounds induce DNA damage, gene mutation, sister chromatid exchange, and chromosomal aberrations
- Cr (III) is expelled from the body mainly in the urine

6.1 INTRODUCTION

Discovered in 1797 by Vauquelin, chromium has diverse industrial applications that have made it an important economic element. Chromium was discovered when French chemist Nicholas Louis Vauquelin analysed the bright red mineral called "*crocite*," in a Siberian gold mine. Captivated by the range of colours radiated by this mineral in solution, Vauquelin named it chromium, derived from the Greek word *chroma* meaning colour. Trivalent chromium is an essential trace element in humans and animals. Pure chromium has no adverse effects but a minor toxic effect is ascribed to its trivalent form if present in large quantities. Chromium is a steel-grey hard transition metal naturally occurring in the Earth's crust in the form of ore ($FeCr_2O_4$). It belongs to d-block, group (VI) and is the first element of the periodic table. It occurs in diverse oxidative stages ranging from Cr (II) to Cr (VI) and out of these oxidation states, Cr (III) and Cr (VI) are highly stable and occur naturally. The forms of Cr differ not only in their oxidation states, but also in their chemical properties and toxicity. Cr (VI) is the most stable form, and rarely occurs in nature, being largely formed during industrial processes. Compounds of Cr (VI) cause acute and chronic toxicity. Cr is detected in minute quantities in the atmosphere, water bodies, and soil. Owing to its wider anthropogenic use, chromium's environmental contamination has increased and it has become an important concern in the last few years. In plants, Cr at low concentrations promotes growth and increase yield, but it is not considered essential to plants. Consequently, the accumulation of chromium in edible plants may present a potential hazard to animals and humans.

Cr (0), being the metallic form, is produced in industry and is used for the manufacturing of steel and other alloys. Cr (VI) and Cr (III) possess different chemical, toxicological, and epidemiological characteristics. Cr (III) serves as a nutritional supplement, and plays a role in glucose and lipid metabolism. Cr (III) in the forms of oxides, hydroxides, and sulphates is less toxic as it is relatively insoluble in water. It has low mobility, and is generally bound to organic matter in soil and aquatic environments. Furthermore, the Cr (III) form tends to make hydroxide which precipitates with Fe at typical groundwater pH values. At high concentrations of oxygen or Mn oxides, Cr (III) can be oxidized to Cr (VI).

Cr (VI) in the form of chromate, dichromate, and CrO_3 is regarded as toxic, owing to its high oxidizing potential, high solubility, and mobility across membranes in living organisms and the environment. Cr (VI) is believed to be highly toxic and induces a range of cellular injuries. Cr (VI) is a powerful epithelial irritant and is regarded as a human carcinogen. Cr (VI) is toxic to many plants, aquatic animals, and microorganisms.

Cr (VI) and Cr (III) also differ in their ability to enter cells, which is the basis of the difference in their toxicities. Cr (VI) is basically similar to sulphate and phosphate anions; hence, cells freely take it up via non-specific anion transporters. Once inside the cell, Cr (VI) undergoes fast metabolic reduction and gets converted to Cr (III). In contrast, Cr (III) compounds cannot enter cells by any transport mechanism.

6.2 SOURCES AND ROUTES OF EXPOSURE

Chromium naturally occurs in rocks, animals, plants, soil, volcanic dust, and gases (Figure 6.1). Cr (III) is essential for normal metabolism of glucose, protein, and fat, possibly by potentiating the action of insulin, and is considered an essential dietary element. Chromium picolinate, a trivalent form of chromium complexed with picolinic acid, is used as a dietary supplement as it is believed to accelerate the metabolism and may have anti-diabetic effects. The body has numerous ways of reducing Cr (VI) to Cr (III). Detoxification of Cr (VI) increases the level of Cr (III). Trivalent chromium is generally found in air emissions, mostly in form of small particles or aerosols. The main sources of chromium in the atmosphere are:

- Automobile brake lining and catalytic converters for automobiles
- Cement-producing industries
- Chemical and refractory processing units
- Industrial processes linked with the production of ferrochrome
- Leather tanneries and chrome pigments
- Ore refining

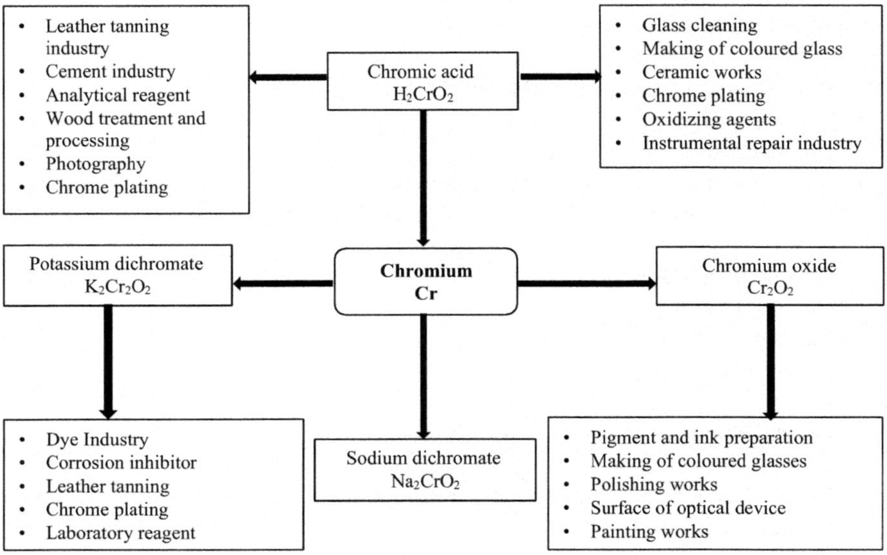

FIGURE 6.1 Common forms of chromium and their applications.

The main source of Cr oral exposure for non-occupational human populations is food and drinking water, for instance, the level of Cr in food ranges from <10 to 1,300 μg/kg, with the highest amounts in meat, fish, fruits, and vegetables. The general population is exposed to Cr (III) by:

- Consuming contaminated food and water
- Inhaling air containing chemicals

The Cr level in unpolluted water is very low, about 1–10 μg/L in rivers and lakes, 0.2–1 μg/L in rainwater, with an average concentration of 0.3 μg/L in ocean water. However, increased industrial applications have resulted in increased release of Cr into soil, groundwater, and air. A raised level of Cr (VI) in drinking water has been reported, posing an important threat of its presence in drinking water. However, there is no specific drinking water standard for hexavalent chromium. The main industrial sources of hexavalent chromium are:

- Chromate pigments in dyes, paints, inks, and plastics or chromate added as an anti-corrosive agent to surface coatings
- Chrome plating using a solution of chromic acid
- Fumes from welding stainless steel or nonferrous chromium alloys
- Impurities present in Portland cement
- Particles released during the smelting of ferrochromium ore

The route of entry of Cr in the human body is through inhalation, ingestion, and dermal absorption. Occupational exposure usually occurs through inhalation and dermal contact, while the general population is usually exposed to it by the ingestion of chromium in soil, food, and water. Workers can inhale airborne hexavalent chromium as a dust, fume, or mist while:

- Carrying out chrome electroplating
- Manufacturing chromate pigments and powders, chromic acid, chromium catalysts, dyes, and coatings
- Smearing and detaching chromate-containing paints and other surface coatings
- Welding or working with stainless steel, high-chrome alloys, and chrome-coated metal
- Chromate production
- Chrome plating
- Stainless-steel production
- Working in tanning industries

Occupational exposure can occur at two levels:

- People living in the vicinity of chromium waste disposal sites or chromium manufacturing and
- Processing plants with greater probability of elevated chromium exposure than the general population

Ingestion: The commonly induced ingestion pathway in the human body is through food consumption and drinking water. Possibly, the ingestion of contaminated soil by children or swallowing contaminated water while swimming or showering could also be taken into consideration. Once Cr is ingested, only 2–3 per cent is absorbed by the gastrointestinal tract, which rapidly reduces Cr (VI) to Cr (III). The average daily intakes from air, water, and food are estimated to be less than 0.2–0.4 micrograms (μg), 2.0 μg, and 60 μg, respectively.

Dermal Contact: Possible non-occupational dermal exposure to Cr includes swimming, bathing, showering, and contact with contaminated soil. Workers in the Cr industry (plating and chemical manufacturing) are more likely to attain some degree of dermal exposure to hexavalent chromium. Due to the fact that Cr (VI) compounds are more soluble in water than Cr (III) compounds, Cr (VI) can penetrate the skin more efficiently than Cr (III). However, the rates of penetration of Cr (VI) and Cr (III) are almost identical if lesions are present on the skin. Skin exposure can occur during direct handling of hexavalent chromium-containing solutions, coatings, and cements. Due to the extremely high boiling point of chromium, gaseous chromium is rarely encountered. Rather, chromium in the environment occurs as particle-bound chromium or chromium dissolved in droplets. Cr (VI) trioxide (chromic acid) and soluble Cr (VI) salt aerosols produce different health effects as compared to insoluble particulate compounds. For example, exposure to Cr (VI) trioxide results in marked damage to the nasal mucosa and perforation of the nasal septum, whereas exposure to insoluble Cr (VI) compounds damages the lower respiratory tract. Dermal exposure to chromium may occur while using products containing chromium, for instance, wood treated with copper dichromate or leather tanned with chromic sulphate.

Inhalation: The respiratory system is the primary target organ for the carcinogenic effects of Cr (VI). The inhalation of chromium-containing aerosols is a major concern. The individual uptake may vary depending on related exposure to other relevant factors, e.g., tobacco smoking, and the particle size distribution in the inhaled aerosol. The concentration of chromium in water bodies varies based on the type of surroundings and nature of the underlying soil. Cr in gaseous form is seldom encountered because of the very low equilibrium vapour pressure. However, the inhalation of dust, fumes, and aerosol containing Cr is an important exposure intake mode. Once inhaled, Cr (VI) gets transferred to blood from particles in the lung three times more readily than Cr (III).

6.3 AETIOLOGY

Acute Exposure: Acute poisoning is likely to occur through the oral route, whereas chronic poisoning is mainly from inhalation or skin contact. Serious exposure to Cr (VI) compounds can be unintentional or intentional, and is rarely occupational or environmental. Acute Cr (VI) poisonings are often fatal regardless of the therapy used.

Chronic Exposure: Recurrent skin contact with chromium dusts can lead to eczematous dermatitis with oedema. Chromate dusts can cause:

- Gingivitis
- Irritation of the conjunctiva
- Irritation of the mucous membrane
- Keratitis
- Nasal ulcers and perforation
- Periodontitis

Contact with chromate solution penetrating skin lesions mainly in the fingers, knuckles, and forearms produces chrome holes or chrome ulcers. These ulcers can enter deep into the soft tissue which becomes the location of secondary infection. Lung cancer is the most life-threatening long-term effect of chromium. Apart from the carcinogenic effect, prolonged exposure to Cr can result in bronchitis, rhinitis, sinusitis, or nasal mucosal polyps. Besides the lungs and intestinal tract, the liver and kidneys are other target organs for chromate toxicity. Acute chromium poisoning generally occurs through the oral route, while chronic poisoning is mostly from inhalation or dermal contact. In an occupational environment, contact dermatitis, irritation, and ulceration of the nasal mucosa are the most commonly reported effects of chronic chromium exposure. Various occupations that may involve exposure to different forms of chromium include:

- Abrasive blasting workers [Cr (III) and Cr (VI)]
- Battery makers [Cr (VI)]
- Candle makers [Cr (III) and Cr (VI)]
- Cement workers [Cr (III) and Cr (VI)]
- Dye makers [Cr (III)]
- Painters [Cr (III) and Cr (VI)]
- Printers [Cr (III) and Cr (VI)]
- Rubber makers [Cr (III) and Cr (VI)]
- Workers involved in the maintenance and servicing of copying machines and the disposal of some toner powders from copying machines [Cr (VI)]
- Workers involved in welding, cutting, brazing, soldering, torching, and other hot work operations [Cr (III) and Cr (VI)]

6.4 PATHOPHYSIOLOGY AND CLINICAL MANIFESTATIONS

Ingestion may cause acute haemorrhagic gastroenteritis, and the resulting massive fluid and blood loss may cause shock and oliguric renal failure. Haemolysis, hepatitis, and cerebral oedema have been reported. The oral intake of Cr (VI) compound may cause the following general symptoms:

- Circulatory collapse
- Diarrhoea

- Epigastric pain
- Fever
- Haemorrhagic diathesis
- Intense gastrointestinal irritation or ulceration and corrosion
- Intravascular haemolysis
- Liver damage
- Muscle spasms
- Nausea
- Renal failure
- Toxic nephritis
- Vertigo
- Vomiting
- Acute multisystem organ failure
- Coma, and even death, depending on the dose

Typical clinical presentations of patients with Cr (VI) compound exposure include:

- Allergic and irritant dermatitis
- Asthma
- Bronchitis
- Lung cancer
- Puncturing of the nasal septum
- Respiratory irritation
- Sinusitis
- Skin ulcers

Acute inhalation can cause upper respiratory tract irritation, wheezing, and non-cardiogenic pulmonary oedema. Chronic exposure to hexavalent compounds may lead to pulmonary sensitization, asthma, and cancer. Acute contact may result in severe corneal injury, deep skin burns, and oral or oesophageal burns. Hypersensitivity dermatitis and nasal ulcers may occur following chronic exposure.

6.5 MECHANISM OF CHROMIUM TOXICITY

The oxidation state and solubility are among the major characteristics governing chromium toxicity. Cr (VI) compounds appear to be much more toxic systemically than Cr (III) compounds because of powerful oxidizing agents in Cr (VI) which tends to be more irritating and corrosive than those in Cr (III) in similar amounts and solubility. Although the mechanism of biological interaction is not certain, the variation in chromium toxicity is related to the ease with which Cr (VI) can pass through cell membranes and its subsequent intracellular reduction to reactive intermediates. Since trivalent chromium Cr (III) is poorly absorbed by any route, the toxicity of chromium is mainly attributable to the Cr (VI) form. It can be absorbed by the lungs and gastrointestinal tract, and even to a certain extent by intact skin.

- Cr (VI) is more toxic than Cr (III), therefore Cr (VI) enters cells more readily than Cr (III) compounds and is ultimately reduced to Cr (III). This process is believed to have a protective effect against the detrimental consequences of Cr (VI)

- Cr (VI) enters cells and reacts spontaneously with intracellular reductants, such as ascorbate, glutathione (GSH), cytochrome reductase, and superoxide dismutase, to generate short-lived intermediates like Cr (V) and/or Cr (IV), free radicals, and ultimately Cr (III)
- Cr (III) is retained inside the cell and interacts with the DNA due to its limited membrane permeability
- Further, Cr (V) undergoes a one-electron redox cycle to regenerate Cr (VI) and simultaneously generates ROS that easily interacts with DNA protein multiplexes, triggering multiple possible apoptotic signalling pathways in various cell types
- Furthermore, Cr (VI) can alter redox homeostasis by creating ROS and depleting physiological antioxidants via the Fenton reaction

High levels of ROS attack lipid and genomes, causing oxidative damage and DNA–protein cross-links, resulting in cellular necrosis. The uptake of chromium from the gastrointestinal tract is comparatively low and depends on numerous factors, including:

- Gastrointestinal transport time
- Chemical form, for instance, organic chromium is more readily absorbed than inorganic chromium
- Water solubility of the compound
- Valence state, for example, Cr (VI) is more readily absorbed than Cr (III)

Once entered in the bloodstream, Cr (VI) is quickly taken up by erythrocytes and after absorption it is reduced to Cr (III) inside the RBCs. Cr (III) does not freely cross the membrane of RBCs, but binds directly to transferrin, an iron-transporting protein in plasma (Figure 6.2).

Reduction of Cr (VI) in the RBCs occurs by the action of glutathione. Since RBC membrane is permeable to Cr (VI) but not Cr (III), Cr (III) formed by the reduction of Cr (VI) gets stuck in the RBCs. Ultimately, the diffusion of Cr (VI), the reduction to Cr (III), and complexing to nucleic acids and proteins in the cell change the concentration equilibrium.

Irrespective of the source, Cr (III) is extensively dispersed in the body and accounts for most of the chromium in plasma or tissue. The highest uptake of Cr (III) as a protein complex is through bone marrow, lungs, lymph nodes, spleen, kidneys, and liver. Autopsies reveal that chromium levels in the lungs are usually higher than in other organs.

Metabolic Pathway (Figure 6.3): After oral exposure, initial resistance against Cr (VI) is through the reduction of Cr (VI) to Cr (III) in the gastric environment where gastric fluid and ascorbate play an important role. It can be absorbed by the lungs and gastrointestinal tract and to a certain extent by skin. Approximately 10 per cent of an absorbed dose is removed by biliary excretion, while a smaller amount is expelled

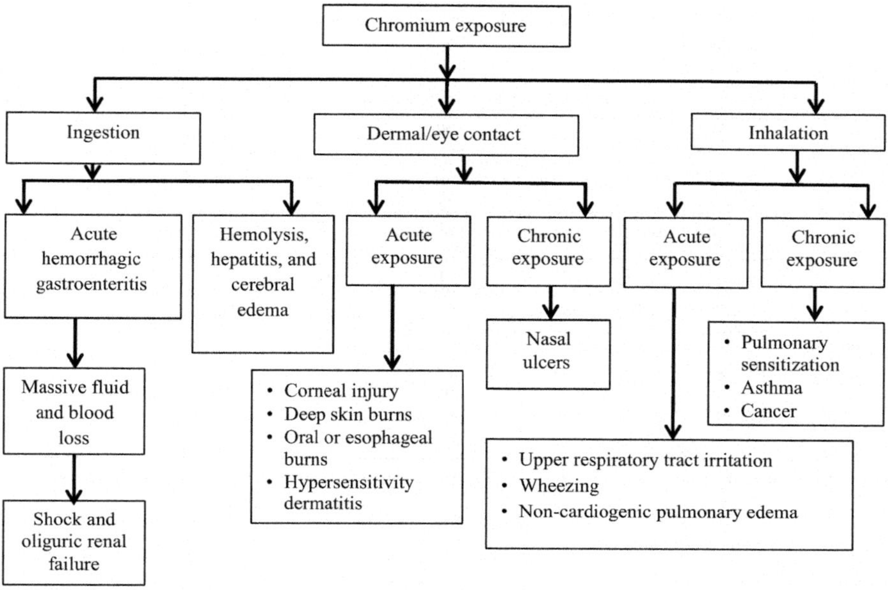

FIGURE 6.2 General impact of chromium.

through the hair, nails, milk, and sweat. Clearance from plasma is generally rapid (within hours), whereas removal from tissue is slower. Since Cr (III) is inadequately taken in by any route, the toxicity of chromium is generally ascribed to the Cr (VI) form. Excretion of absorbed chromium occurs mainly via urine. In humans, the kidneys expel about 60 per cent of an absorbed Cr (VI) dose in the form of Cr (III) within eight hours of ingestion.

The reduction of Cr (VI) is believed to act as a detoxification process when it happens at a distance from the target site for toxic or genotoxic effects, while reduction of Cr (VI) may trigger chromium toxicity if it occurs in or near the cell nuclei of target organs. If Cr (VI) is reduced to Cr (III) extracellularly, Cr (III) is not readily transported into cells and hence toxicity is not seen. The equilibrium that occurs between extracellular Cr (VI) and intracellular Cr (III) is what eventually controls the amount and rate at which Cr (VI) enters cells and exhibits its toxic effects. Cr (VI) enters diverse cell types and under physiological conditions can be reduced by hydrogen peroxide (H_2O_2), glutathione (GSH) reductase, ascorbic acid, and GSH to produce reactive intermediates, including Cr (V), Cr (IV), thiyl radicals, hydroxyl radicals, and finally, Cr (III). Any of these species could attack DNA, proteins, and membrane lipids, thus upsetting cellular functions and integrity.

6.6 TOXICOLOGICAL EFFECTS OF CHROMIUM

Chromium causes different effects on different parts of the body (Figure 6.3), which vary from one system to another.

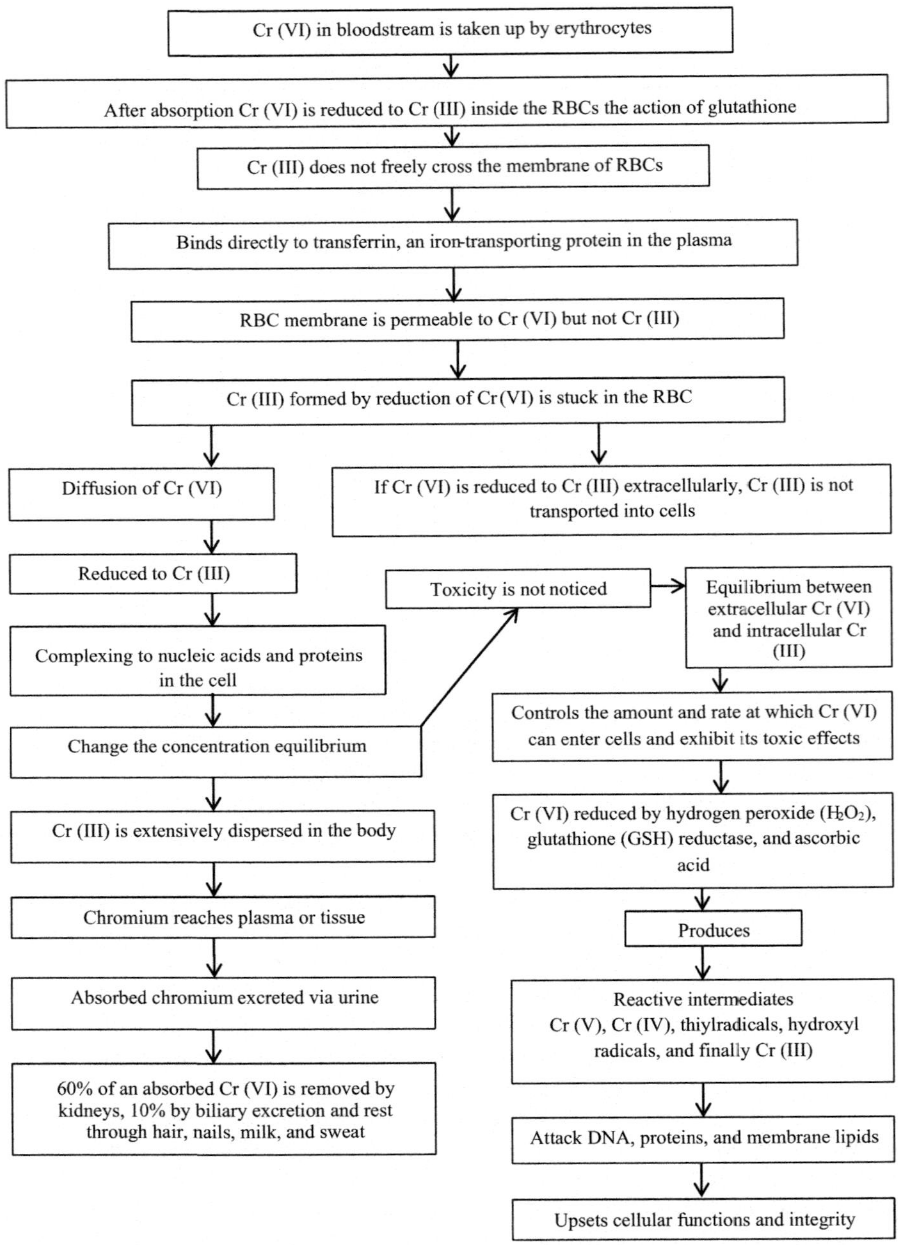

FIGURE 6.3 Mechanism of chromium toxicity in the human body.

Toxicological Effects on the Respiratory System: Occupational exposure often includes mixed exposure to both Cr (III) and Cr (VI). Human occupational exposure clearly shows that, when breathed in, chromium compounds act as respiratory tract irritants, causing airway irritation, obstruction, and later lung, nasal, or sinus cancer. The dose, duration of exposure, and compounds involved determine chromium's effects. Pulmonary irritant effects after breathing of chromium dust include:

- Asthma
- Chronic bronchitis, irritation, pharyngitis, rhinitis
- Congestion and hyperemia
- Polyps of the upper respiratory tract
- Tracheobronchitis
- Ulceration of the nasal mucosa with possible septal perforation

Toxicological Effects on the Skin: Dermal exposure to chromium is known to produce irritant and allergic contact dermatitis. Irritant dermatitis is mainly linked with the direct cytotoxic properties of chromium; while allergic contact dermatitis is an inflammatory response intervened by the immune system. Allergic contact dermatitis is a cell-mediated immune response that occurs in a two-step process. This mechanism involves two steps, the first step being induction, where chromium is absorbed into the skin and triggers sensitization; the next step is basically an immune response. When exposed to chromium above a threshold level, sensitized individuals show an allergic dermatitis response. Localized erythematous or vesicular lesions at points of contact or generalized eczematous dermatitis suggest sensitization. Chromium allergic dermatitis is characterized by:

- Dryness
- Erythema
- Fissuring
- Papules
- Scaling
- Small vesicles
- Swelling

Solubility and pH are primary determinants of an allergic response. Penetration into skin causes painless erosive ulceration "chrome holes" with prolonged healing. These usually occur on the fingers, knuckles, and forearms. Additionally, occupational exposure to Cr (VI) compounds has been linked with outcomes on the skin, nasal septum, and eardrums. Chromium is one of the most common skin sensitizers that has a skin-sensitizing effect. A probable source of chromium exposure is waste dumps for chromate-producing plants, resulting in local air or water pollution.

Carcinogenic Effects: Occupational exposure to Cr (VI) compounds in an industrial setup is linked with increased risk of cancer, mainly of the respiratory system. Studies of labourers working in the chromium pigment, chrome-plating, and ferrochromium industries present a statistically substantial link between exposure to Cr (VI) and lung cancer among workers. Along with lung cancer, there is an increased risk of nasal and sinus cancers. The WHO has classified Cr (VI) as a human carcinogen. Carcinogenicity is related with inhalation of the less soluble or insoluble Cr (VI) compounds. Toxicology of Cr (VI) does not reside with the elemental form. The cellular toxicity of chromium may result from:

- Damage to cellular components during the reduction process from hexavalent to trivalent chromium
- Generation of free radicals, including DNA damage

Renal Effects: Renal effects followed by inhalation or oral exposure to Cr (VI) compounds have been reported. Though glomerular injury is observed in chromium workers, the main renal injury is tubular, with low doses acting especially on proximal convoluted tubules. Injury to brush border membrane is a feature of chromate nephropathy. Severe toxicity can lead to acute tubular necrosis and acute renal failure. Low-dose chronic Cr (VI) exposure normally produces transient renal effects.

Hepatic Effects: Cr (VI) is known to cause severe liver effects in workers exposed to chromium trioxide in the chrome plating industry. Reported liver effects include:

- Deranged liver cells
- Elevated level of liver enzymes
- Hepatic failure
- Hepatomegaly
- Increased number of Kupffer cells
- Lymphocytic and histiocytic infiltration
- Necrosis

Gastrointestinal Effects: The effects on the gastrointestinal tract include:

- Abdominal pain
- Gastrointestinal haemorrhage
- Gut mucosal necrosis
- Hypertrophic gastritis
- Pancreatitis
- Ulcer formation
- Vomiting

Cardiovascular Effects: The main cardiovascular impacts include:

- Cardiogenic shock
- Cardiopulmonary arrest
- Cardiovascular collapse and shock
- Haemorrhage
- Hypotension

- Metabolic acidosis
- Severe respiratory distress
- Ventricular arrhythmias

Haematological Effects: The haematological effects of Cr (VI) manifest as:

- Decreased haemoglobin and haematocrit
- Increased total WBC counts, reticulocyte counts, and plasma haemoglobin
- Intravascular haemolysis
- Thrombocytopenia

Cellular, Genotoxic, and Mutagenic Effects: Chromium results in oxidative stress, which causes cytotoxicity, cell proliferation, and spontaneous mutations, which can cause tumours and cell death (Figure 6.4). It seems to exert its genetic effects by binding directly to DNA, which results in:

- Breaking of DNA strands
- DNA oxidative damage
- DNA–DNA cross-links
- DNA–protein cross-links
- Increase in sister chromatid exchanges in lymphocytes
- Increased frequencies of chromosome aberrations in lymphocytes

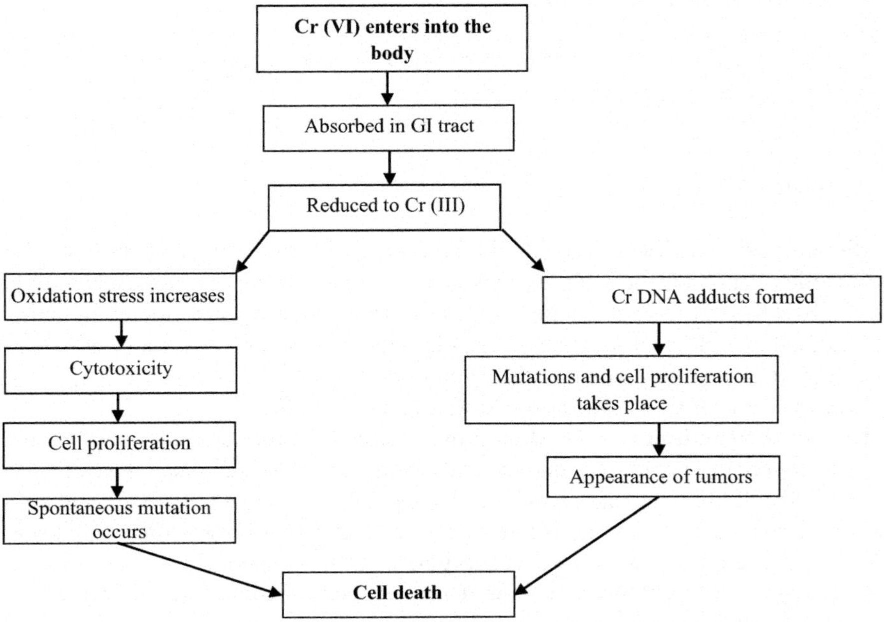

FIGURE 6.4 Mechanism of chromium toxicity at the cellular level.

- Mutagenic effect
- Production of stable DNA–chromium complexes

Other Effects: People working in chromium industries, especially chrome plating plants with poor exhaust facilities, inhale chromium trioxide fumes and may experience:

- Dizziness
- Erosion and discolouration of the teeth
- General weakness
- Headache
- Papillomas of the oral cavity and larynx
- Severe corneal injury

6.7 DIAGNOSIS

The diagnosis of chromium toxicity is determined by the history of exposure and clinical manifestations. Chromium can be determined in the blood and urine; however, hair or nail analysis has no clinical relevance. Urinary chromium excretion is a suitable index of exposure in occupational settings. However, it reveals exposure over the previous 1 or 2 days only. The following tests can help in the diagnosis of Cr (VI) exposure:

- B_2-microglobulin
- Blood and urine chromium levels
- Complete blood count
- Liver function tests (SGOT or AST, SGPT or ALT, and bilirubin)
- Specialized Screening Tests
- Urinalysis

Specialized Tests

Blood or Serum Chromium Levels: To analyse chromium in different biological media, a high-performance anion-exchange liquid chromatography technique is used where Cr (VI) and complexes of Cr (III) are rapidly determined in plasma and other biological specimens. Cr (VI) enters RBCs but Cr (III) does not; therefore, it is possible to distinguish sources and types of exposure Cr (VI) versus other forms of Cr by measuring RBC versus serum Cr.

Urinary Chromium Levels: Monitoring of urinary chromium is limited by rapid depletion of the body's burden and variations in metabolism. Excretion of urinary chromium generally reflects absorption over the previous couple of days. A urinary chromium concentration of 40 µg/L to 50 µg/L right after a work shift confirms exposure to air levels of soluble Cr (VI) compounds.

Urinary β_2-Microglobulins Levels: Chrome platers exposed to Cr (VI) show a higher value of urinary β_2-microglobulins than in unexposed controls. Hair or nail analyses can rarely be used because it is impossible to distinguish chromium bound within the hair during protein synthesis from chromium deposited on the hair from dust, water, or other external sources. The normal range of hair levels without exposure to chromium is between 50 parts per million (ppm) to 100 ppm. Levels of chromium and its compounds in biologic complexes can be determined using chromatographic and colourimetric techniques while chromium sensitivity is diagnosed by patch testing and lymphocyte proliferation testing.

Other Measurements: When a patient has confirmed Cr (VI) inhalation exposure, a chest radiograph and pulmonary function test can be initiated along with other tests.

6.8 TREATMENT AND MANAGEMENT OF CHROMIUM POISONING

A. Emergency and Supportive Measures

1. **Inhalation.** Supplemental oxygen can be given followed by treatment of wheezing and monitoring the victim closely for delayed-onset noncardiogenic pulmonary oedema.
2. **Ingestion**
   a. Immediate dilution with water followed by treatment of haemorrhagic gastroenteritis with aggressive fluid and blood replacement. Endoscopy may be done to assess the extent of oesophageal or gastric injury.
   b. Haemoglobinuria resulting from haemolysis can be treated with alkaline diuresis if it occurs.

B. Specific drugs and antidotes

1. Chelation therapy is not very effective.
2. Ascorbic acid, which converts hexavalent to less toxic trivalent compounds, is suggested.

C. Decontamination

1. **Inhalation**: The victim can be taken away from the exposure location and supplemental oxygen if available can be given.
2. **Skin**: Removal of contaminated clothing followed by immediately washing the exposed areas with soap and water can minimize the dermal effects. A 10 per cent EDTA ointment may help in the removal of chromate scabs. A 10 per cent topical solution of ascorbic acid is suggested to enhance the conversion of hexavalent chromium to a less toxic trivalent state.
3. **Eyes**: Copious moistening of the eyes with lukewarm water or saline followed by fluorescein examination can prevent corneal injury.
4. **Ingestion**: Giving milk or water can reduce the corrosive effects of Cr. For large ingestions, gastric lavage is performed.

6.9 IMPACT OF CHROMIUM ON THE ENVIRONMENT

Chromium in the Environment: Chromium enters natural waters by erosion of Cr-containing rocks, leaching of soils, and direct release from industrial operations, among others. In aquatic environments Cr may undergo reduction, oxidation, sorption, desorption, dissolution, and precipitation. The aqueous solubility of Cr (III) is a function of the pH of the water. Under neutral to basic pH, Cr (III) precipitates, while under acidic pH it solubilizes. Cr (VI) chromate and dichromate are extremely soluble under all pH conditions. The concentration of Cr in soil varies significantly based on the natural composition of the rocks and sediments that constitute them. The chromium level in soil increases mostly due to anthropogenic activities, including removal of chromium-bearing liquids and solid wastes. Usually, Cr in soil occurs as a mixture of both Cr (III) and (VI). As in an aquatic environment, Cr in soil or sediment undergoes a variety of transformations, such as oxidation, reduction, sorption, precipitation, and dissolution. Oxidants present in the soil can oxidize Cr (III) to Cr (VI).

Chromium in Plants

Chromium Uptake: The pathway of Cr uptake in plants is not yet clear. Being a non-essential element, Cr does not have any specific uptake mechanism and is dependent on Cr speciation. Cr (III) uptake by plants is a passive process, i.e., it does not involve any energy expenditure. The uptake of Cr (VI) is believed to be an active mechanism accomplished by carriers for the uptake of essential elements such as sulphate. Cr also competes with Fe, S, and P for carrier binding. Cr (VI) has higher solubility and thus its bioavailability is more toxic at lower concentrations than Cr (III), which tends to form stable complexes in soil. Cr toxicity is dependent on metal speciation, which is a determining factor for its uptake, translocation, and accumulation.

Chromium Accumulation and Translocation: Cr accumulates mainly in roots and shoots; however, roots accumulate the major part, with usually only a small part being translocated to the shoots. Studies have shown that Cr is mainly accumulated in roots, followed by stems and leaves; however only small amounts of Cr are translocated to leaves. Plants with the potential for phytoremediation of chromium contamination can tolerate metals through chelation with appropriate high-affinity ligands, biotransformation with reductants, and compartmentalization in the cytoplasm or vacuole. Thus, Cr immobilization in vacuoles in plant root cells may represent an important mechanism of Cr detoxification by plants.

Growth and Development: The presence of Cr in the medium compromises several processes in plants, for instance, plant germination. The ability to germinate in the presence of Cr may indicate a degree of tolerance to Cr. A decreased rate of germination is one of the key responses to HMs exposure, such as Cd, Pb, and Hg, which can be related to decreased activity of α- and β-amylase under Cr stress. Amylase hydrolysis of starch is required for sugar supply to the developing embryo. Decreased amylase activity due to Cr reduces the availability of sugar to the developing embryo, which inhibits seed germination. Besides germination, root growth is often affected by HMs. Decreased root growth in the presence of Cr (VI) is due to inhibition of root cell division and/or elongation, which occurs because of tissue collapse and subsequent inability of roots to absorb water and nutrients from the medium combined with an extended cell cycle. A reduced root surface in Cr (VI) stressed plants may decrease the capacity of plants to search for water in the soil, contributing to water stress.

Stem growth is usually affected by Cr exposure, which in turn is due to reduced root growth and subsequent decreased nutrients and water transport to higher parts of the plant. Additionally, Cr transport to the aerial parts of the plant immediately affects the cellular metabolism of shoots, adding to the reduced plant height. Likewise, leaf area generally decreases in response to an increased concentration of Cr, which is attributed to a decrease in the number of cells in the leaves inhibited by salinization or decreased cell size. With continued Cr supply, the lamina of old leaves become necrotic, perpetually wilted, dry, and shed. Plant yield depends on leaf growth, leaf area, and leaf number. As Cr affects most of the biochemical and physiological process in plants, productivity and yield are also affected. Cr also affects plant photosynthesis, leading to decreased yield and eventually death. Both Cr (III) and Cr (VI) can cause ultrastructural changes in chloroplasts, inhibiting photosynthesis.

Decreased chlorophyll content at a high Cr concentration suggests that chlorophyll synthesis and/or chlorophyllase activity are affected by a high concentration of Cr. Chromium, being similar to other essential elements, affects plant mineral nutrition. The decrease in Fe concentration in leaf tissue in response to Cr toxicity is suggestive of Cr (VI) interference in the availability of Fe, which impairs Fe metabolism. Different Cr concentrations affect the activities of antioxidant enzymes, for instance, peroxidase, catalase (CAT), glucose-6-phosphate dehydrogenase, and superoxide dismutase (SOD).

While evaluating the effect of Cr (VI) on root cell growth and division, it has been seen that the mitotic index decreases with increased concentration of Cr (VI). Furthermore, Cr (VI) also affects chromosome morphology and increases the frequency of c-mitosis, chromosome bridges, anaphase bridges, and chromosome stickiness.

BIBLIOGRAPHY

Avudainayagam, S., et al. (2003) Chemistry of chromium in soils with emphasis on tannery waste sites. *Reviews of Environmental Contamination and Toxicology* 178: 53–91.

Cifuentes, F. R., Lindemann, W. C., and Barton, L. L. (1996) Chromium sorption and reduction in soil with implications to bioremediation. *Soil Science* 161: 233–241.

Cohen, M. D. and Kargacin, B. (1993) Mechanisms of chromium carcinogenicity and toxicity. *Critical Reviews in Toxicology* 233: 255–281.

Costa, M. and Klein, C. B. (2006) Toxicity and carcinogenicity of chromium compounds in humans. *Critical Reviews in Toxicology* 36: 155–163.

Dayan, A. D. and Paine, A. J. (2001) Mechanisms of chromium toxicity, carcinogenicity and allergenicity: Review of the literature from 1985 to 2000. *Human & Experimental Toxicology* 20: 439–451.

Gibb, H. J. and Lees, P. S. (2000) Lung cancer among workers in chromium chemical production. *American Journal of Industrial Medicine* 38(2): 115–126.

Holmes, A. L., Wise, S. S., and Wise, J. P. (2008) Carcinogenicity of hexavalent chromium. *Indian Journal of Medical Research* 128: 353–372.

Kimbrough, D. E., et al. (1999) A critical assessment of chromium in the environment. *Critical Reviews in Environmental Science and Technology* 291: 1–46.

Oliveira, H. (2012). Chromium as an environmental pollution: Insights on induced plant toxicity. *Journal of Botany* 1: 1–8.

Panda, S. K. and Choudhury, S. (2005) Chromium stress in plants. *Brazilian Journal of Plant Physiology* 171: 95–102.

Ramachandran, V., D'Souza, T. J., and Mistry, K. B. (1980) Uptake and transport of chromium in plants. *Journal of Nuclear Agriculture and Biology* 94: 126–128.

Shahinasi, E., Devolli, A., and Mariola, K. (2019) Inorganic toxicity: Chromium. In: *Inorganic Toxicity: Environment & Human Health*, Avinash Kumar (ed.), pp. 98–114. Saarbrücken, Germany: Lambert Academic Publishing.

Shanker, A. K., Cervantes, C., Loza-Tavera, H., and Avudainayagam, S. (2005) Chromium toxicity in plants. *Environment International* 315: 739–753.

Sharma, P., Singh, S. P, Parakh, S. K., and Tong, Y. W. (2022) Health hazards of hexavalent chromium Cr VI and its microbial reduction. *Bioengineered* 13(3): 4923–4938.

Shekhawat, K., Chatterjee, S., and Joshi, B. (2015) Chromium toxicity and its health hazards. *International Journal of Advanced Research* 3(7): 167–172.

Suranjana, A. R. and Manas, K. R. (2009) Bioremediation of heavy metal toxicity with special reference to chromium. *Al Ameen Journal of Medical Science* 22: 57–63.

Yao, H., et al. (2008) Oxidative stress and chromium VI carcinogenesis. *Journal of Environmental Pathology, Toxicology and Oncology* 27: 77–88.

Zayed, A. M. and Terry, N. (2003) Chromium in the environment: Factors affecting biological remediation. *Plant and Soil* 2491: 139–156.

Zhitkovich, A. (2011) Chromium in drinking water: Sources, metabolism, and cancer risks. *Chemical Research in Toxicology* 24: 1617–1629.

7 Heavy Metal Toxicity
Mercury

Learning is like mercury, one of the most powerful and excellent things in the world in skilful hands; in unskilful, the most mischievous.

Alexander Pope

After going through this chapter, the reader will learn about the sources and routes of exposure to mercury, its aetiology, pathophysiology, and clinical manifestations, the mechanisms of mercury toxicity, diagnosis of mercury toxicity, and the treatment and management of mercury poisoning

FACT FILE

- The element named "*mercury*" was known of before 1500 BCE and was used as a representation of the planet Mercury
- The name mercury also represents the Roman God named Mercury, known for his swiftness
- Hg is the chemical symbol of mercury and the name is derived from the Latin word *hydrargyrum* meaning "liquid silver" and is sometimes also known as quicksilver
- It has a shiny silvery liquid appearance and is the only metallic compound other than bromine that is liquid at room temperature and pressure
- The atomic number of Hg is 80, while its atomic weight is 200.592 gm and its density is 13.534 g/cm^3
- Mercury is assigned as a transitional or post-transitional metal
- It is the most rare element found in the Earth's crust and accounts for only about 0.08 ppm by weight
- It is found in the mineral called cinnabar (mercuric sulphide, which is the source of red pigment called vermilion)
- Mercury is a relatively poor conductor of heat and a mild conductor of electricity
- It exhibits +1 or +2 oxidation state but sometimes it has a +4 oxidation state
- Mercury exists in three forms: elemental or metallic mercury (Hg^0), inorganic mercury (Hg^+, Hg^{2+}), and organic mercury (commonly methyl, ethyl, or phenyl mercury)

DOI: 10.1201/9781003343721-7

- Organic mercury compounds such as methylmercury (Me-Hg) or ethyl mercury (Et-Hg) are more toxic than the inorganic compound
- The order of increasing toxicity related to different forms of mercury is: $Hg^0 < Hg^{2+}, Hg^+ < CH_3\text{-}Hg$
- Natural mercury is composed from seven isotopes: Hg^{196} (0.15 per cent), Hg^{198} (9.97 per cent), Hg^{199} (16.87 per cent), Hg^{200} (23.10 per cent), Hg^{201} (13.18 per cent), Hg^{202} (29.86 per cent), and Hg^{204} (6.87 per cent)
- Mercury is used in fluorescent lamps, thermometers, barometers, batteries, float valves, and dental amalgams
- Mercuric chloride ($HgCl_2$) has been used as a preservative and for developing photographic film
- The disinfectant mercury compound thimerosal is an organomercury compound found in vaccines, tattoo inks, contact lens solutions, and cosmetics
- Mercury has no physiological role in human metabolism and hence the human body has no mechanism to excrete mercury actively
- The vascular effects of mercury include oxidative stress, inflammation, thrombosis, VSM proliferation and migration, endothelial dysfunction, dyslipidemia, immune dysfunction, and mitochondrial dysfunction

According to the World Health Organization:

- Mercury is a naturally occurring element found in air, water, and soil that is highly toxic and carcinogenic in nature
- It has toxic effects on the nervous, digestive, and immune systems, lungs, kidneys, skin, and eyes
- Mercury is released either naturally from volcanic eruptions and weathering processes or from anthropogenic activities such as coal-fired power stations, and the mining of mercury, gold, and other metals
- The maximum contaminant level (MCL) for mercury in drinking water is 0.002 mg/L or 2 parts per billion (ppb)
- A major source of organic methylmercury exposure in the general population is fish consumption
- Mercury is one of the top ten chemicals or groups of chemicals of major public health concern

7.1 INTRODUCTION

Mercury is a heavy metal (HM) known for its toxicity and is a liquid at room temperature and pressure. It is sometimes called quicksilver and is easily alloyed with other metals such as gold and silver. Although mercury had been used for centuries both as a medicine and as a poison, most of its medicinal uses have been discontinued and it is currently used for commercial and industrial purposes only. Mercury occurs naturally in the Earth's crust and is released into the environment through volcanic activity, weathering of rocks, and as a result of human activity. Anthropogenic activities that release mercury into the atmosphere are mainly from coal-fired power stations, industrial processes, waste incinerators, and mining of metals. Mercury is a widespread environmental toxicant which induces severe changes in body tissues and causes a wide range of health problems. Both humans and animals are exposed to various chemical forms of mercury in the environment. These include elemental mercury vapour (Hg^0), inorganic mercurous (Hg^{+1}) and mercuric (Hg^{+2}), and the organic mercury compounds, mainly as methyl, ethyl, and similar groups attached to mercury. Due to its ubiquitous nature in the environment, humans, plants, and animals have difficulty in avoiding exposure to some forms of mercury. Mercury is also utilized in:

- Dentistry (dental amalgams)
- Electrical industry (switches, thermostats, batteries)
- Industrial processes such as the production of caustic soda
- Nuclear reactors
- As a solvent for reactive and precious metals, antifungal agents for wood processing, and a preservative of pharmaceutical products

There are various sources of mercury and it is emitted in different forms into the atmosphere. During the environmental recycling of mercury under diverse environmental processes, one form of mercury is usually changed to another form which increases its toxicity level depending on the nature of the mercury element (see Figure 7.1). The exposure to and bioaccumulation of mercury, generally organic

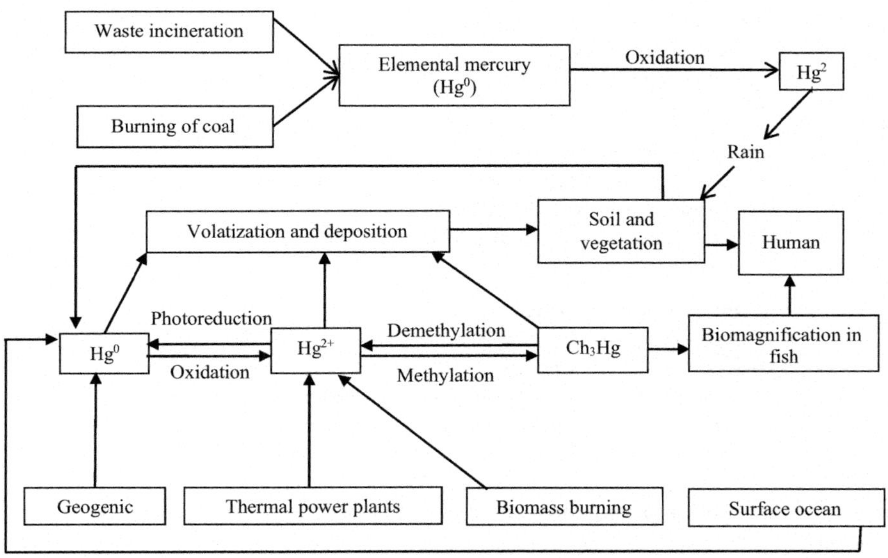

FIGURE 7.1 Environmental recycling of mercury from atmospheric emissions, deposition, exposure, and bioaccumulation.

Note: Hg [0]: (elemental mercury), Hg [II]: (inorganic mercury), and CH_3Hg [II]: (methylmercury).

mercury, in the human body affect the CNS and result in multiple chronic disorders and even death in some cases.

7.2 SOURCES AND ROUTES OF EXPOSURE

Exposure of humans to mercury is through environmental pollution, food contamination, dental care and other preventive therapeutic practices, agricultural and industrial operations, and occupational operations. The most important sources of mercury exposure (see Table 7.1) generally include the following:

- Atmospheric exposures from outgassing from rocks or through volcanic activity
- Inhalation of elementary mercury vapour via occupational or dental amalgam
- Breaking of fever thermometers in the mouth causing its ingestion
- Consumption of contaminated fish
- Mining (mercury and gold in particular)
- Burning of coal

TABLE 7.1
Sources and Routes of Exposure for Different Forms of Mercury (Hg)

Forms of Hg	Sources of Exposure	Exposure Route
Elementary mercury	**Environmental** • Volcanic explosions • Rock weathering • Degassing **Anthropogenic (inadvertent)** • Combustion: Fossil fuels (coal), waste incineration • Industrial: Gold/silver mining, batteries, switches, fluorescent lights, thermometers, sphyngo-manometers **Anthropogenic (intentional)** • Dental amalgams • Ritual and folk medicine use • Anthropogenic activities (combustion of fossil fuel; gold or silver mining; waste incineration; batteries; fluorescent lights; semi-conductor solar cells; relays; thermometers)	• Inhalation (mostly) • In a few cases, ingestion and dermal
Inorganic mercury	• Industrial products (disinfectants; antimicrobials; pharmaceuticals, cosmetics; photography, alternative medicines)	• Skin absorption • In a few cases, ingestion
Organic mercury	• Environmental conservation (fish and shellfish; bio-magnification) • Industrial processes (fungicides; bactericides) • Vaccine preservatives (thiomersal)	• Absorption in gastrointestinal tract • Parenteral absorption • Trans-placental route

- Bio-magnifications (mercury enters into water and is transformed into methylmercury which makes its way through the food chain into fish, shellfish, and eventually to humans)

Route of Exposure: A diagrammatical representation of the routes of exposure to inorganic and organic mercury is shown in Figure 7.2.

7.3 AETIOLOGY

Mercury generally exists in organic and inorganic forms. The inorganic form can further be sub-divided into elemental mercury (liquid mercury) and mercury salts. Elemental exposures are typically derived from mercury-containing devices. Organic mercury is also sub-divided into short-chain and long-chain compounds. Organic mercury exposure usually occurs after ingestion of contaminated seafood, paints containing mercury, or thimerasol (flu vaccine). Personnel potentially exposed to mercury include, but are not limited to, the following:

- Chlorine and caustic soda production workers
- Cosmetic producers
- Dental personnel
- Electroplaters
- Explosives manufacturers
- Felt makers and leather tanners
- Grinding machine operators
- Hazardous waste site personnel
- Ink manufacturers
- Laboratory personnel
- Manufacturers of batteries, fluorescent lamps, mercury vapour lamps, switches, rectifiers
- Metallurgists
- Miners and processors of cinnabar (hgs), gold, silver, copper, zinc
- Paint and pigment manufacturers
- Painters
- Paper millers
- Pesticide/fungicide production and application workers
- Pharmaceutical producers
- Plumbers

In addition, foetuses, infants, and children are potentially at a higher risk of adverse effects due to mercury exposure in the body. Methyl and dimethyl mercury (organic mercury) generally originate from biological sources such as fresh- or salt-water fish. Upon exposure, methylmercury readily crosses the placenta during the prenatal stage and damages the most sensitive part of the body, i.e., the nervous system. Due to the fact that methylmercury concentrates in breast milk, nursing infants can be adversely affected. In addition, infants and children are also at high risk of inhaling elemental mercury from broken thermometers or other home-based sources. Because mercury

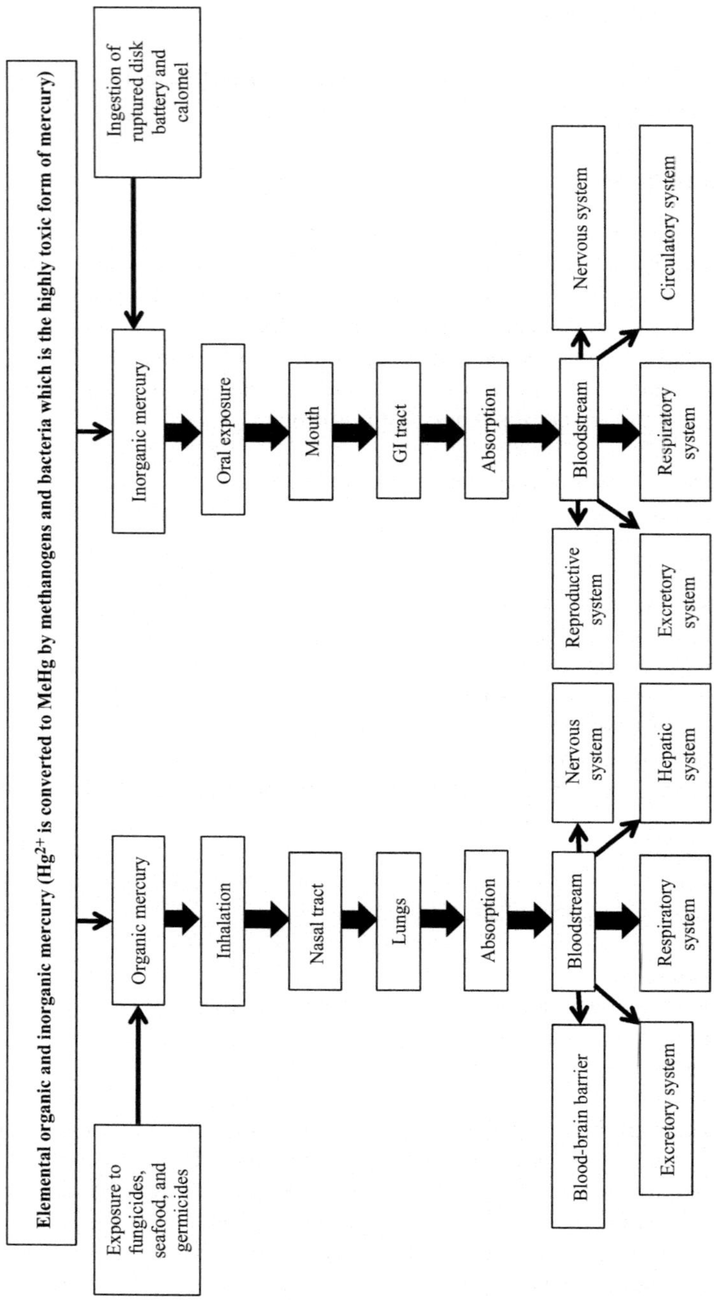

FIGURE 7.2 Effects of inorganic and organic Hg in the human body.

vapour is heavier than air, it tends to settle on the ground, initiating inhalation. Occupational exposure to mercury is highly significant in adults. People working in coal-burning power plants, mining processes of gold and mercury, battery producers, pharmaceutical companies, etc. are more susceptible to mercury exposure. Burning hazardous wastes, breaking-off mercury products, producing chlorine, and improper treatment and disposal of products or wastes containing mercury, can release mercury into the environment. Exposure to metallic mercury in human comes from mercury vapour outgassing from amalgam fillings. Moreover, consumption of fish to lower the blood cholesterol level increases the risk of exposure to mercury through the diet.

7.4 PATHOPHYSIOLOGY AND CLINICAL MANIFESTATIONS

The risk of mercury toxicity is dependent on the form of mercury and its route of exposure. The most common route of exposure to metallic mercury is due to inhalation of mercury vapour. Generally, mercury vapour has a low concentration in the atmosphere and it is not a common route of exposure. However, mercury vapour shows potential impact in occupational exposures such as dentistry for tooth restoration, gold mining, and manufacturing of scientific instruments and electrical devices. However, the organic form of mercury is more susceptible to exposure in the environment as well as in the human body.

Inorganic Mercury

Elemental or Metallic (Hg^0) Mercury: Elemental mercury is mainly absorbed during breathing and its vapours are highly lipophilic in nature, therefore these are effectively absorbed through the lungs and tissue lining the mouth. Elemental mercury is totally absorbed when inhaled, but poorly absorbed when ingested. Acute symptoms caused due to inhalation of elementary mercury vapours include chills, cough, fever, and shortness of breath, and gastrointestinal problems such as nausea, vomiting and diarrhoea, weakness, headaches, and visual disorders. Long-term inhalation of elemental mercury may result in a lack of intellectual functioning, problems in visual judgments, memory loss, anxiety, and psychomotor retardation. Once Hg enters the body, it shows great affinity for sulfhydryl groups and forms a bond with sulphur-containing amino acids. It is transferred to the brain and is either dissolved in serum or adheres to red cell membranes. Metallic mercury gets deposited in the foetal brain by bypassing the placenta, causing damage to the central nervous system. However, metallic mercury oxidizes quickly and forms highly reactive mercuric mercury. Once it enters the bloodstream, the breakdown of metallic mercury occurs.

In addition to the brain, metallic mercury is also deposited in the liver, kidneys, skin, breasts, lungs, myocardium, adrenals, pancreas, salivary glands, testes, and prostate and may be linked with dysfunctioning of these organs. Clinical manifestations of intoxication cause chest pain, dry cough, hypoxemia, dyspnoea, and ventilatory patterns. Moreover, mercury has the affinity to bind to the surface of T-cells which influences their functioning.

Mercurous (Hg^{2++}) Mercury: Mercurous salts are generally colourless, white, or light-yellow powders and include mercurous acetate (antibacterial agent), mercurous chloride or calomel (cathartic, diuretic, antiseptic, and antisyphilitic agent), mercurous nitrate (used to blacken brass), and mercurous oxide (used to make electric batteries). Mercurous mercury salt or calomel is poorly soluble in water and its absorbance in the intestine is not significantly severe. However, some of it is expected to undergo oxidation to more readily absorbable forms. Other than the transitional form between mercuric and metallic mercury, the chances of mercurous mercury surviving in the body are minimal. According to some, with the absorption of mercurous mercury in the body, calomel is occasionally linked with pink disease or acrodynia.

Mercuric (Hg^{++}) Mercury: Mercuric salts are usually colourless or white crystals or intensely yellow- or red-coloured powders. These salts include mercuric nitrate (used in working with felt), mercuric cyanide and mercuric oxide (topical antiseptics), and mercuric chloride (antiseptic and disinfectant). Generally, mercuric salts are more soluble and produce more serious poisonings than mercurous salts. Mercuric mercury either adheres to sulfhydryl groups on erythrocytes, metallothionein, or glutathione, or remains suspended in plasma. However, it fails to cross the blood–barrier but accumulates in foetal tissue or the placenta. It also shows a significant amount of accumulation in the liver, choroidal plexus, testes, and much less in epithelial tissue. Once administered, mercuric mercury is excreted mostly through urine and stool or through sweat, breast milk, and saliva. The symptoms of acute inorganic mercury inhalation include:

- Chest pain
- Dry cough
- Tightness
- Dyspnoea
- Acute chemical pneumonitis
- Bronchiolitis

The clinical findings of mercury toxicity include:

- Carotid obstruction
- Hypertension
- Coronary heart disease (CHD)
- Increases in carotid intimal medial thickness (IMT)
- Myocardial infarction (MI)

Chronic exposure to inorganic mercury can cause:

- Nephrotic syndrome with haematuria and anuria
- Immune complex nephritis, especially membranous nephropathy

Organic Mercury Compounds

Organo-mercury compounds are readily absorbed in the body by inhalation, dermal contact, and ingestion. Most available data on organic mercury compounds refer to methylmercury, which is a major source of human mercury exposure and is commonly found in fish. Methylmercury is distributed uniformly to all tissues and is concentrated more in blood, unlike elemental mercury or mercury ions. A high concentration of methylmercury in the red blood cells is significant where it is metabolized to mercury ions at a slow rate. Despite excretion, the half-life of one-third of ethyl mercury behaves in a similar way as methylmercury. Moreover, organo-mercurials and phenylmercury compounds are absorbed less efficiently by the gastrointestinal system as compared to methylmercury compounds. Dimethylmercury is also efficiently absorbed through the skin, and at times can cause death.

Methylmercury is concentrated mainly in the brain, kidneys, liver, placenta, bone marrow, and foetus, especially the foetal brain. Furthermore, methylmercury attaches itself to the targeted area and slowly undergoes a demethylation process to convert itself into inorganic mercury. Once methylmercury enters the bloodstream, it adheres to sulfhydryl groups, mainly to those in cysteine. The adherent sulfhydryl group in cysteine acts as a transporter and is accountable for the further distribution of intoxicants to other peripheral tissues in the body.

The route of exposure, duration, intensity, and form of mercury influence its clinical manifestation. The primary organ system affected by chronic exposure to organo-mercury compounds and elemental mercury is usually the nervous system, while that of mercury salts is the kidneys. Methylmercury does not produce severe symptoms on mild exposure but high exposure leads to acute gastro-intestinal symptoms, regional destruction of neurons, and delayed neurotoxicity. The majority of the effects of mercury toxicity are found primarily in the CNS and kidneys, where it accumulates after exposure.

7.5 MECHANISMS OF MERCURY TOXICITY

Mercury toxicity is induced by different mechanisms such as:

- Altering cell membrane integrity
- Modifying intracellular calcium balance and membrane potential
- Unsettling immune functions
- Disturbing or inhibition of enzymes
- Inducing oxidative stress
- Inhibition of protein and DNA synthesis
- Interruption of microtubule formation

The molecular mechanism of mercury toxicity (Figure 7.3) is usually based on its chemical activity and biological features, which indicate mostly the involvement of oxidative stress. Through this oxidative stress, mercury follows the mechanism of sulfhydryl reactivity, as described below:

1. Mercury binds to phosphoryl, carboxyl, and amide groups in biological molecules
2. Methylmercury is responsible for inducing oxidative stress because of which its free radicals have great potential to cause neurotoxicity

3. In addition, the accumulation of serotonin, aspartate, and glutamate also contribute to inducing methylmercury neurotoxicity
4. Oxidative stress has an effect on calcium homeostasis
5. Activation of phospholipase A2 results in an increase in reactive oxygen species due to increase production of arachidonic acid
6. The alteration of chemical homeostasis is induced both by organic and inorganic mercury but through a different mechanism. In the case of organic mercury, there is increase in intercellular calcium due to the influx of calcium from the extracellular medium and mobilizing intracellular stores, whereas inorganic mercury (Hg^{2+}) compounds tend to increase the intracellular calcium stores only through the influx of calcium from the extracellular medium
7. Methylmercury is also converted to the inorganic form in the CNS and adheres to sulfhydryl-containing molecules
8. The primary site for the production of reactive oxygen species (ROS) is mitochondria through normal metabolism. Inorganic mercury has been shown to increase the production of these reactive oxygen species, causing defects in oxidative phosphorylation and electron transport at the ubiquinone–cytochrome b5 step
9. Inorganic mercury and methylmercury compounds tend to bind to thiol containing protein, e.g., glutamine, cysteine, albumin, etc.
10. Binding to the endogenous thiol-groups facilitates the distribution of mercury compounds in the body. It also protects the compounds from binding to other proteins, thus providing a protective mechanism. A diagrammatical representation of mechanism of mercury toxicity is presented in Figure 7.3.

7.6 TOXICOLOGICAL EFFECTS OF MERCURY

The clinical manifestation of an individual exposed to mercury toxicity depends on the dose, and length and form of exposure. Acute toxicity is generally linked with inhalation and ingestion of elemental mercury and inorganic mercury, respectively, whereas chronic toxicity is more commonly linked with exposure to organic mercury compounds. Acute exposure is caused by the inhalation of elemental mercury which generally leads to pulmonary symptoms. The initial signs and symptoms include:

- Chills
- Fever
- Metallic taste
- Pleuritic chest pain
- Shortness of breath

Other possible symptoms can also appear, such as:

- Stomatitis
- Lethargy
- Confusion
- Vomiting

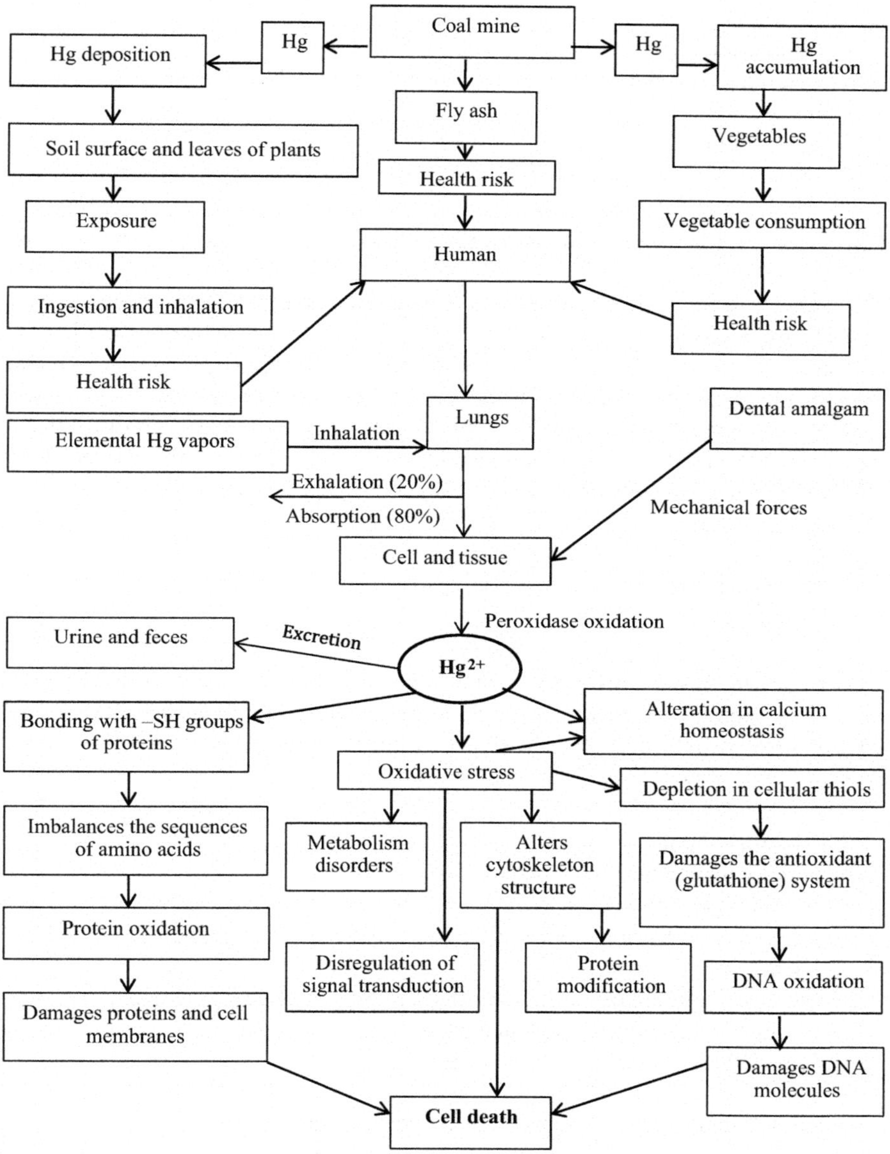

FIGURE 7.3 Mechanisms of mercury toxicity.

- Fatigue
- Muscular weakness
- Weight loss

The classic triad found in chronic toxicity is tremors, gingivitis, and neuropsychiatric disorders such as:

- Anorexia
- Depression
- Emotional instability
- Insomnia
- Memory loss
- Uncontrolled perspiration

Effects on the Cardiovascular and Pulmonary Systems: Mercury accumulation in the heart contributes to multiple cardiovascular problems, particularly cardiomyopathy. Indeed, mercury levels in the heart tissue of individuals dying from idiopathic dilated cardiomyopathy were found to be 22,000 times higher in heart tissue than mercury levels in other deaths caused by heart disease. Endothelial dysfunction, mitochondrial dysfunction, and hypertension are other health implications of mercury toxicity.

Toxic vapours formed due to the burning of mercury or mercury vaporization enter the respiratory system and from there readily enters into the circulation. Even chronic inhalation of low concentrations of mercury (0.7–42 $\mu g/m^3$) can cause:

- Impaired cognitive skills
- Sleep disturbances
- Tremors

Effects on the Gastrointestinal System: The ingestion or absorbance of mercury through epithelial cells results in various digestive disturbances as it inhibits the production of digestive enzymes. The effects of mercury on the gastrointestinal system typically present as:

- Abdominal pain
- Bloody diarrhoea
- Inflammatory bowel disease
- Inhibition of the production of trypsin, chymotrypsin, and pepsin in the body
- Ulcers
- Increased amount of undigested food products in the bloodstream
- Prompting of immune-mediated reactions
- Reduced resistance to pathogenic infection

Exposed individuals may experience common symptoms such as:

- Anorexia

Effects on the Central Nervous System: The most devastating effect of mercury on the nervous system is cellular detoxification, which can cause cells to either die or live in a state of chronic malnutrition. Mercury poisoning in the CNS leads to:

- Damage to the blood–brain barrier
- Penetration of the brain by other toxic metals and substances

- Primary sensory neuropathy
- Autism
- Depression
- Drowsiness
- Fatigue
- Hallucinations
- Insomnia
- Memory loss
- Paranoia
- Tremors of the hands, head, lips, tongue, jaw, and eyelids
- Weight loss

Mercury is known to have various effects on other special sensory systems such as optic neuropathy, retinopathy, blindness, hearing loss, a reduced sense of smell, and abnormal touch sensation.

Effects on the Skin: The most common symptoms of contact dermatitis after exposure to mercury compounds include:

- Erythema
- Irritation
- Mild swelling
- Scaling
- Urticaria
- Vesiculation

Topical and systemic exposure cause a mercurial reaction on the skin followed by allergy with mild pain. *Acrodynia* or *pink disease*, a well-known mercury allergy, results in:

- Diarrhoea
- Oedema of the hands and feet
- Erythema of the palms and soles
- Hypertension
- Poor muscle tone

Effects on the Renal System: Research studies have claimed that exposure to mercury in the renal system can cause many kidney injuries. Mercury can potentially produce substantial damage to the kidneys and produce chronic renal diseases such as:

- Acute tubular necrosis
- Glomerular disease (glomerulonephritis)
- Membranous glomerulonephritis
- Nephrotic syndrome
- Nephrotic-range proteinuria

- Renal cancer
- Sub-acute-onset nephrotic syndrome
- Tubular dysfunction

Effects on the Endocrine System: Mercury is believed to impair the functioning of the endocrine system by reducing hormone–receptor binding which generally affects the insulin, oestrogen, testosterone, and adrenaline hormones. Mercury can inhibit catecholamine degradation, which results in the accumulation of epinephrine and hyperhidrosis, tachycardia, and ptyalism (hypersalivation).

Low levels of pituitary function due to mercury toxicity lead to depression and suicidal thoughts, and can cause frequent urination as well as high blood pressure. Like the pituitary gland, the thyroid can also accumulate mercury, as a result of which thyroid hormone production is blocked by occupying iodine-binding sites with these inhibiting or altering hormone production, impairing body temperature control, and causing hypothyroidism and thyroid inflammation. Similarly, the pancreas is also susceptible to the toxic effects of mercury and produces effects such as interference with normal biological function and dysregulation of blood glucose levels.

Effects on the Reproductive and Foetal Systems: Mercury can initiate pathophysiological changes that may affect reproductive function by altering the circulating levels of follicle-stimulating hormone (FSH), luteinizing hormone (LH), oestrogen, progesterone, and the androgens. In males, mercury can cause spermatogenesis, reduced epididymal sperm count and testicular weight, and erectile dysfunction, while in females, mercury affects the oestrogen and progesterone levels leading to ovarian dysfunction, painful or irregular menstruation, premature menopause, and tipped uterus.

Along with reproductive issues, mercury is linked with foetal toxicity and can cause miscarriages, spontaneous abortions, and low birth weights. Mercury can inhibit the development of the foetal brain by crossing through the placenta and cause cerebral palsy and psychomotor retardation in later stages of development. Table 7.2 summarizes the acute and chronic symptoms of mercury compounds causing multiple disorders.

Historical Case Studies Illustrating the Health Impacts of Mercury: Two epidemics that have occurred due to methylmercury poisoning events are worth mentioning. The first case occurred in the Japanese villages of Minamata Bay (1953) and the second occurred in rural Iraq in 1971–1972. One of the major instances of organic mercury toxicity in history is Minamata disease (1956), which is sometimes called "*dancing cat fever*." In Minamata Bay, Japan, a chemical factory called Chisso discharged methylmercury into water which was subsequently ingested by fish. Methylmercury became bio-accumulated in the food chain from plankton, then microorganisms, up to fish and shellfish. Later, when the methylated fish was consumed by local villagers, they started exhibiting signs of neurologic damage, such as visual loss, extremity numbness, hearing loss, and ataxia. Babies exposed to the methylmercury in utero were most severely affected, and had effects such as low birth weight, seizure disorders, profound developmental delay, incomplete visual loss

TABLE 7.2
Acute and Chronic Effects of Mercury Toxicity in the Human Body

Target Area	Effects on Human Body (Acute Exposure)	Effects on Human Body (Chronic Exposure)
Cardiovascular	• Heart palpitations • Hypertension • Hypovolemic shock • Oxidative stress • Thrombosis	• Endothelial dysfunction • Hypertension • Immune dysfunction • Mitochondrial dysfunction • Tachycardia • Vascular smooth muscle (VSM) proliferation and migration
Pulmonary	• Cough • Emphysema • Interstitial fibrosis, • Pleuritic chest pain • Pneumatocele • Pneumonitis • Respiratory distress syndrome (RDS) • Shortness of breath	• Respiratory distress syndrome (RDS)
Gastrointestinal tract	• Bloody stool • Diarrhoea, nausea, vomiting, and abdominal pain	• Constipation • Diarrhoea • Generalized distress
Central nervous system	• Convulsions • Decreased reflexes • Electrocardiogram (EEG) anomalies • Hearing loss • Irritability • Lethargy • Nerve conduction • Psychomotor and tremors	• Anorexia • Ataxia • Depression • Dysarthria • Insomnia • Memory loss • Paresthesias • Peripheral neuropathy • Shyness • Tremor • Unsteady gait • Vasomotor disturbances
Skin and keratinized tissues	• Buccal pain • Burning and bleeding • Contact dermatitis • Erythematous and pruritic skin rash, • Alopecia • Mucosal inflammation (stomatitis) and greyish membranes	• Acrodynia (pink disease) • Alopecia • Gingivitis • Presence of thin blue lines on gums
Endocrine system	• Depression • Dysregulation of blood glucose level • Hyper-salivation • Hypertension	• Adrenal hyperplasia • Hyper salvation • Hyperhidrosis • Hypothyroidism

(*Continued*)

TABLE 7.2 (Continued)
Acute and Chronic Effects of Mercury Toxicity in the Human Body

Target Area	Effects on Human Body (Acute Exposure)	Effects on Human Body (Chronic Exposure)
		• Tachycardia • Thyroid inflammation
Renal	• Anuria • Haematuria • Oliguria • Proteinuria failure	• Albuminuria • Polydipsia • Polyuria
Reproductive/ foetal	• Deafness • Speech problems • Spontaneous abortion	• Blindness • Erectile dysfunction • Foetal brain damage • Incoordination • Ovarian dysfunction • Painful or irregular menstruation • Retardation • Seizures and paralysis • Spermatogenesis • Spontaneous abortion
Musculoskeletal	• Lumbar pain	• Loss of muscle tone • Muscle weakness • Paralysis • Tremor
Others	• Fever and chills • Foul breath • Loosening of teeth • Metallic taste • Photophobia	• Blushing • Perspiration • Salivation • Weight loss

(including tunnel vision) or total blindness, and hearing loss. Officially, over 2265 victims were recognized, while thousands more people were affected.

The second epidemic of severe methylmercury intoxication was in 1971–72 in rural Iraq where the hospitalization of about 7,000 people and deaths of 460 individuals were reported. Methylmercury was ingested as a result of bread prepared and eaten from wheat seed treated with a mercury-based fungicide. Both of these methylmercury poisoning incidents not only caused deaths but produced multiple and long-lasting intoxication symptoms that included mental retardation, cerebral palsy, blindness, deafness, and dysarthria, especially in children exposed in utero.

7.7 DIAGNOSIS

Mercury generally exists in several physical and chemical statuses and may undergo biotransformation. A diagnosis of mercury toxicity is often confirmed in the following ways:

Physical Examination: Abdominal and rectal examinations, with stool guaiac testing, and skin examination

Urine: Excretion of mercury in urine fairly well indicates the presence of inorganic and elemental mercury exposure while it is unreliable for organic mercury (methylmercury) because it is mostly removed in the faeces. A quantity of more than 100 μg/L produces neurological signs, while a concentration greater than 800 μg/L is often fatal

Blood: The concentration of methylmercury in RBCs is very high in acute toxicity but varies in chronic toxicity. The whole blood mercury concentration is often less than 10 μg/L, but it may reach 20 μg/L. Generally, the mercury blood levels are helpful for recent exposure to toxicity but they do not serve as a guide to therapy

Hair: While blood levels are more accurate for acute exposures, long-term exposures are best reflected in hair mercury measurements. Hair has a high sulfhydryl content so that mercury forms covalent bonds with sulphur which is then found in abundance in hair samples. This results in accurate toxicity measurements in the case of organic mercury (methylmercury). In general, mercury concentrations in the hair do not exceed 10 mg/kg

Cerebrospinal Fluid: The cerebrospinal fluid (CSF) mercury concentration is measured by applying a mass spectroscopy technique. Increased values of CSF mercury levels have been found mostly in workers exposed to mercury vapours

Other Diagnostic Evaluation

Diagnosis tests depend on the clinical situation, which may include complete blood cell count, and renal and liver function tests in acute elemental mercury vapour.

Chest X-ray: A flat plate radiograph of the abdomen is visualized to identify ingested elemental mercury, which appears radiopaque

Magnetic Resonance Imaging (MRI): This helps to confirm clinical and pathologic findings related to mercury toxicity

Single-photon emission computed tomography (SPECT): This can be used to demonstrate right cingulate hyper-metabolism and inattention following exposure to inorganic mercury

Electrophysiology: Electrophysiologic studies have demonstrated sensorimotor neuropathy, typically axonal, in some workers exposed to elemental mercury or mercury vapours

Along with these procedures, other diagnostic evaluations include electrocardiography (ECG) pulmonary function test (PFT), cardiovascular monitoring, electroneuromyography, and neuropsychologic tests.

7.8 TREATMENT AND MANAGEMENT OF MERCURY POISONING

Immediate Considerations and Decontamination: Treatment of mercury toxicity usually consists of evacuating the patient from the source of exposure followed by supportive care and chelating therapy. Patients having emotional squeal and cognitive disorders may require psychotropic medications. Before treatment, some immediate considerations and decontamination are required to be carried out, as follows:

- The primary step is to remove the source of contamination
- Monitoring of proper functioning of vital organs is necessary in primary management of acute exposure
- After pulmonary aspiration, bronchial lavage should not be done so that mercury particles cannot disperse further into the lungs
- Chest X-rays must be done to determine the extent of dispersion
- Inorganic mercury produces severe corrosive injuries such as oropharyngeal oedema and upper airway obstruction, for which endoscopic examination is required. To overcome obstruction of the airway, IV fluid therapy and endotracheal intubation and/or tracheostomy are required
- Skin should be washed with soap and water in case of direct contact with mercury
- Gastrointestinal decontamination should be implemented for inorganic mercury salt because of systemic absorption

Chelation Therapy: Treatment of inorganic mercury toxicity generally involves the use of chelating agents. Chelating agents may contain sulfhydryl groups, which tend to bind the mercury ions and facilitate their excretion through urine and faeces. These agents are specialized in removing a portion of the body's mercury stores. Different kinds of chelating agents used include the following:

Succimer (Chemet): This is used in inorganic and organic mercurials and is a chelator of choice in cases of chronic or mild toxicity

Dimercaprol (BAL): This kind of chelating agent is used for treating acute inorganic mercury toxicity. Dimercaprol is used only in acute ingestion and is administered intramuscularly every four hours, mixed in a peanut oil base. It is then excreted through the urine and bile. Dimercaprol is usually provided to patients suffering from renal failure

D-penicillamine (Depen, Cuprimine): D-penicillamine is an oral, thiol-based chelator for treatment of acute or chronic toxicity. It forms a complex with mercury and is excreted in the urine, which makes it non-useful in the case of renal failure. Succimer has now completely replaced penicillamine, due to its strong metal-mobilizing capacity and lower side-effects

Combination Therapy with Chelating Agents

Nowadays, combination therapy has surpassed single-therapy methods of treating HM toxicity. Co-administration of DMSA with MiADMSA has turned out to be more effective than mono-therapy with MiADMSA by not only controlling the lipid peroxidation but also balancing the decreased catalase activity. It provides better clinical recoveries and minimizes the possible side effects by reducing the chelating agent dose.

Management of Mercury Toxicity: Workplace hygiene is vital, while careful monitoring and disposal of industrial waste must be considered indispensable for averting mercury poisoning. Once exposed to mercury, washing of exposed skin and hair with mild soap and water must be done. Hypothermia must be avoided while decontaminating children or the elderly. In cases of ingestion, emesis must not be induced. Elemental mercury takes time to absorb in the gastrointestinal tract and acute toxicity through the route of exposure is minimal. Washing the mouth with a small quantity of borax after ingestion and immediately providing saline purgative is also recommended. Pregnant women and nursing mothers must avoid consuming larger fish due to the fact that the concentration of mercury is higher in large fish as compared to smaller fish. People vacuuming or sweeping mercury spills in an enclosed space can inhale elemental mercury. That is why concerned authorities must handle any spillages with suitable mercury decontamination kits and procedures.

7.9 IMPACT OF MERCURY ON THE ENVIRONMENT

Atmospheric Environment: Elemental mercury generally evaporates into the air, which acts as source for inhalation. A larger amount of inhalation of contaminated air is a greater risk to health. Mercury and methylmercury exposure to ultra-violet (UV) light emitted by the sun can detoxify its effects on the environment. UV light breaks down methylmercury into inorganic or elemental mercury, which can leave the aquatic ecosystem and again re-enter into the atmosphere as gas. Some sources of mercury compounds in air are household products such as thermostats, barometers, and switches in large appliances. Fluorescent bulbs also contain mercury powder or dust which may spill metallic mercury in the home either deliberately or accidently.

Aquatic Environment: Inorganic mercury is deposited from the atmosphere, which upon methylation gets converted into organic methylmercury. This methylmercury has high toxicity and potential for accumulation in aquatic organisms. In addition, some water bodies also contribute to mercury deposition from direct discharge of industrial wastes, mining wastes, or naturally occurring mercury minerals. Methylmercury is generally responsible for causing toxicity in aquatic organisms. Methylmercury readily crosses the biological membranes and causes toxicity, which consequently increases its concentration while moving up the food chain. In other words, small organisms containing mercury are eaten by the tiny ocean fish and then larger fish eat smaller fish, and as a result the mercury becomes accumulated in larger fish. This bioaccumulation of mercury in top predator fish, where the concentration of mercury can be extremely high, is responsible for causing harmful effects when consumed by humans and fish-eating animals.

Terrestrial Environment: Mercury is widely known as a toxic, persistent, and mobile contaminant that does not degrade in the environment. Different kinds of natural processes are responsible for depositing atmospheric mercury. The deposition of mercury mainly occurs in oxidized form (Hg^{2+}), followed by transformation through the biological and chemical processes of methylation. Generally, the deposition of mercury pollutants with low vegetation is about three to five times lower on the ground than in forests. The largest proportion of mercury is emitted from tropical regions followed by the temperate zone, and the lowest levels are from the polar regions. The mercury content in soil and plants is much higher in areas with high pollution, such as areas near Hg mines, as compared to other areas. Mercury contamination in the soil can affect the growth of crops or even kill plants, which eventually affects human health through bioaccumulation. Soil not only acts as a sink for mercury but also acts as a source of mercury deposition. Soil receives the mercury content from the environment and then re-emits the mercury back to the atmosphere, water, and plants.

Wildlife: Wildlife are at greater risk of mercury contamination because of its ability to magnify in ecosystems. Bio-magnification and bioaccumulation of mercury, mainly methylmercury, have a significant influence on various animal and bird species. In the case of bio-magnification of methylmercury, initially inorganic mercury undergoes methylation with the help of anaerobic microorganisms in wetlands, lakes, and reservoirs. Once methylmercury is released by microbes, it instantly enters the food chain and bio-magnifies easily in predatory species of animals and birds. In mammals, mercury toxicity mostly impacts the CNS, causing lethargy, limb paralysis, and behavioural impairments. In addition, mercury poisoning can cause reduced food take, difficulty in flying, poor muscle coordination, and progressive weakness in many bird species. In the case of bioaccumulation, animals excrete mercury slowly, which results in the accumulation of a high amount of mercury among the top-level predators as they consume a high number of contaminated prey items. Once methylmercury reaches a top-level predator, such as a largemouth bass (freshwater) or king mackerel (saltwater), it may have increased by as much as a million times from bacteria to fish.

BIBLIOGRAPHY

Azevedo, B., et al. (2012) Toxic effects of mercury on the cardiovascular and central nervous systems. *Journal of Biomedicine & Biotechnology* 1: 1–11.

Berglund, M., et al. (2005) Inter-individual variations of human mercury exposure biomarkers: A cross-sectional assessment. *Environmental Health* 4(20): 1–11.

Björkman, L., Lundekvam, B. F., and Lagreid, T. (2007) Mercury in human brain, blood, muscle and toenails in relation to exposure: An autopsy study. *Environmental Health* 6(30): 1–14.

Clarkson, T. W. and Magos, L. (2006) Toxicology of mercury and its chemical compounds. *Critical Reviews in Toxicology* 36: 609–662.

Clarkson, T. W., Magos, L., and Greenwood, M. R. (1972) Transport of elemental mercury into fetal tissues. *Biology of the Neonate* 213: 239–244.

Crespo-López, M.E., Macêdo, G. L., and Pereira, S. (2009) Mercury and human genotoxicity: Critical considerations and possible molecular mechanisms. *Pharmacological Research* 604: 212–220.

Dantzig, P. I. (2003) A new cutaneous sign of mercury poisoning? *Journal of the American Academy of Dermatology* 49: 1109–1111.

Davidson, P. W., Myers, G. J., and Weiss, B. (2004) Mercury exposure and child development outcomes. *Pediatrics* 113: 1023–1029.

Davis, B. J., et al. (2001) Mercury vapour and female reproductive toxicity. *Toxicological Science* 592: 291–296.

Guzzi, G., Grandi, M., and Cattaneo, C. (2006) Dental amalgam and mercury levels in autopsy tissues: Food for thought. *American Journal of Forensic Medicine and Pathology* 271: 42–45.

Haffner, H. T., et al. (1991) Morphological and toxicological findings after intravenous injection of metallic mercury. *Deutsche Medizinische Wochenschrift* 116(36): 1342–1346.

Heyer, N. J., et al. (2004) Chronic low-level mercury exposure, BDNF polymorphism, and associations with self-reported symptoms and mood. *Toxicological Science* 81(2): 354–363.

Holmes, P., Hames, K. A. F., and Levy, L. S. (2009) Is low-level mercury exposure of concern to human health? *Science of Total Environment* 408: 171–182.

Hua, M. S., Huang, C. C., and Yang, W. J. (1995) Chronic elemental mercury intoxication: follow-up case study. *Brain Injury* 10: 377–384.

Hursh, J. B., Clarkson, T. W., and Cherian, M. G. (1976) Clearance of mercury Hg-197, Hg-203 vapour inhaled by human subjects. *Archives of Environmental Health* 31(6): 302–309.

Klaassen, C. D. (2007) *Casarett & Daull's Toxicology: The Basic Science of Poisons*. New York: McGraw Hill.

Kungolos, A., Aoyama, I., and Muramoto, S. (1999) Toxicity of organic and inorganic mercury to *Saccharomyces cerevisiae*. *Ecotoxicology and Environmental Safety* 43(2): 149–155.

Larry, A. B., Catherine, A. H. S., and Ruth E. W. (2002) Toxicology of mercury. *Laboratory Medicine* 33(8): 614–625.

Li, R., et al. (2017) Mercury pollution in vegetables, grains and soils from areas surrounding coal-fired power plants. *Scientific Report* 7: 1–9.

Lund, B. O., Miller, D. M., and Woods, J. S. (1991) Mercury induced H_2O_2 formation and lipid peroxidation in vitro in rat kidney mitochondria. *Biochemistry Pharmacology* 42: 181–187.

Matsuo, N., Suzuki, T., and Akagi, H. (1989) Mercury concentration in organs of contemporary Japanese. *Archives of Environmental Health* 44(5): 298–303.

Mostafalou, S. and Abdollahi, M. (2013) Environmental pollution by mercury and related health concerns: Renotice of a silent threat. *Achieves of Industrial Hygiene and Toxicology* 64: 179–181.

Nelson, L. S., et al. (2011) *Goldfrank's Toxicologic Emergencies*. New York: McGraw Hill.

Rafati-Rahimzadeh, M., Rafati-Rahimzadeh, M., Kazemi, S., and Moghadamnia, A. A. (2014) Current approaches of the management of mercury poisoning: Need of the hour. *DARU Journal of Pharmaceutical Science* 22: 46.

Raj, D. and Maiti, S. (2019) Sources, toxicity, and remediation of mercury: An essence review. *Environmental Monitoring and Assessment* 191(9): 1–22.

Reinhardt, J. W. (1992) Side-effects: Mercury contribution to body burden from dental amalgam. *Advances in Dental Research* 6: 110–113.

Sakamoto, M., Nakamura, M., and Murata, K. (2018) Mercury as a global pollutant and mercury exposure assessment and health effects. *Nihon Eiseigaku Zasshi* 73(3): 258–264.

Sarkar, B. A. (2005) Mercury in the environment: Effects on health and reproduction. *Reviews on Environmental Health* 20: 39–56.

Suda, I. and Takahashi, H. (1992) Degradation of methyl and ethyl mercury into inorganic mercury by other reactive oxygen species besides hydroxyl radical. *Archives of Toxicology* 66(1): 34–39.

Suzuki, T., Takemoto, T. I., Shishido, S., and Kani, K. (1977) Mercury in human amniotic fluid. *Scandinavian Journal of Work, Environment and Health* 31: 32–35.

Tchounwou, P. B., Ayensu, W. K., Ninashvilli, N., and Sutton, D. (2003) Environmental exposures to mercury and its toxicopathologic implications for public health. *Environmental Toxicology* 18: 149–175.

WHO. (1991) *Inorganic Mercury: Environmental Health Criteria.* Geneva, Switzerland: World Health Organization.

Wu, M. F., Ison, J. R., Wecker, J. R., and Lapham, L. W. (1985) Cutaneous and auditory function in rats following methyl mercury poisoning. *Toxicology and Applied Pharmacology* 79(3): 377–388.

Yoshida, M. (2002) Placental to foetal transfer of mercury and fetotoxicity. *Tohoku Journal of Experimental Medicine* 196(2): 79–88.

Zahir, A., Rizwi, S. J., Haq, S. K., and Khan, R. H. (2005) Low dose mercury toxicity and human health. *Environmental Toxicology Pharmacology* 20: 351–360.

8 Heavy Metal Toxicity
Cadmium

Scarily, cadmium is not even the worst poison among the elements. It sits above mercury, a neurotoxin.

Sam Kean

After going through this chapter, the reader will learn about the sources and routes of exposure to cadmium, its aetiology and clinical manifestations, the mechanism of cadmium toxicity, the diagnosis, treatment, and management of cadmium poisoning, and the environmental impacts of cadmium.

FACT FILE

- Cadmium was named after the Greek mythological character Cadmus, the founder of Thebes
- Pure cadmium metal was first discovered by the German chemist Friedrich Stromeyer in 1817
- Cadmium (Cd) is a soft, malleable, ductile, silver-white divalent metal that is placed with zinc and mercury in the Periodic Table in group II B of D block
- Cd was discovered in contaminated zinc compounds sold in pharmacies in Germany, with Germany remaining an important producer of this metal
- The mineral form of Cd is greenockite or cadmium sulphide
- Cd occurs in association with sulphide ores of zinc, lead, and copper, and is most often isolated during the production of zinc; many attributes of cadmium are similar to those of zinc, however it forms complex compounds
- Cd has relatively low melting (320.9°C) and boiling (765°C) points and a relatively high vapour pressure
- Cd is resistant to corrosion and is used as a protective plate on other metals
- In air, Cd is rapidly oxidized into cadmium oxide
- In the presence of reactive gases, for instance CO_2, SO_2, water vapour, sulphur trioxide, or hydrogen chloride, Cd vapuors react to produce cadmium carbonate, hydroxide, sulphite, sulphate, or chloride, respectively, and these compounds are generally formed in chimney stacks and released into the environment

DOI: 10.1201/9781003343721-8

- Inorganic compounds of Cd, namely, acetate, chloride, and sulphate, are quite soluble in water, while cadmium oxide, carbonate, and sulphide are almost insoluble
- Cd is a relatively rare element and is not found in nature in the pure state but in tiny amounts in air, water, soil, and food
- All soils and rocks, including coal and mineral fertilizers, contain some Cd
- Besides its applications in electroplating, nuclear fission, televisions, anticancer drugs, and semi-conductor devices, Cd is also used in batteries, mostly in rechargeable nickel-cadmium batteries
- Cadmium has no role in higher organisms and is considered toxic
- The earliest reported cases of cadmium toxicity were in Belgium in 1858 where workers inhaled Cd dust through polishing silver with cadmium carbonate
- Public awareness about Cd toxicity started with the post-World-War-II outbreak of "*itai-itai*" disease ("ouch-ouch" disease) in Japan, which was believed to have been caused by the release of cadmium-contaminated water from the Kamioka mine
- Cd accumulates in the kidneys, and it damages kidney tissue leading to high blood pressure, kidney and heart disease

8.1 INTRODUCTION

Cadmium is the rarest element in the Earth's crust. It spreads through wind during natural processes such as erosion, or through the burning of cadmium-containing ores in volcanic emissions. Airborne specks of cadmium settle on the soil and watercourses as dust. Though surface water contains some dissolved Cd, its concentration tends to be low as the metal is quickly absorbed by sea life, particularly shellfish. Humans play an important role in generating Cd and discharging it into the environment through anthropogenic and industrial activities, such as mining, smelting, and refining metal ores – mainly zinc, lead, and copper. Another important source of contamination of soil is the use of chemical fertilizers derived from rock phosphate and sewage sludge. Until the mid-1900s, Cd was mainly used as a leather tanning agent or as a pigment in dyes. Today, Cd is mainly used in rechargeable batteries, usually in combination with nickel or silver oxides. Cd plays a significant role in numerous leading-edge technologies such as solar cells. Though trace amounts of many metals are vital for the wellbeing of living organisms, there is no scientific evidence revealing cadmium's nutritional role. Cd is grouped and placed among the transition elements. Cadmium, being a transition metal with a chemical structure similar to silicon and germanium, possesses some similar semi-conducting characteristics. Humans can be harmed by a single large exposure to Cd, and by long-term exposure to higher-than-normal concentrations. Until the mid-1900s, cadmium had few industrial uses and people were seldom exposed to Cd, therefore it was not placed in the category of metals of health concern. But, with growing demand for Cd in industrial processes, its worldwide use and the toxic effects began to be recognized.

8.2 SOURCES AND ROUTES OF EXPOSURE

Volcanic activity is one of the natural sources for increasing environmental cadmium concentrations. Cd was first produced commercially in the twentieth century. It is a by-product of the zinc industry; its production is therefore determined basically by that of zinc. Before the First World War, Cd was not recovered frequently from zinc plants or other non-ferrous metals plants, which brought about uncontrolled Cd pollution of the environment for decades. Cd is now used in television screens, lasers, batteries, paint pigments, cosmetics, photodiodes, phototransistors, photovoltaic solar cells, transparent electrodes, anti-reflection coatings, and in galvanizing steel, as a barrier in nuclear fission, and it was used along with zinc to weld seals in lead water pipes prior to the 1960s.

Most of the Cd that occurs in air is linked with particulate matter in the respirable range. Cd is mostly released to the atmosphere as elemental cadmium and cadmium oxide, and from industrial sources as cadmium sulphide, a result of coal combustion and non-ferrous metal production, or cadmium chloride, a result of refuse incineration. Cd subsists in the air for a relatively short period of time (days to weeks), but this is long enough to permit long-range transport in the atmosphere. Nowadays, Cd is used in a variety of other products, for instance, for chemically stabilizing some plastics. It is also used to produce special solder alloys that melt at comparatively low temperature. In the paint industry, cadmium sulphide gives a yellow shade, while cadmium selenide produces colours ranging from yellow to red. Owing to its ability to absorb emitted particles, control rods and shields of nuclear reactors are made of cadmium. Cd is also used to make phosphor us compounds that glow when bombarded with electrons – this technology produces the image in a television tube. Phosphate fertilizers also contribute a large cadmium load.

Drinking water contains a very low concentration of Cd, generally in the range 0.01–1 μg/litre. For non-smokers, food constitutes the principal environmental source of cadmium. The lowest concentrations are found in milk (around 1 μg/kg). The cadmium concentration is in the range 1–50 μg/kg in meat, fish, and fruit, and 10–300 μg/kg in staple foods such as wheat, rice, and potatoes. The highest cadmium levels (100–1000 μg/kg) are found in the internal organs (kidneys and livers) of mammals and aquatic animals. Contamination of drugs and dietary supplements is a source of contamination that can produce long-term health effects. When cultivated on cadmium-polluted soil, some crops, such as rice, can accumulate considerable amounts of cadmium. Human exposure to Cd mainly occurs through inhalation or ingestion. Based on particle size, 10–50 per cent of inhaled Cd dust is absorbed. Absorption through skin contact is insignificant. Depending on the particle size, about 5–10 per cent of consumed Cd is absorbed. Intestinal absorption is greater in persons with iron, calcium, or zinc deficiency. Cigarette smoking is another significant source of cadmium exposure. The Cd levels in blood and kidneys are greater in smokers than non-smokers. Industrial exposure can be huge in occupational settings, for example, welding or soldering, and can cause chemical pneumonitis. After absorption, Cd is transported all over the body, generally joined to a sulfhydryl group-containing protein like metallothionein. The half-life of cadmium in the blood is estimated to be

75–128 days, but this half-life mainly denotes the deposition in organs, not clearance from the body. Consequently, blood, hair, and urine Cd levels are poor substitutes for body burden and mainly reveal fresh exposures, as with other heavy metals.

8.3 AETIOLOGY

On account of the fast industrial development and technological advancements, Cd as a pollutant has been introduced into the environment. It is taken up in high amounts from contaminated water, food, and air. High concentrations of Cd are present in crustaceans, molluscs, oysters, cephalopods, and crabs; it is also found in offal products, such as liver and kidney, oil seeds, cocoa beans, and a few wild mushrooms. Foodstuffs obtained from crops grown in contaminated soil contain a high concentration of Cd as compared to meat, egg, milk, and dairy products. Among these, rice, wheat, green leafy vegetables, potatoes, carrots, etc. have a high Cd concentration. Vegetarians and shellfish consumers are exposed to higher cadmium intakes than omnivores. Cd presents specific hydro-chemical features, affecting its potential mobility in groundwater. At near neutral pH (<6.5) Cd remains in solution, contrary to the typical fixation of other heavy metals. Thus, heavy metal contamination of potable water and rivers with Cd may occur in areas closer to factories or mines. Another environmental source of Cd exposure is tobacco use.

8.4 PATHOPHYSIOLOGY AND CLINICAL MANIFESTATIONS

Inhaled cadmium causes smoking-related respiratory diseases, such as pulmonary disease and lung cancer. Cd is widely used in electroplating and industrial paints, and represents a hazard when sprayed. Atmospheric deposition of airborne Cd, mining, and the use of fertilizers containing cadmium phosphate and sewage sludge on farms may contaminate soil and increase its absorption by crops and vegetables. Starting in 1912, exposure to cadmium was a serious problem in Toyama Prefecture (Japan), where many people consumed rice cultivated using Cd-contaminated irrigation water. Cd was released into the Jinzu river basin by mining companies that were prosecuted for the damage. It resulted in the outbreak of *itai-itai* disease, with patients showing a wide range of symptoms, for instance:

- Anaemia
- Coughing
- High rate of fractures and distortion of long bones
- Intense bone pain
- Kidney failure, leading to death
- Low bone mass
- Low-grade bone mineralization
- Micro-architectural deterioration of bone tissue
- Osteomalacia
- Osteoporosis
- Severe skeletal decalcification

Workers engaged in soldering or welding metals containing cadmium, and smelting and refining metals are exposed to Cd in the air, in addition to those involved in making Cd products, such as batteries, coatings, or plastics. Moreover, artists working with cadmium pigments, containing cadmium sulphide and cadmium sulphoselenide, can accidentally ingest paint pigments. Being one of the global health problems that affect many organs, Cd poisoning has been reported from different parts of the world. Long-term exposure to cadmium through air, water, soil, and food manifests as cancer and organ system toxicity, such as in the skeletal, urinary, reproductive, cardiovascular, and respiratory systems, and also the central and peripheral nervous systems. Human exposure to Cd compounds has serious health implications.

8.5 MECHANISMS OF CADMIUM TOXICITY

Cadmium toxicity is demonstrated in several organs. Cadmium induces:

- Tissue injury through oxidative stress
- Epigenetic changes in DNA expression
- Inhibition or upregulation of transport pathways, mainly in the proximal S1 segment of the kidney tubule
- Competitive interference with the physiologic action of Zn or Mg
- Inhibition of haem synthesis
- Impairment of mitochondrial functioning, possibly bringing about apoptosis
- Depletion of glutathione
- Structural distortion of proteins because of Cd binding to sulfhydryl groups

The clinical toxicity of cadmium depends on the route, dose, and duration of exposure. In humans, the main impacted organ is the kidney, where the S1 segment of the proximal tubule is a major target of Cd accumulation, with clinically evident defects in protein, amino acid, glucose, bicarbonate, and phosphate reabsorption as a consequence of Cd-induced oxidative damage to transport proteins and mitochondria. This in turn induces apoptosis of tubular cells. Effective antioxidant therapies are being searched for and there is *in vitro* evidence that selenium and zinc may partly antagonize the toxic effects of cadmium. Approximately 30 per cent of cadmium in the body is deposited in the kidneys' tubular regions. Diabetic patients are more vulnerable to renal tubular damage from Cd exposure.

Excretory and Skeletal Systems: Cd exposure may harm vitamin D metabolism in the kidney, with a harmful impact on bone. This effect, together with direct Cd-induced damage to the gut absorption of calcium and instability of collagen metabolism produces osteomalacia and/or osteoporosis. The most extreme example of this process is *itai-itai* disease in Japan. Cd toxicity in bone stimulates fibroblast growth factor 23, which prompts phosphaturia and decreases the uptake of phosphate, leading to osteomalacia. Cd is toxic to MC3T3 osteoblasts and encourages osteoclasts, thereby inducing osteoporosis. These factors combine to induce calciuria, increase bone resorption, and decrease bone mineral density. Epidemiological studies have

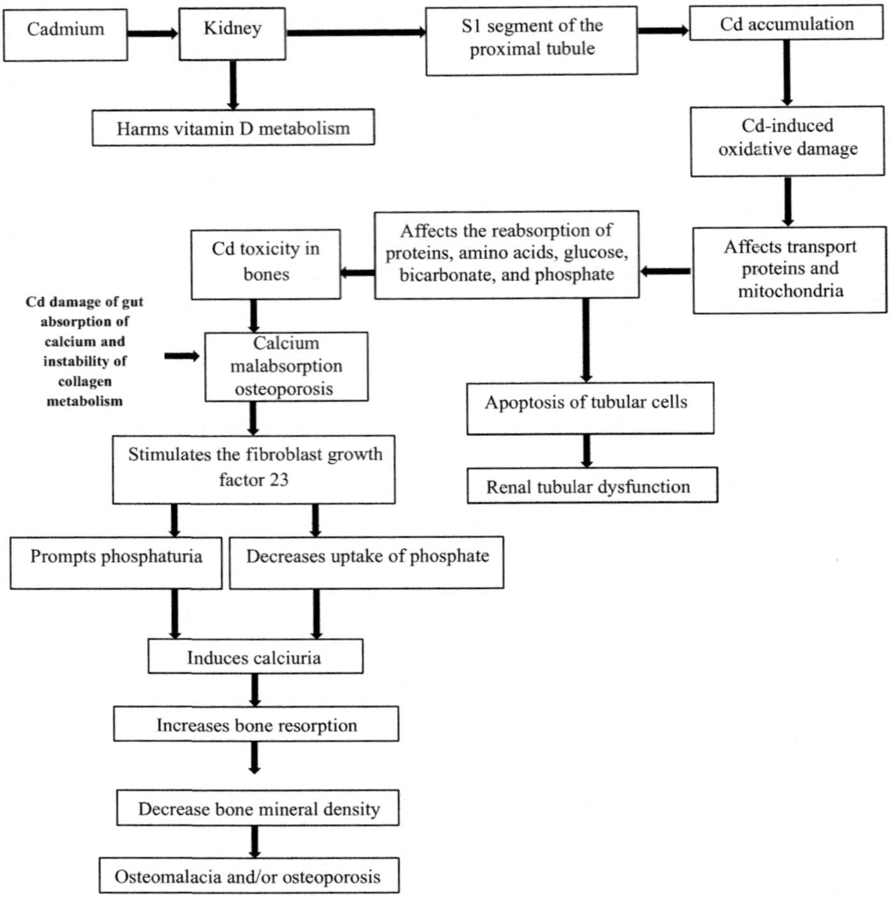

FIGURE 8.1 Mechanism of Cd toxicity in the kidneys and bones.

revealed that exposure to Cd promotes the development of musculoskeletal diseases, such as osteoporosis, rheumatoid arthritis (RA), and osteoarthritis (OA) (Figure 8.1).

Cardiovascular System: Cadmium affects the cardiovascular system in several ways, such as (Figure 8.2):

- Cd induces hypertension and diabetes
- Direct impact on gene transcription in vascular endothelium
- Sudden cardiac death
- Peripheral arterial disease
- Increased vascular intima media thickness

Immune System: The immune system is also affected by Cd-induced damage at several levels. Prenatal Cd exposure may impair post-natal T-cell production and

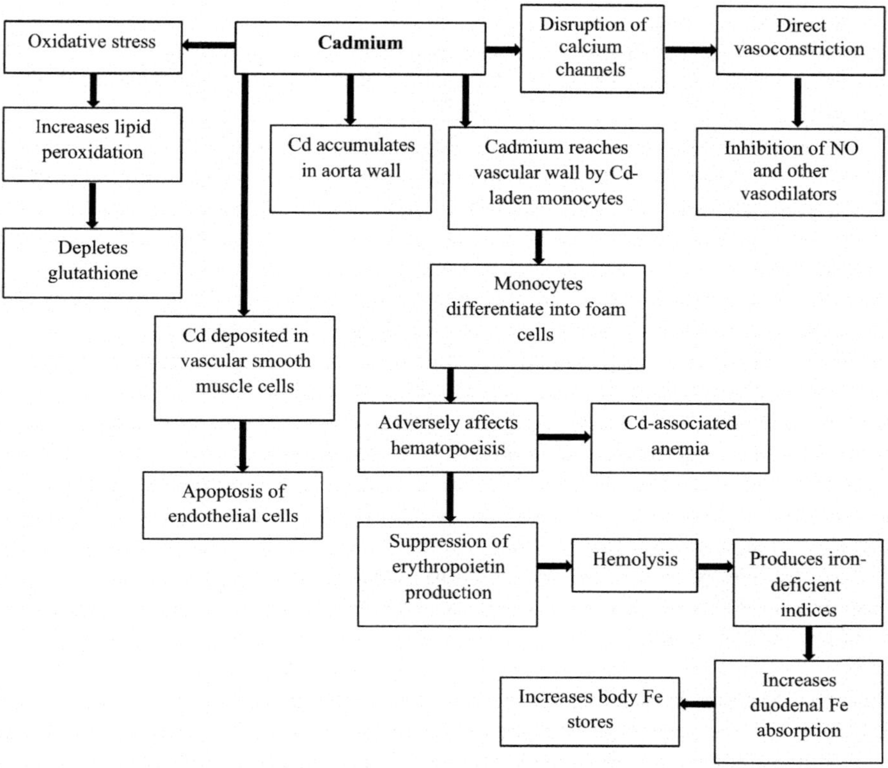

FIGURE 8.2 Mechanism of Cd toxicity in the circulatory system.

the response to immunization, as well as dysregulated thymocyte development. Post-natal Cd exposure arrests the cell cycle and apoptosis in splenocytes. Cd induces:

- Increased rates of auto-immunity
- Increased production of nonspecific antibodies
- Decreased production of antigen-specific antibodies
- Lymphocyte proliferation
- Suppression of natural killer cell activity
- Myocardial infarction/direct myocardial structural damage

Endocrine System: Cd considerably disrupts the capacity of the endocrine system, apparently dysregulating the pituitary hormones. Studies have shown that an elevated level of Cd in blood suppresses TSH production, while increased urine Cd is related to elevated levels of serum T_3 and T_4. Cd exposure is a known risk factor for developing insulin resistance. The Cd effect on insulin resistance may be reduced by supplementation with Fe, Ca, Mg, and Zn.

Reproductive System: Studies in animals have shown that Cd exposure induces male infertility as a result of damage to the blood–testis barrier, decreasing germ cell adhesion, leading to germ cell loss, reduced sperm count, and subfertility or infertility.

Along with strong teratogenic and mutagenic effects, cadmium shows adverse effects at low doses on both male and female human reproduction and affects pregnancy and/or its outcome. This is due to changes in the expression of many genes in the embryo, leading to the aberrant methylation of them within the placenta and embryo. Cd induces the modification of steroidogenesis, which manifests in the form of disorders of the menstrual cycle and reproductive hormones, delays in puberty and menarche, pregnancy loss, premature birth, and reduced birth weight.

Cd has been adversely linked to bone mineral density in post-menopausal women. Women have a higher cadmium body burden than men, showing a higher concentration of Cd in the blood, urine, and kidneys. Cd toxicity requires immediate medical treatment. Emesis or a gastric lavage may be beneficial soon after exposure via ingestion. Detoxification of Cd with EDTA (ethylenediamine tetraacetic acid), BAL (British Anti-Lewisite, dimercaprol), DMSA (2,3-dimercapto-succinic acid), and DMPS (2, 3-dimercapto-1-propane sulfonic acid) is also possible. MiADMSA (monoisoamyl DMSA) is a water-soluble lipophilic chelating agent that can access different endogenous ligands, which is desirable to its parent compound.

Digestive System: Depending on the dose and nutritional composition, the uptake of Cd through the gastrointestinal tract is about 5 per cent of the ingested amount of cadmium in humans. An average German citizen has a daily intake of 30–35 μg cadmium; 95 per cent of this is taken up with food and drinks. An average smoker has an additional intake of 30 μg per day. Several factors can increase this amount, such as low intake of vitamin D, calcium, and trace elements like zinc and copper. It has been shown that a high-fibre diet increases the dietary cadmium intake. Cd ingested in high doses irritates the gastric epithelium. Acute poisoning via Cd ingestion occurs after consuming acidic food or beverages improperly stored in containers with a Cd glaze. Symptoms of severe cadmium ingestion are:

- Nausea
- Vomiting
- Abdominal cramps and pain
- Diarrhoea
- Tenesmus

Respiratory System: Severe, often deadly, respiratory disease can occur after short-term inhalation exposure to a high concentration of Cd compounds; however, such exposures are now very uncommon. These types of exposures are more common in occupational settings, such as cadmium alloy production, welding involving Cd-coated steel, and Cd smelting and refining. The initial symptoms resemble flu-like illness, chills, fever, and myalgias. Later symptoms include chest pain, cough, and dyspnoea, followed by bronchospasm and haemoptysis. Histologic findings in the lungs after Cd exposure include:

- Fibroblastic proliferation
- Hyperaemia of the trachea and bronchi
- Hyperplasia of the alveolar lining cells
- Intra-alveolar haemorrhage
- Pulmonary oedema
- Thrombosis of small blood vessels

Biologically active compounds, namely, polyphenolic substances, melatonin, carotenoids, quercetine, resveratrol, vitamin E, vitamin C, L-carnitine, etc., occurring naturally in plants are useful for averting the harmful health effects of exposure to heavy metals, including cadmium. Among these, polyphenols and antioxidants are present in vegetables, fruits, and spices, and in drinks like green tea, red wine, and cocoa. Owing to the high affinity of Cd ions to hydroxyl groups and the capacity to increase the expression of intestinal metallothioneins (MTs), polyphenols decrease the absorption of this xenobiotic metal from the gastrointestinal tract. Bio-elements, for example, Mn, Se, Zn, Mg, Cl, and Fe, decrease Cd absorption from the gastrointestinal tract and cause its accumulation in the body. Moreover, some of the elements present at active sites of antioxidant enzymes protect the weakening of anti-oxidative potential induced by Cd and oxidative stress development. Melatonin, a natural antioxidant, may protect biological organisms from oxidative stress.

8.5.1 Mechanisms of Cd Toxicity at the Cellular Level

Cd is known to cause oxidative stress and histologically evident membrane disturbances in the CNS, with reduced acetylcholinesterase activity, increased oxidative stress markers, depletion of glutathione, superoxide dismutase 2, and other antioxidants, and depletion of catalase, glutathione peroxidase, and glutathione-S-transferase (Figure 8.3). These changes lead to the apoptosis of cortical cells in the CNS, probably owing to phosphorylation of calcium/calmodulin-dependent protein kinase II. Cd can also inhibit influx through calcium channels. Clinically, humans with raised blood or urine Cd exhibit decreased attention level and memory. Furthermore, those with high urinary Cd levels had considerably decreased low-frequency hearing.

At the cellular level, Cd^{2+}-induced apoptosis and mitochondrial effects are very significant. The mitochondrial respiratory chain plays a vital role in maintaining energy homeostasis through oxidative phosphorylation, producing energy in the form of adenosine triphosphate (ATP), which is necessary for life. Moreover, mitochondria are involved in the synthesis of amino acids, lipids, and phospholipids, ion homeostasis, motility, and apoptosis.

The numbers and roles of mitochondria differ in animal tissues and cells relative to energy needs and in response to physiological or environmental modifications.

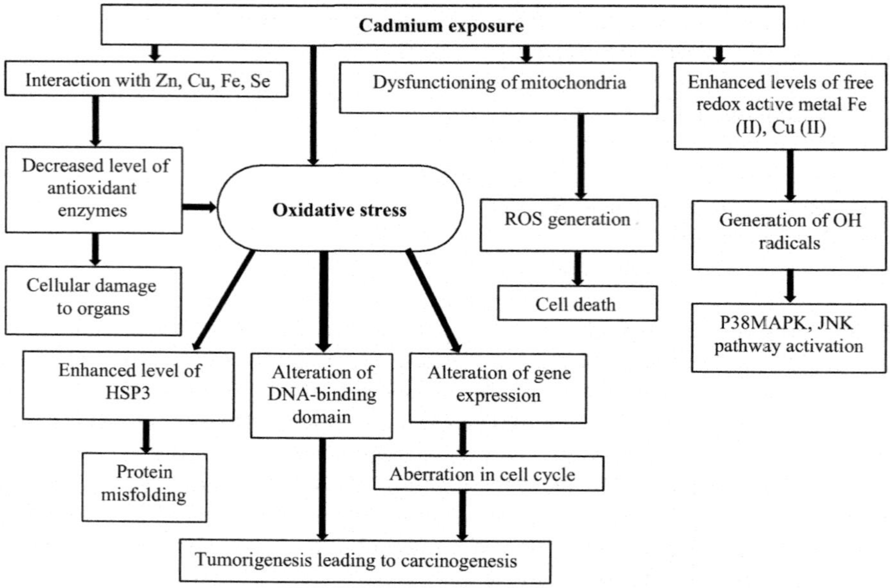

FIGURE 8.3 Mechanism of Cd toxicity at the cellular level.

Malfunctioning of mitochondria is linked to aging and many diseases, including cancer. Mitochondria consist of four sub-compartments:

- Outer membranes
- Inner membranes
- Inter-membranes space
- Matrix

The mitochondrial morphology apparatus is also controlled through post-translational modifications, such as phosphorylation, ubiquitination, and sumoylation. Mitochondria contain the genome, made of a small double-stranded circular DNA molecule (mtDNA). The human mitochondrial chromosome contains 37 genes, including 13 that encode proteins of the respiratory chain; with the remaining 24 genes encoding for rRNA and tRNA, which are required for the protein-synthesizing apparatus of mitochondria. Nuclear genes encode for other mitochondrial proteins. Generally, mitochondria contain five to seven genomes that are stabilized by DNA-binding proteins and that are different from the nuclear histones. Given the restricted activities of DNA repair and high ROS production, mammalian mtDNA is more prone to oxidative damage and is exposed to a higher mutation rate than nuclear DNA.

Important targets of Cd are the thiol groups (–SH) of cysteines present in proteins. Inactivation of the sulfhydryl groups of enzymes produces several functional deficits in nuclei, the endoplasmic reticulum, and mitochondria. The toxic effect linked with Cd is mostly because of the blocking of the mitochondrial electron-transfer chain by damaging electron flow through complex III. Cd-induced changes in mitochondria include:

- Inhibition of ADP- and uncoupler-stimulated respiration
- Increased ion penetrability in the inner mitochondrial membrane by opening of a mitochondrial permeability transition pore
- Modified activity of mitochondrial proteins (enzymes and transporter systems across the outer and inner membranes) by:
  - ✓ Preventing respiratory chain enzymes
  - ✓ Collapsing membrane potential
  - ✓ Mitochondrial inflammation
  - ✓ Inhibition of respiration
- Increased permeability and decreased mitochondrial membrane potential release cytochrome C and activate the caspase pathway
- Inhibition of ATPase, lactate dehydrogenase (LDH), superoxide dismutase (SOD), and glutathione peroxidase (GPx) activities
- Enhanced levels of ROS and lipid peroxidation
- Cd is involved in Fenton reactions. Even though Cd does not act as a catalyst in the Fenton reaction, it can produce ROS by indirectly displacing an endogenous Fenton metal (e.g., Fe^{2+}) from proteins, thereby increasing the amount of free redox-active metals
- Cd may alter the cellular redox status by reacting with exogenous and endogenous antioxidants, such as glutathione (GSH). Cd may damage mitochondria and not only interferes with Ca^{2+} signalling but also increases mitochondrial ROS formation

Cd ions act as xenobiotics by inhibiting the activity of complex II (succinate dehydrogenase) and complex III (cytochrome bc1 complex) of the ETC more than acting on complex I (NADH dehydrogenase or NADH: ubiquinone oxidoreductase) and complex IV (cytochrome oxidase). The principal site of ROS production resides in complex III, and ROS amassing affects the mitochondrial membrane potential and triggers a sequence of events, including apoptosis. There are two different apoptotic pathways:

- The extrinsic or death receptor-mediated pathway (triggers apoptosis in response to external stimuli)
- The intrinsic or mitochondrial-mediated pathway (triggers apoptosis in response to internal stimuli, such as DNA damage)

These two pathways are interrelated to each other and the molecules in one of the two pathways influence those of the other. The mitochondrial pathway is related to toxic stimuli like ROS, UV radiation, ionizing radiation, Ca^{2+}, and Cd^{2+}, either indirectly or through increases in Ca^{2+} and ROS.

The intrinsic pathway (Figure 8.4) is a result of the cellular stress signal as DNA damage triggers caspase-9. In addition to Cd toxicity, apoptosis can be induced by caspase-independent events or by Ca^{2+}–calpain coupled processes. Excessive ROS

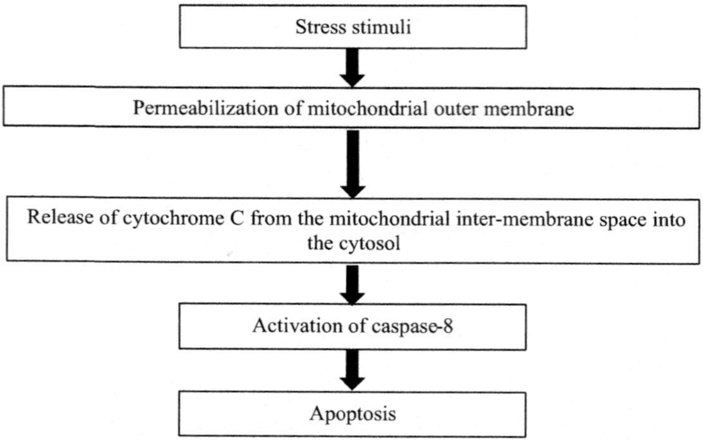

FIGURE 8.4 Extrinsic or death receptor-mediated pathway.

production leads to a free radical attack of mitochondrial membrane phospholipids and mitochondrial membrane depolarization, which is an important step in the intrinsic pathway of apoptosis. The implementation pathway affects the typical cyto-morphological features, such as:

- Apoptotic bodies
- Cell shrinkage
- Chromatin condensation
- Formation of cytoplasmic blebs

Moreover, apoptotic cells show protein cleavage, protein cross-linking DNA break-down, and phagocytic recognition. Besides apoptosis, excessive ROS production can lead to:

- Depolarization and disturbed integrity of the mitochondrial membrane
- Free radical attack on phospholipids
- Macromolecule oxidation
- mtDNA mutation

Major elements of mitochondrial apoptosis are represented by:

- Generation of ROS
- Modification of the mitochondrial membrane potential
- Activation of caspase-9

In vivo, Cd induces apoptosis in several organs, including the liver and kidneys. In cultured cells, it causes a cellular stress response that ends in activation of the mito-chondrial apoptosis pathway. In intact cells, Cd^{2+}-induced mitochondrial damage with mitochondrial apoptosis is a late event, which occurs at least 15–24 h after Cd^{2+} exposure. Cd has been categorized as a carcinogenic agent when inhaled, while there is insufficient evidence in support of oral ingestion to classify Cd as a carcinogenic xenobiotic. Major Cd-induced carcinogenic mechanisms include:

- Abnormal DNA methylation
- Alterations in gene expression
- Lessening of apoptosis
- Cell proliferation
- Damage to DNA
- Decreased DNA repair capacity
- Epigenetics
- Induced inflammatory processes
- Oxidative stress
- ROS

Zinc finger motifs in steroid hormone receptors and other DNA-binding molecules offer a target for Cd binding and modification in their biological function. The chemical resemblance of Cd to Zn is responsible for the toxic effects of Cd. Since oxidative stress is vital in Cd toxicity, it is known to encourage tumour development through mutagenesis and impacts the cell cycle. A functional DNA repair system removes errors due to metabolism and environmental carcinogens; however, insufficient repair mechanisms permit the amassing of damaged DNA, which stimulates cancer. This way, Cd can inhibit DNA repair, including nucleotide and base excisions and mismatch. Loss of the DNA repair mechanism allows amassing of cells with impaired DNA, which after cell division brings about carcinogenic mutations.

8.6 DIAGNOSIS

The diagnosis of Cd toxicity after acute exposure involves the recording of history, physical examination, and symptoms. In chronic exposure, a urinary Cd test can determine the total body Cd burden. Fluids and tissues may be tested for Cd toxicity. This can be done with:

- Blood tests
- Hair or nail analyses
- Urine tests

8.7 TREATMENT AND MANAGEMENT

Cd fumes pose a significant risk to the lungs; hence, airway management is important and so supplemental oxygen is applied. The ingestion of Cd raises somewhat different concerns. Activated charcoal and gastric lavage are believed to be the best options for treatment. Besides monitoring of gastrointestinal injury, renal and hepatic dysfunction are also important. Chelators, though advised, have not been absolutely proven to be beneficial in managing Cd toxicity. In the case of chronic poisoning, the patient's dysfunction is usually ruled out after routine screening. The first intervention is removal of the patient from the site of exposure. To reduce the chances of Cd toxicity, it is important to:

- Avoid the use of Cd-containing fungicides
- Consult a doctor for advice on quitting smoking, as this is the main source of Cd intake
- Follow occupational health and safety measures at workplace
- Identify potential sources of Cd in and around the living place, at work, and playing areas
- Keep nickel–cadmium batteries out of the reach of children
- Make an inventory of Cd-containing products
- Eat a balanced diet that provides enough Cl, Fe, Zn, and proteins

8.8 IMPACT OF CADMIUM ON THE ENVIRONMENT

Among metals, cadmium is considered to be highly toxic to living organisms including humans, and it has biological activity in terrestrial and aquatic organisms. Developments in agriculture and industries have resulted in a higher concentration of Cd in agricultural soil. Cd enters ecosystems via many anthropogenic activities and emissions to the environment. The deposition of Cd in plants grown in Cd-polluted soil poses serious problems to the health of animals and humans due to its high mobility in polluted soils.

Cd pollution in groundwater and soil is witnessed across the globe. Cadmium, though bio-persistent, has toxicological effects and remains for many years after being consumed by organisms. Anthropogenic sources of Cd in groundwater and soil include the metal industry, traffic, landfills, mining, combustion emissions, and sewage sludge. Likewise, the utilization of synthetic phosphate fertilizers containing Cd is a common contamination cause of the increasing Cd concentration in groundwater and soil. Cd is a main component of refined products of petroleum as an impurity. It has potential to enter the food chain and may be toxic to biota. Generally, landfills, Cd-Ni batteries, and solid municipal wastes are the main sources of Cd pollution. Leachate from compost and sludge accounts for 2–5 per cent of Cd deposition in soil. Leachates from various sources cause Cd pollution in groundwater and leachate from farmyard manure, atmospheric deposition, inorganic fertilizers, and municipal solid waste which contribute 10–25 per cent, 15–50 per cent, and 30–5 per cent, and 2–5 per cent, respectively, of Cd in soil. A Cd coating provides superior resistance to corrosion for vehicles, mostly in highly corrosive environments such as aerospace and marine. Cd pollution from Zn smelters in soil via atmospheric deposition and solid waste leaching releases Cd into soil. Similarly, groundwater contamination occurs instantaneously from various sources and diverse routes, which may inhibit the detection of Cd sources, routes, and geogenic phenomena.

Cd is a non-nutritive metal that is considered harmful to the environment and humans, mostly affecting the kidneys and skeleton. If inhaled, Cd acts as a carcinogen. Cd is used in batteries, paints, plastics, electroplating, etc. It is released into the atmosphere from metal production and the burning of fossil fuels. Cd is naturally present in the Earth's crust and oceans but can be added to the soil through natural and anthropogenic activities such as irrigation water, manure, and fertilizers. The presence of Cd in the environment, increased by activities such as long-term use of coal and fossil-fuel and mining, poses serious health hazards. The accumulation of Cd in soil can affect crops and eventually animals and human b's dependent on them. Cd is a water-soluble metal pollutant and is non-essential for plant growth. Cd is known to:

- Trigger the synthesis of reactive oxygen species
- Hinder the utilization, uptake, and transport of essential nutrients and water
- Modify photosynthetic machinery, thus resulting in plant tissue death

The effects of Cd depend on the dose as well as the plant species, with some plants showing Cd tolerance through the following cellular responses:

- Affecting plant metabolism
- Changes in gene expression
- Compartmentalization of ligand–metal complexes
- Generation of enzymatic and non-enzymatic antioxidants
- Hormonal and/or nutritional balances
- Discontinuing carbon fixation
- Decreasing bio-synthesis of chlorophyll content and photosynthetic activity
- Metal ion homeostasis
- Production of metal-detoxifying peptides
- Reduced photosynthesis, growth, and productivity
- Synthesis of osmolytes
- Disturbed plant–water relations

Chlorosis and shunted growth are easily identifiable symptoms of Cd toxicity. Higher toxicity prevents plant growth and results in plant necrosis. Exposure to Cd in soil induces osmotic stress in plants, and minimizes leaf relative water content, stomatal conductance, and transpiration, resulting in plant physiological damage. Cd toxicity causes the over-production of ROS, which damages plant membranes leading to the destruction of cell biomolecules and organelles. Cd also reduces plant uptake of Fe and Zn, resulting in leaf chlorosis. Usually, Cd interferes with the transport and uptake of Ca, P, Mg, K, and Mn in plants.

BIBLIOGRAPHY

Achmad, R. T., Budiawan, B., and Ibrahim, A. E. (2017) Effects of chromium on human body. *Annual Research and Review in Biology* 13: 1–8.

Bernhoft, R. A. (2013) Cadmium toxicity and treatment. *The Scientific World Journal* 7: 1–8.

Cannino, G., Ferruggia, E., Luparello, C., and Rinaldi, A. M. (2009) Cadmium and mitochondria. *Mitochondrion* 9(6): 377–384.

Chunhabundit, R. (2016) Cadmium exposure and potential health risk from foods in contaminated area, Thailand. *Toxicology Research* 32(1): 65–72.

Cuypers, A., Plusquin, M., Remans, T., et al. (2010) Cadmium stress: An oxidative challenge. *BioMetals* 23(5): 927–940.

Eum, K. D., Lee, M. S., and Paek, D. (2008) Cadmium in blood and hypertension. *Science of the Total Environment* 407: 147–153.

Friberg, L. (1983) Cadmium. *Annual Review of Public Health* 4: 367–373.

Godt, J., Scheidig, F., Grosse-Siestrup, C., et al. (2006) Toxicity of cadmium and resulting hazards for human health. *Journal of Occupational Medicine and Toxicology* 1: 22.

Hasan, S. A., Fariduddin, Q., Ali, B., Hayat, S., and Ahmad, A. (2009) Cadmium: Toxicity and tolerance in plants. *Journal of Environmental Biology* 30: 165–174.

Il'yasovam, D. and Schwartz, G. G. (2005) Cadmium and renal cancer. *Toxicology and Applied Pharmacology* 207: 179–186.

Jali, P., Pradhan, C., and Das, A. B. (2016) Effects of cadmium toxicity in plants: A review article. *Scholars Academic Journal of Biosciences* 4: 1074–1081.

Jarup, L. (2002) Cadmium overload and toxicity. *Nephrology Dialysis Transplantation* 17: 35–39.

Jarup, L., Rogenfelt, A., and Elinder, C. G. (1983) Biological half-time of cadmium in the blood of workers after cessation of exposure. *Scandinavian Journal of Work, Environment and Health* 94: 327–331.

Joseph, P. (2009) Mechanisms of cadmium carcinogenesis. *Toxicology and Applied Pharmacology* 238(3): 272–279.

Kirkham, M. B. (2006) Cadmium in plants on polluted soils: effect of soil factors, hyperaccumulation and amendments. *Geoderma* 137(1–2): 19–32.

Klaassen, C. D., Liu, J., and Diwan, B. A. (2009) Metallothionein protection of cadmium toxicity. *Toxicology and Applied Pharmacology* 238(3): 215–220.

Kubier, A. Wilkin, R.T., and Pichler, T. (2019) Cadmium in soils and groundwater: A review. *Applied Geochemistry* 108: 1–16.

Liao, M. and Xie, X. M. (2004) Cadmium release in contaminated soils due to organic acids. *Pedosphere* 14: 223–228.

Menke, A., Muntner, P., Silbergeld, E. K., Platz, E. A., and Guallar, E. (2009) Cadmium levels in urine and mortality among U.S. adults. *Environmental Health Perspectives* 117(2): 190–196.

Nawrot, T., Geusens, P., Nulens, T. S., and Nemery, B. (2010) Occupational cadmium exposure and calcium excretion, bone density, and osteoporosis in men. *Journal of Bone and Mineral Research* 25: 1441–1445.

Nordberg, G. F. (2004) Cadmium and health in the 21st century historical remarks and trends for the future. *Biometals* 17: 485–489.

Park, R. M., Stayner, L. T., Petersen, M. R., et al. (2012) Cadmium and lung cancer mortality accounting for simultaneous arsenic exposure. *Occupational and Environmental Medicine* 69(5): 303–309.

Qin, S., Liu, H., Nie, Z., Rengel, Z., Gao, W., Li, C., and Zhao, P. (2020) Toxicity of cadmium and its competition with mineral nutrients for uptake by plants: A review. *Pedosphere* 30: 168–180.

RafatiRahimzadeh M., RafatiRahimzadeh M., Kazemi S., and Moghadamnia A. A. (2017) Cadmium toxicity and treatment: An update. *Caspian Journal of Internal Medicine* 83: 135–145.

Rani, A., Kumar, A., Lal, A., and Pant, M. (2014) Cellular mechanisms of cadmium- induced toxicity: A review. *International Journal of Environmental Health and Research* 24: 378–399.

Seregin, I. V. and Ivanov, V. B. (2020) Physiological aspects of cadmium and lead toxic effects on higher plants. *Russian Journal of Plant Physiology* 48: 523–544.

Song, Y., Jin, L., and Wang, X. (2017) Cadmium absorption and transportation pathways in plants. *International Journal of Phytoremediation* 19: 133–141.

Thévenod, F. (2003) Nephrotoxicity and the proximal tubule: Insights from cadmium. *Nephron* 93(4): 87–93.

Waalkes, M. P. (2003) Cadmium carcinogenesis. *Mutation Research* 533: 107–120.

Wahid, A., Ghani, A., and Javed, F. (2008) Effect of cadmium on photosynthesis, nutrition and growth of mungbean. *Agronomy for Sustainable Development* 28: 273–280.

Wolf, M. B. and Baynes, J. W. (2007) Cadmium and mercury cause an oxidative stress-induced endothelial dysfunction. *BioMetals* 201: 73–81.

9 Radioactive Metals and Radiation Toxicity

It is easier to denature plutonium than to denature the evil spirit of man.

Albert Einstein

After going through this chapter, the reader will learn about: radioactive elements, a list of radioactive elements, nuclear stability, radioactive decay and its types, the measurement of radioactivity, sources and routes of exposure, its aetiology, nuclear disasters in history, its pathophysiology and clinical manifestation, mechanisms of toxicity, toxicological effects, diagnosis, the treatment and management of radioactive metals poisoning, and the impact of radioactive metals on the environment.

FACT FILE

- Klaproth named uranium after the planet Uranus
- Romans used uranium ore as a yellow colour for colouring their pottery
- Uranium has a lower density than gold and tungsten
- Uranium-enriched gun type fission weapon, "*Little Boy*," the atomic bomb, which was dropped over Hiroshima on 6th August 1945 was the first nuclear weapon ever used
- Radium was the first synthetically made radioactive element
- 14,000 times more caesium 137 than the Hiroshima atomic bombing was released following the Fukushima incident
- Marie Curie discovered the element Polonium while experimenting with pitchblende
- Marie Curie died because of her exposure to ionizing radiation resulting in aplastic anaemia, a rare blood disorder caused by failure of the bone marrow to form new blood cells
- Astatine is the rarest naturally occurring radioactive element on Earth
- Radon is a naturally occurring radioactive gas which is present in high concentration in indoor environments, such as homes and workplaces

DOI: 10.1201/9781003343721-9

- A radioactive energy drink called "*Radithor*" was sold in the US market for a decade in the 1920s
- Radium 226 and 228 isotopes were used to cure impotency and other diseases
- X-rays, being the oldest and most useful technology for medical diagnosis, were discovered accidently when German scientist Wilhelm Conrad Roentgen was experimenting with vacuum tubes in 1895
- Tritium is the lightest radioactive isotope
- Bananas are slightly radioactive
- Radiotherapy is used to treat cancer patients
- Survivors of nuclear bombings of the Hiroshima and Nagasaki are called "*Hibakusha*," and were victims of extreme discrimination due to belief that radiation poisoning was highly contagious
- The Manhattan Project secretly tested the effects of radiation on its own citizens, including secretly injecting pregnant women with radioactive mixtures and feeding children radioactive oatmeal. In the end, about 1,400 children died due to these experiments
- During the Manhattan Project, a man was injected with plutonium without his knowledge or consent and he survived for 20 years, eventually surviving the highest radiation dose known for any human
- Cosmic rays refer to high-speed particles that travel the universe
- Atoms and molecules can be ionized by using ionizing radiation, which has a level of energy of more than 10 eV
- Heating is considered to be the primary effect of radiation when the tissues interact with non-ionizing radiation, which includes radio waves, microwaves, infrared light, and visible light
- Chest X-rays give off about 30 millirems of radiation
- On December 27, 2004, the Earth was hit by the largest blast of radiation in recorded history. It came from a *Giant Flare*, a magnetar about 50,000 light years away

9.1 INTRODUCTION

A large number of toxicants exist in our surrounding environment and contaminate the four pillars of Earth, i.e., the atmosphere, hydrosphere, lithosphere, and biosphere. Heavy metals (HMs), metalloids, radioactive elements, pesticides, and pollutants pose a serious threat to living organisms. HMs are naturally occurring elements that have a high atomic weight and a density at least five times higher than that of water. Their multiple industrial, domestic, agricultural, medical, and technological applications have led to their wider distribution in the environment, raising concerns over their potential effects on human health and the environment. With the assumption that heaviness and toxicity are interrelated, HMs also include metalloids, such as arsenic, that are able to induce toxicity at a lower level of exposure. In recent years, there has been an increasing global ecological and public health concern linked with environmental contamination by HMs along with radioactive toxic metals. Radioactive metals are natural as well as synthetic that produce alpha (α), beta (β), and gamma (γ) rays. Radioactive materials release ionizing radiation, a form of energy. When a person comes into contact with radiation, the energy may be absorbed by the body. Humans use radioactive metals to their advantage, such as for electricity generation, energy for nuclear power plants, fuel for rockets, and treatment of carcinogenic tumours. Uranium and thorium are the dominant radioactive metals which are the main elements of radioactive mineral rocks. Uraninite, thorianite, monazite, carnotite, brannerite, and davidite are important radioactive minerals. The total time taken by a radioactive element to disappear is referred to as its half-life or decay rate. Initially, radioactive material decays quickly, but after sometime, the rate will slow down until only tiny amounts are left. Radioactive metal that decays faster is believed to be more radioactive. Uranium and thorium are major radioactive metals that induce both radiological and chemical toxicity. Human exposure has increased considerably as a consequence of the exponential increase in their industrial, medical, and technological applications. Some important milestones in the research into radioactive metals and radiation are listed in Table 9.1.

The environment surrounding living organisms is exposed to radiation. Radiation is the transmission of energy in the form of waves that travel through space or other mediums. The word *radiation* gives an idea about the atomic radioactive particles, nuclear energy, and radioactivity, but radiation has many other forms. Sound, heat, light, ultraviolet rays, infrared rays, microwaves, radio, television, and wireless communications are all forms of radiation. Sunlight, which is essential for photosynthesis, is received in the form of radiation. People are exposed to radiation from cosmic rays, as well as from radioactive materials found in the soil, water, food, air, and also inside the human body. Anthropogenic sources of radiation are widely used

TABLE 9.1
Important Milestones in the Research on Radioactive Metals and Radiation

Year	Scientist	Discovery
1775	Percivall Pott	Relationship between exposure to chimney soot and the incidence of squamous cell carcinoma of the scrotum among chimney sweeps
1800	William Herschel	Infrared rays
1801	Johann Wilhelm Ritter	UV rays
1863	Rudolph Virchow	Identified WBCs in cancerous tissue
1895	Wilhelm Roentgen	X-rays while working on cathode tube
1896	Henri Becquerel, the father of radioactivity	Accidental discovery of radioactivity
1897	J. J. Thomson	Electron
1897	Ernest Rutherford	Alpha, beta, and gamma particles
1899	Tor Stenbeck and Tage Sjogren	Radiation therapy to cure cancer
1900–1908	Marie Curie and Pierre Curie	Polonium and radium
1902	Theodor Boveri	Proposed that cancerous tumours arise from single cells that have experienced chromosome damage
1932	James Chadwick	Neutron
1942	Enrico Fermi's team	World's first sustained and human-controlled nuclear chain reaction at Chicago Pile 1 (CP-1)

in medicine, industry, nuclear power plants, and academic and research applications. In ancient times, doctors diagnosed broken bones, head injuries, or tumours using their best guess and physical examinations. The diagnosis or treatments of such injuries remained painful and sometimes patient died during the treatment, such as in the case of injuries caused by embedded objects like bullets. Radiology originated as a medical sub-specialty in the first decade of the twentieth century after the discovery of X-rays by Roentgen and developed at a rapid pace until the Second World War. The extensive use of X-ray imaging and the advent of digital computers and new imaging modalities like ultrasound and magnetic resonance imaging have combined to create an explosion of diagnostic imaging techniques in the past few decades.

9.2 RADIOACTIVE ELEMENTS

Radioactive elements are natural as well as synthetic toxic elements that are made up of atoms having unstable nuclei and which release atomic radiation to attain stability. Radiation transforms radioactive elements into other atoms, which may be radioactively stable or unstable which undergo further decay. The stability of the nuclei of an atom can be calculated by taking into account the ratio of neutrons to protons. Elements of the periodic table up to atomic number 20 are stable and contain an equal number of neutrons and protons. As the atomic number increases, the repulsion between various protons increases and the process requires more protons and neutrons. Thus, with increasing atomic number, the neutron-to-proton ratios of stable nuclei also increases. All radioactive elements are part of the modern periodic table created by Mosley. They either have no stable naturally occurring isotope, or else are entirely artificial. These radioactive elements undergo radioactive decay i.e., alpha (α), beta (β), and gamma (γ) decay, to release neutrons.

9.2.1 Classification of Radiation

Living organisms have evolved in an environment which is surrounded by natural and man-made radiation. There are broadly two forms of radiation (Figure 9.1), as described below.

Natural Radiation: Radiation found in the atmosphere, lithosphere, hydrosphere, biosphere, and even in space is known as natural radiation. This type of radiation is

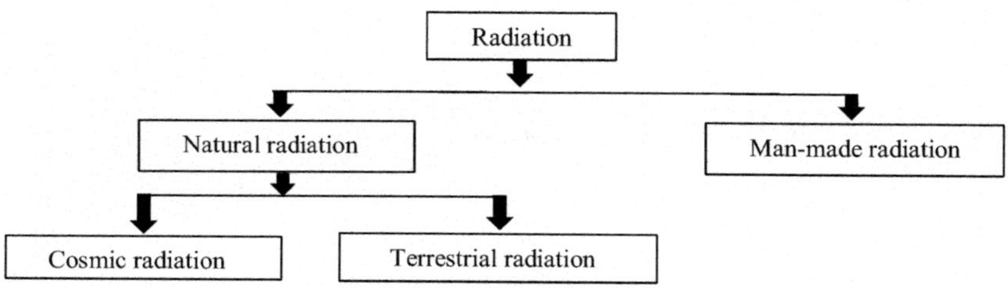

FIGURE 9.1 Types of radiation.

also known as natural background radiation and is always around organisms whether indoors or outdoors, sleeping or working, eating or traveling. Natural radiation can either be cosmic or terrestrial.

- **Cosmic Radiation**: This type of radiation is produced in space as cosmic rays from celestial bodies and reaches the Earth's surface from space. Cosmic radiation is the dominant form of natural radiation on Earth. Sunlight, UV rays, and gamma rays are examples of cosmic radiation.
- **Terrestrial Radiation**: This type of radiation is produced in soil, air, rock, water, or vegetation due to the decay of radioactive elements already present on Earth. Solar radiation is absorbed by the Earth and extra radiation is reradiated back to space by the Earth's surface. Uranium and potassium decay are examples of terrestrial radiation.

Man-made Radiation: Radiation produced from anthropogenic sources such as nuclear tests, nuclear weapons, TV, mining, etc. is known as man-made radiation and its exposure to the public is mainly during medical procedures, such as diagnostic X-rays, nuclear medicine, and radiation therapy.

Electromagnetic Spectrum

The electromagnetic spectrum is a collection of frequencies, wavelengths, and photon energies of electromagnetic waves that travel at the speed of light in a vacuum (Figure 9.2). It is a graphical representation of electromagnetic waves arranged according to their wavelength. It is a complete spectrum of all forms of light. An electromagnetic wave consists of electric and magnetic fields and consists of ionizing as

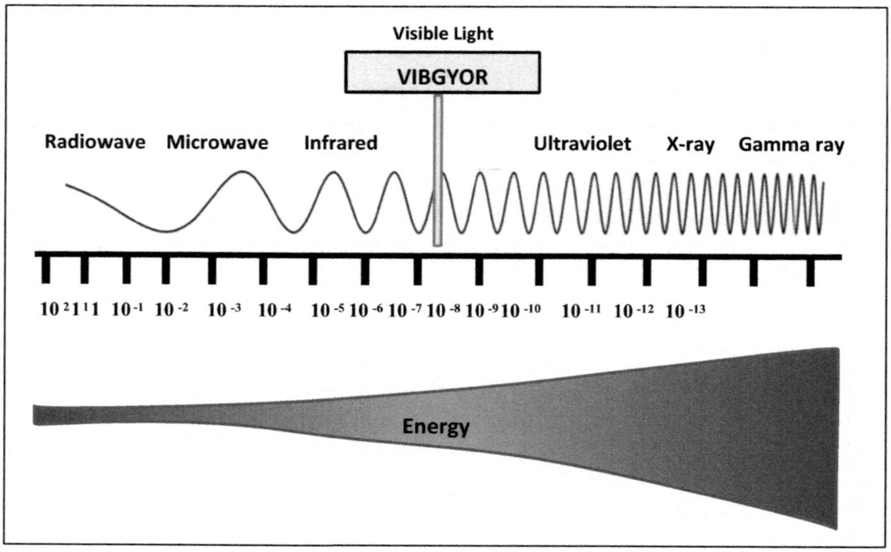

FIGURE 9.2 Electromagnetic spectrum.

well as non-ionizing radiation. Based on the energy of radiated particles, the radiation energy that travels through space is broadly classified into two broad categories, as described below.

Ionizing Radiation: This is radiation with a relatively large amount of energy that can remove tightly bound electrons from the orbit of an atom, causing the atom to become charged or ionized. These types of radiation are highly penetrating and can easily damage body tissue. The energy released by atoms travels in the form of electromagnetic waves (gamma or X-rays) or particles (neutrons, beta, or alpha). This spontaneous disintegration of atoms is called radioactivity. Excess energy is emitted in the form of ionizing radiation. The ionization process results in the formation of two charged ions, i.e., a molecule with a net positive charge and a free electron with a negative charge. The ionizing radiations of concern in most radiological work are alpha particles, beta particles, gamma rays, and neutron particles.

Gamma Rays (γ): Gamma rays are produced by radioactive and nuclear processes such as radioactive decay or positron annihilation. Gamma rays have a wavelength of less than 10 picometres with the highest frequency and shortest wave. Gamma radiation is similar to X-rays in that both easily pass through living materials. These rays have sufficient energy to ionize matter and hence can damage living cells. The damage produced in the cell or tissue is related to the number of ionizing paths created in the absorbing material. These rays are similar to light but they have much higher energy and can travel great distances through air or space. Unprotected gamma rays are harmful inside and outside the body and can more easily penetrate the skin than bones. Protection from gamma radiation is best provided by very dense materials, such as lead, concrete, or steel. Owing to its high penetration power, gamma radiation can result in radiation exposure to the entire body rather than just a small area of tissue near the source. Gamma radiation is considered an external hazard.

X-rays: These are electromagnetic waves or photons, and have no electrical charge but can ionize matter as a result of direct interactions with orbital electrons and they also have a very high penetrating power. X-rays are produced when a metallic anode is bombarded by very high-energy electrons. Gamma rays and X-rays are physically identical, with the only difference being the place of origin. X-rays are characterized by their probability of interacting in a given material. An X-ray beam travels through the air and comes in contact with the body tissues, and produces an image on a metal film. These rays are capable of traveling in a vacuum and are not affected by an electromagnetic field. They can propagate independently, i.e., without a medium, and are diverging rays, i.e., they cannot be focused on a single point and can cause photoelectric emissions.

Alpha particles (α): Alpha particles are emitted during the decay of certain radioactive materials. Compared to other types, the alpha particle has a relatively large mass and consists of two protons and two neurons. Being a highly charged particle, it is emitted from the nucleus of an unstable atom. The positive charge causes the alpha particle (+) to strip electrons (–) from nearby atoms when it passes through the material, thus ionizing the atoms. The alpha particle deposits a large amount of energy in a short distance of travel. Most alpha particles are stopped by a few centimetres of air, a sheet of paper, or a dead layer (outer layer) of human skin. Alpha particles are not regarded as an external radiation hazard, as they are easily stopped by a dead layer of skin. If alpha-emitting radioactive material is inhaled or ingested, it becomes a source of internal exposure. Internally, the source of alpha radiation is in close contact with body tissue and can deposit a large amount of energy in a small volume of body tissue.

Beta particles (β): A beta particle is an energetic electron emitted during radioactive decay. Compared to an alpha particle, a beta particle is nearly 8000 times less massive and has half the electrical charge. The beta particle causes ionization by the same forces at work with alpha radiation – mostly electrical interactions with atoms which it encounters as it travels. However, since it is not highly charged, the beta particle does not cause ionization. It travels further before giving up all its energy and finally coming to rest. The beta particle has limited penetrating ability. Its typical range in air is up to about 10 feet. In human tissue, the same beta particle would travel only a few millimetres. Beta particles are easily protected by comparatively thin layers of plastic, glass, aluminium, or wood. Dense materials such as lead must be avoided while shielding beta radiation due to an increased production of X-rays in a shield. Beta particles are potentially hazardous to the skin and eyes and cannot penetrate deep tissue, e.g., bone marrow or other internal organs.

Neutron Particles (n): Neutron particles are ejected from the nuclei of atoms. A neutron has no electrical charge. Due to their neutral charge, neutrons do not interact directly with electrons in matter. A direct interaction occurs as a result of a collision between a neutron and the nucleus of an atom. A charged particle causing ionization may be emitted during these interactions, which is known as indirect ionization. Neutrons do not experience electrostatic forces, but have a relatively high penetrating ability and are difficult to stop. Like gamma radiation, the range is not absolutely defined. The distance they travel depends on the probability of an interaction in a particular material. Moderate- to low-energy neutron radiation is best shielded by a material with a high hydrogen content, such as water or polyethylene plastic. High-energy neutrons are best protected by more dense materials such as steel or lead. Sometimes, a multi-layered shield can be used to first slow down very "fast" neutrons, and then absorb the "slow" neutrons. Like gamma radiation, owing to their high penetrating ability, neutrons produce an external whole-body hazard.

Non-Ionizing Radiation: Non-ionizing radiation does not have the amount of energy required to ionize an atom. Non-ionizing radiation is described as a series of energy waves composed of oscillating electric and magnetic fields traveling at the speed of light. These do not cause ionization. They generally emit heat, which at times is so intense that it causes burns. These are further of different types, as discussed below.

Ultraviolet Rays: These rays have a shorter wavelength as compared to visible light and hence they are invisible to the naked eye. Ultraviolet rays are present in the wavelength range from around 10 nm to around 400 nm. UV radiation is divided into four regions, i.e., near (400–300 nm), middle (300–200 nm), far (200–100 nm), and extreme (below 100 nm). The three types of UV radiation are classified accordingly in respect of their wavelength. On interaction of UV radiation with biological materials, it is divided into three divisions, i.e., UV-A (400–315 nm), UV-B (315–280 nm), and UV-C (280–100 nm). They differ in their biological activity, to the extent that they can penetrate the skin. The shorter the wavelength of light, the more harmful is the UV radiation. However, short-wavelength UV radiation is less capable of penetrating the skin. Short-wavelength UV-C is by far the most damaging form of UV radiation. However, UV rays are completely filtered by the atmosphere, and generally do not reach the Earth's surface. UV-B is biologically active but it cannot penetrate beyond the superficial skin layer and is responsible for the delayed tanning and burning of the skin. Most solar UV-B is filtered by an atmospheric layer. The intensity of UV is higher when it is closer to the equator.

Visible Light: This part of the electromagnetic spectrum can be seen and distinguished optically by the human eye. Visible light has a wavelength of around 400 nm to 700 nm. White light is a combination of various colours where every colour can be linked with a particular wavelength of light. When white light passes through the prism, it scatters into seven colours of the visible light spectrum, namely violet, indigo, blue, green, yellow, orange, and red (Table 9.2). The range of visible spectrum in the electromagnetic spectrum ranges from the IR to UV region.

Microwaves: Microwaves are high-frequency radio-waves and are part of the electromagnetic spectrum. Microwaves are reflected, transmitted, or absorbed by materials in their path in a similar manner as the visible light. Microwaves are mostly used for TV broadcasting, radar for air travel, sea navigational aids, and telecommunications including mobile phones. They are used in industry for processing materials, in medicine for diathermy treatment, and in kitchens for cooking food. Metallic materials completely reflect microwaves, while

TABLE 9.2
Visible Light Spectrum

Color	Wavelength of Light	Frequency of Light
Violet	380–450 nm	680–790 THz
Indigo	450–485 nm	620–680 THz
Blue	485–500 nm	600–620 THz
Green	500–565 nm	530–600 THz
Yellow	565–590 nm	510–530 THz
Orange	590–625 nm	480–510 THz
Red	625–740 nm	405–480 THz

TABLE 9.3
Different Types of Infrared Rays

Type of Infrared	Wavelength	Use
Near-infrared	0.75 to 1.4 micrometres	• Materials science • Fibreoptic communication • Medical applications
Short wavelength infrared	1.4 to 3 micrometres	• Telecommunications • Military purposes
Medium infrared	3 to 8 micrometres	• Chemical industry • Astronomy
Long-wavelength infrared	8 to 15 micrometres	• Astronomical telescopes • Optical fibre communications
Far infrared	15 to 1000 micrometres	• Cancer therapy

non-metallic materials such as glass and some plastics are mostly transparent to microwaves.

Infrared Rays: This is electromagnetic radiation with wavelengths longer than those of visible light. The wavelength varies from 700 nanometres to 1 millimetre. Most of the thermal radiation emitted by objects near room temperature is infrared. Table 9.3 highlights different types of infrared rays, their wavelengths, and uses.

Hot objects and molecules produce infrared waves. Infrared is used extensively in remote sensing. Astronomers use optical devices such as mirrors, solid-state digital detectors, and lenses to study objects from space with the help of infrared waves.

Radio waves: These are the building block of radio communications with the longest wavelength and lowest frequency in the electromagnetic spectrum. Radio waves are naturally produced by lightning or astronomical objects. Like all other electromagnetic radiations, radio waves travel at the speed of light in a vacuum. Like waves on a pond, a radio wave is a series of repeating peaks and valleys. The entire wave pattern, before it repeats itself, is called a cycle. A radio wave is generated by a transmitter and then detected by a receiver. An antenna allows a radio transmitter to send energy into space and a receiver to pick up energy from the space.

Sources of Radiation: The major sources of naturally occurring radiation are listed below.

Cosmic radiation: This comes from the sun or outer space and consists of positively charged particles, as well as gamma radiation. At higher elevations the amount of atmosphere shielding cosmic rays decreases and thus the dose increases.

Terrestrial radiation: These are natural sources of radiation in the ground, rocks, building materials, and drinking water supplies. Radium, uranium, and thorium contribute to terrestrial radiation.

Internal sources: Human bodies (radionuclides like potassium-40).

Human sources: This consists of anthropogenic sources. The following are man-made radiation sources.

Medical radiation sources: X-rays are identical to gamma rays; however, they originate outside the nucleus. X-rays produce an ionizing radiation hazard. Along with X-rays, radioactive isotopes are used in medicine for diagnosis and therapy.

Consumer products: TVs, old luminous dial watches, smoke detectors, and lantern mantels are some examples.

Atmospheric testing of nuclear weapons: Fallout from atmospheric nuclear weapons testing.

Industrial uses: X-ray machines and radioactive sources (radiography) to test pipe welds, bore-holes, etc.

Application of Radiation: Radiation applications are listed in Table 9.4.

9.2.2 Nuclear Stability

Nuclear stability refers to the steadiness of a nucleus of an atom and is basically a state in which the nucleus of an atom remains in its current state unless an outside agent interacts with it. Radioactivity and nuclear stability are interrelated concepts. Let's discuss these concepts with a short story.

TABLE 9.4
Applications of Radiation

Medicine	Radiation Therapy	For treating leukaemia
	Ultrasonography	To get an image of foetus in mother's womb
	Diagnostic radiology	Imaging modalities in diagnosis of diseases
	Positron emission tomography	Diagnosis of brain cells
	Argon laser	Eye surgery
Agriculture	Crop improvement through mutation breeding	Developing crop mutant with desirable agronomic traits
	Soil quality improvement	Isotope of Cs-137 is used to estimate soil erosion rates
	Pest control	Sterile insect techniques
	Improvement of animal production	Improving breeding and nutritional management of dairy cattle through radioimmunoassay
Industry	Smoke detector	Ionization smoke detector uses alpha radiation
	Thickness gauge	Beta radiation is used to control thickness of paper or tires in factories
	Sterilization (food irradiation)	Beta radiation is used to kill microbes in food items, to prevent sprouting in potatoes, onions, and garlic, and to increase the shelf life of fruits, such as mangoes, strawberries, and papaya
	Surgical sterilization	Sterilizing cardiac pacemakers and syringes using radiation
	Detection of leakage	Detecting leaks in underground pipes using radiation or radioactive elements

In a village named Nucleus, three brothers named Proton, Neutron, and Electron live happily in their three respective houses. They live in stability with each other. One day a foreign sub-atomic particle reached Nucleus village and forcefully tried to invade the houses of Electron, Proton, and Neutron. This created instability in Neutron village. Thus, the three brothers forcefully removed the extra foreign neutron and some energy was released during this process.

This story beautifully explains that a stable nucleus is one that remains in its current state forever, until an outside sub-atomic particle interacts with it. Generally, stable nuclei have an almost equal number of neutrons as protons, and a strong excess of one or the other will result in an unstable nucleus. The ratio of neutrons to protons in a stable nucleus is thus around 1:1 for small nuclei ($Z < 20$). This ratio increases slowly with atomic number up to about 1.58 at high Z. There are only two stable nuclei with $Z > N$ (more protons than neutrons): ^{1}H and ^{3}He has one more proton than it has neutrons. There are some tendencies, or rules, that stabilize nuclei. By definition, $A = N + Z$, i.e., it is the total number of nucleons in a nucleus. For any Z there is a range of stable N. At low Z, $N \approx Z$; as Z increases, the value of N/Z for which stability occurs gradually rises to around 1.58 for the heaviest stable nuclei, namely, those of lead ($Z = 82$). For the most part, alpha and beta decays tend to occur in a way that moves a nucleus toward the range of N/Z.

9.2.3 Radioactive Decay

Radioactive decay is a spontaneous nuclear transformation that is unaffected by pressure, temperature, chemical form, etc. This insensitive extranuclear condition allows characterization of radioactive nuclei by their decay period, mode, and energy of decay without taking into consideration their physical or chemical condition. Radioactive decay is exhibited by atoms of larger atomic number with less nuclear stability. In atoms of smaller atomic number, little to no radioactive decay is observed, while atoms with even-numbered protons and neutrons generally show a degree of stability as compared to atoms with odd numbers of protons or neutrons. During radioactive decay, unstable radionuclides become stable by a spontaneous nuclear transformation from one nuclear species to another with a different mass or atomic number or both.

9.2.4 Types of Radioactive Decay

Alpha Decay: In this type of decay an element with an unstable nucleus transforms to another element by emitting two protons and two neutrons. Emitted particles are known as alpha particles. Alpha particles result in extensive ionization in matter due to the energy of the particle. Alpha particles are practically not influenced by collision with electrons, since the mass of the electron in contrast with the mass of alpha particles is negligible. However, an alpha particle can collide with a nucleus resulting in deflection of the alpha particle or the nucleus captures the particle to induce a nuclear reaction. Alpha decay can be generally represented by:

$$^{A}_{Z}X \rightarrow {}^{A-4}_{Z-2}X + {}^{4}_{2}He$$

Where X = radioactive element; A= mass number; and Z = atomic number.

Beta Decay: The emission of beta particles from a radioactive nucleus is called beta decay. Beta particles are particles having a pair of positron and neutrino or electron and anti-neutrino. Beta decay is mainly of three types:

- **Beta minus decay (β^-)**: In this, the nucleus emits an electron and the neutron converts to a proton inside the nucleus.
- **Beta plus decay (β^+)**: In this, the nucleus emits a positron. The proton converts to a neutron inside the nucleus.
- **Beta capture:** In this, the nucleus captures an electron from a lower energy level in the atom and the proton gets converted to a neutron inside the nucleus.

Gamma Decay: This is disintegration of a parent nucleus to a daughter through emission of high-energy photons. In gamma decay, a nucleus changes from a higher energy state to a lower energy state through the emission of electromagnetic radiation. Gamma rays are characterized by radionuclides present in the emission of a photon, hence intensities of emitted photons of gamma radiation determine the radionuclides in a sample.

Units of radioactivity are:

Rem: Unit used for equating radiation absorption with biological damage. Occupational radiation exposure is recorded in rems

Roentgen: Unit used for measuring the exposure

Rad: Unit used for measuring absorbed dose in any material

Becquerel: SI unit of radioactivity is represented by the symbol Bq. One Becquerel is the number of radioactive nuclei disintegrations per second

1 Becquerel = One disintegration per second

Curie: The rate of decay is measured in Curies. It is the oldest unit of radioactivity, and is represented by the symbol Ci. One Curie is defined as the radioactivity of one gram of radium element

1 Curie = 37 billion disintegrations per second

9.3 SOURCES OF RADIOACTIVE METALS

Although radioactive metals are naturally occurring elements that are found throughout the Earth's crust, most environmental contamination and human exposure result from anthropogenic activities, such as mining, smelting operations, industrial production, and radiation leakage from nuclear power plants. Reported sources of HMs in the environment include geogenic, industrial, agricultural, pharmaceutical, domestic effluents, and atmospheric sources. Major sources of radioactive metals are (Figure 9.3):

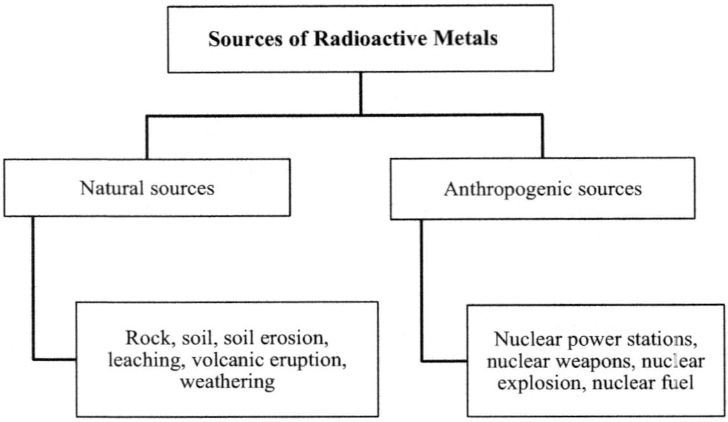

FIGURE 9.3 Sources of radioactive metals.

Natural Sources: Soil and rocks are natural reservoirs of radioactive metals like uranium, thorium, and radium. Natural phenomena such as weathering and volcanic eruptions are also reported to significantly contribute to radioactive metal pollution

Anthropogenic Sources: Man-made sources include mining, nuclear explosions, nuclear fuel, nuclear weapons, etc.

9.4 ROUTES OF EXPOSURE OF RADIOACTIVE METALS

Radioactive metals are highly toxic to living organisms as well as for the environment. They induce radiological and chemical toxicity. They even affect the structures of carbohydrates, proteins, and nucleic acids. Most radioactive toxicants enter living organisms through several routes. These toxic elements produce adverse responses which may be sudden (acute toxicity) or long term (chronic toxicity). The main routes by which radioactive toxicants enter in human body are described below.

Occupational Exposure

Exposure to an internal and external dose of ionizing radiation or radioactive dust received by workers during the course of work is defined as occupational exposure. Occupational radiation exposure is measured in Roentgen Equivalent Man (REM). Occupational exposure involves three modes, i.e., oral, dermal, and inhalation (Figure 9.4).

Oral Exposure: Oral exposure or ingestion is the most common route of exposure for naturally occurring radioactive toxic metals like uranium and thorium. Exposure can occur through the ingestion of food or drinking water contaminated with radioactive metal through mining or waste activities. Uranium gets adsorbed through the roots of plants or can be ingested during the intake of food.

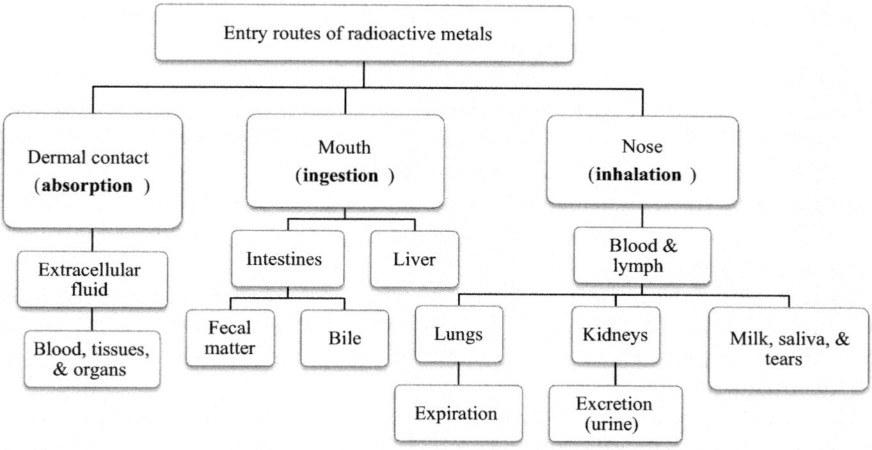

FIGURE 9.4 Routes of entry of radioactive metals.

Inhalation: Inhalation of uranium powder is the primary exposure route for workers. Mining activities increase the air concentration of uranium, thereby exposing mine workers to elevated levels of uranium. Uranium enrichment processing has the potential to release additional uranium into the atmosphere.

Workers inhale the dust from mining, milling, enrichment, recycling, and combustion of uranium-containing coal. The dust even gets accumulated on workers' skin, hair, and clothes, and can then spread to other areas.

Dermal contact: Dermal contact is a possible exposure pathway for naturally occurring radioactive toxic metals. Dermal exposure includes workers exposed to radioactive powdered metals or those coming in contact with radioactive waste. It is not likely that community members would be exposed to increased levels of uranium via direct skin contact, though children playing in contaminated areas may be at risk of this type of exposure.

Incidental and Accidental Exposure: Coming into contact with harmful radioactive elements by chance due to nuclear leakage or during a nuclear disaster is referred to as an incidental or accidental exposure. During the Chernobyl and Fukushima-Daiichi nuclear disasters, thousands of people were accidentally exposed to harmful radioactive metals.

9.5 AETIOLOGY

The unwanted or undesirable presence of radioactive metals in a given ecosystem or environment is referred to as radioactive contamination. These radioactive metals pollute the environment and pose a serious threat to organisms. Exposure to radioactivity can cause structural damage to the body and alters DNA structure, which leads to unpredictable point or non-point mutations. Some major causes of radioactive metal contamination and radiation are discussed below.

Cosmic rays: Cosmic rays containing alpha particles, beta particles, X-rays, and gamma rays reach the Earth's surface and damage human bodies. The intensity of cosmic rays is lowest at the equator and highest at the poles.

Medical Practices: Many radioactive substances are used in medicines as markers. These radioactive substances at times induce a negative effect on an otherwise healthy organ.

Medical Waste: Radioactive substances are used for treating certain diseases such as cancer or tumours. In the case that these hazardous elements are not suitably transported or stored, they can easily contaminate the surrounding environment.

Meteors/Meteorites: Celestial rocks like meteors or meteorites are a source of radioactive metals. When the meteors strike the Earth's surface a large number of radioactive elements are released.

Nuclear Accidents: The Chernobyl and Fukushima-Daiichi nuclear disasters were the most devastating nuclear disasters in human history. Thousands of people were accidentally exposed to harmful radioactive dust or metals during these nuclear disasters. A large number of radionuclides were generated in the atmosphere, which settled down with rain, thereby contaminating the soil and water bodies. Finally, they entered into food chain causing serious problems to living organisms.

Nuclear Power Plants: Nuclear energy is used to produce electricity through controlled nuclear reactions. This method causes less environmental impact but produces radioactive by-products that require adequate storage. Negligence in handling the radioactive elements can cause a longer-lasting impact on environment.

Nuclear Tests: Nuclear tests have been carried out in various uninhabited places around the world. These tests leave a radioactive footprint on those areas, causing considerable damage to the ecosystem and environment. Many floral species are affected by nuclear tests.

Nuclear Waste: Nuclear power plants produce a lot of nuclear radioactive wastes, which acts as a source of radioactive elements that release huge amounts of radiation.

Nuclear Weapons: These are major sources of nuclear radioactivity which release a huge number of radioactive metals into the environment and have a long-lasting effect on the environment. The Hiroshima and Nagasaki nuclear bomb explosions in Japan are well-known examples.

Radiation Therapy: This makes use of high doses of radiation to kill cancer cells and shrink tumours. At low doses, X-rays are used to image broken bones. However, the effect of radiation therapy is not specific to tumours and may produce toxicity in adjacent cells or organs.

Radioactive Elements: Radioactive elements are made up of atoms having unstable nuclei, and during the process of radioactive decay to attaining stability these elements release radiation.

Radioactive Minerals: Minerals containing uranium-235, thorium-232, and plutonium-239 are capable of emitting energetic radiation.

Radionuclides: Unstable radionuclides in the atmosphere can split into smaller parts emitting energetic radiation.

9.6 NUCLEAR DISASTERS IN HISTORY

Human civilization has encountered five main incidents where a large number of radioactive elements have reached the environment and produced devastating effects on humans, flora, and fauna along with different components of ecosystems. The main nuclear incidents in history are listed in Tables 9.5–9.8.

9.7 PATHOPHYSIOLOGICAL AND CLINICAL MANIFESTATIONS

The mechanism of cell killing from radiation is not selective for tumour cells, and optimal delivery of radiation therapy (RT) is a balance between maximization of the tumour dose and minimization of the normal tissue dose (see Figure 9.5). Thus, the study of radiation toxicity is a core component of radiation oncology training, and practices which are guided by an extensive and evolving literature on factors linked with toxicity probability. The study of radiation toxicity is also relevant to nuclear medicine and the evolving field of "theranostics," in which radionuclides linked to targeting agents have both diagnostic and therapeutic utility, although dosimetric studies on efficacy and toxicity have been limited.

TABLE 9.5
Hiroshima and Nagasaki Atomic Bombing

Date	6th to 9th August, 1945
Country	Japan
Location	Hiroshima and Nagasaki cities
Reason	Atomic bombs were dropped by US Air Force during the Second World War
Major Radioactive Metals	Uranium and plutonium
Key Points	• First time nuclear weapon was used • Little Boy (uranium-enriched gun type fission bomb) was dropped on 6th August 1945 over Hiroshima city • Fat Man (Plutonium implosion type atomic bomb) was dropped on 9th August 1945 over Nagasaki city
Causalities	• About 140,000 people in Hiroshima and 74,000 people in Nagasaki
Immediate effects	• 70 per cent of the buildings were destroyed • Ground temperature reached 4000°C • Radioactive rain
Long-term Impact	Leukaemia, radiation sickness, thyroid, breast, lung, and intestine cancers, abnormalities in foetuses, miscarriages, intellectual disabilities, impaired growth and mutations

TABLE 9.6
Chernobyl Nuclear Disaster

Date	26 April, 1986
Country	Former USSR
Location	Chernobyl Nuclear Plant, Pripyat
Reason	Flawed Reactor Design
Major Radioactive Metals	Iodine, strontium, and caesium
Key Points	• Nuclear and radiation accident • Radioactive iodine-131 was the first killer • Strontium-90 and caesium-137 were long-term killers
Causalities	• About 50–100 deaths from the initial explosion
Immediate Effects	Radiation sickness, cataracts
Long-term Impact	Leukaemia, thyroid cancer

TABLE 9.7
Fukushima-Daiichi Nuclear Disaster

Date	11th March, 2011
Country	Japan
Location	Okuma, Fukushima
Reason	Earthquake followed by tsunami
Major Radioactive Metals	Xenon-133, iodine-131, caesium-134, and caesium-137
Key Points	• A massive earthquake with a magnitude of 9.0 generated several powerful tsunami waves in the ocean causing damage to reactors of the nuclear power plant • The earthquake which caused the nuclear disaster was the largest in Japan's history • 14,000 times more caesium-137 was released from Fukushima nuclear disaster than Hiroshima atomic bombing
Causalities	No deaths reported directly but more than 2000 people were affected
Immediate Effects	Radiation level increased in groundwater, ocean, and food
Long-term Impact	Leukaemia, thyroid cancer, anxiety, depression, post-traumatic stress disorder (PTSD), and radiation sickness

Ionizing radiation directly or indirectly (via reactive oxygen species) damages DNA, prompting a cascade of events that may lead to cell death. The degree of cell killing and resistance varies based on properties such as the degree of differentiation, mitotic rate, and cumulative and fractional radiation dose. RT side effects are categorized as acute, sub-acute, or late. Acute effects begin within 1 or 2 weeks after starting RT and are often inflammatory or reflect the depopulation of rapidly growing epithelial cells. The timing of symptoms relates to the turnover and transit time for normal tissue stem cells to repopulate damaged tissue; patients finishing RT are counselled that acute side effects may continue to worsen before recovery. Late

TABLE 9.8
Three Mile Island Nuclear Accident

Date	18 March, 1979
Country	United States
Location	Pennsylvania
Reason	Equipment failure and operator error
Major Radioactive Metals	Xenon-133
Key Points	Reactor No. 2 got damaged and radioactive substances were released into the environment
Causalities	No deaths reported directly
Immediate effects	Nausea
Long-term Impact	Metallic taste, diarrhoea, hair loss, and damage to flora

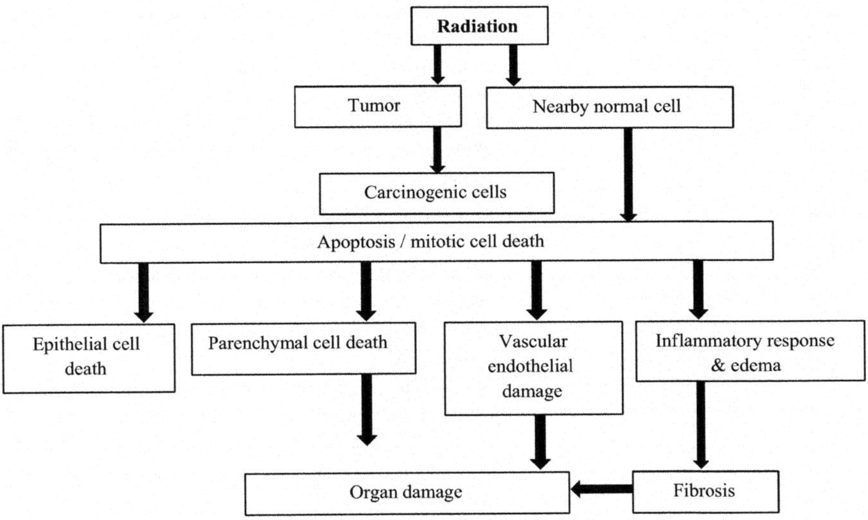

FIGURE 9.5 Pathophysiology of radioactive elements on normal cells.

effects often reflect fibrosis, vascular injury, or other gradual changes in slowly dividing tissues, with end-organ damage possibly manifesting years after the treatment. Residual DNA damage rarely causes delayed carcinogenesis. RT is a loco-regional treatment. Therefore, the anatomic properties of affected organs and tissues influence the pathogenesis and clinical presentation of toxicity. The cellular effects of ionizing radiation on contact with radioactive metals range from cell death to DNA damage, which lead to chromosomal abnormalities or mutations or may cause or predispose to cancer.

Therapeutic radiation, diagnostic radiology, and nuclear accidents/warheads all have carcinogenic potential. Table 9.9 describes the effects, causes, and clinical manifestations of radioactive metals or radiation therapy on the human body.

TABLE 9.9
Effects, Causes, and Clinical Manifestations of Radioactive Metals or Radiation Therapy

Skin, connective tissue, and breast	Radiation dermatitis	• Common and expected in treatment of head and neck, vulvar, anal, skin, extremity (e.g., sarcoma), and breast cancers • Effects are produced by cutaneous epithelial depopulation
	Soft tissue fibrosis and breast cosmesis	• Late fibrosis can develop from high-dose radiation therapy, akin to wound healing after surgery • Fibrosis may adversely affect breast cosmesis
	Lymphedema	• This occurs because of impaired lymphatic drainage from fibrosis and/or direct nodal removal. Symptoms include swelling, pain, and limited range of motion, which may develop immediately or in months to years • Severity and risk are related to the degree of surgery and radiation
	Bone	• Acute effects on bone include hematologic suppression and inflammation • Demineralization and osteoporosis may increase the risk of symptomatic fracture, especially of the ribs, femur, and pelvis
Central nervous system	Acute effects	• RT is frequently used in treating neoplasms in or near brain and spinal cord, including benign brain tumours, malignant gliomas, ocular/CNS lymphoma, and metastases • Patients receiving RT develop mild-to-moderate acute dry mouth and dry eye from dose to parotid and lacrimal gland, respectively
	Neurocognitive	• Neurocognitive changes are potential late effects which manifest from months to years after treatment
	Radiation necrosis	• Increasing use of radiation after treatment in the form of increased intracranial pressure and vascular endothelial damage
Endocrine glands	Pituitary gland	• Radiation striking pituitary gland produces effects on the hypothalamic–pituitary axis resulting in endocrinopathies
	Thyroid gland	• Hypothyroidism in patients exposed to radioactive metals
Hair	Follicle stem cell	• Risk of permanent or severe hair loss is also seen in the case of a high dose
Lungs	• RT-induced lung injury during treatment for lung and oesophageal cancer, which manifests as sub-acute radiation pneumonitis or late fibrosis	
Heart	• Cardio-toxicity includes ischemic heart disease and myocardial infarction	
Gastro-intestinal	Oesophagus	• Acute esophagitis in patients after treatment of lung and oesophageal cancer with epithelial depopulation and inflammation

TABLE 9.9 (Continued)
Effects, Causes, and Clinical Manifestations of Radioactive Metals or Radiation Therapy

	Stomach	• Patients receiving palliative RT for upper abdominal tumours experience toxicity from incidental irradiation of abdominal organs
	Colon, rectum, and anus	• RT for treatment of prostate, vulva, cervical, endometrial, rectal, and anal cancers causes rectal toxicity
Excretory system	Urethra	• Urethral and bladder toxicity are important considerations for prostate and gynaecologic cancers
	Urinary bladder	• Bladder toxicity occurs through direct urothelial damage, along with effects on urinary sphincter and detrusor muscle from fibrosis and vascular ischemia
Reproductive system	Female	• RT and chemotherapy for malignancies in pelvis may adversely affect fertility, hormones, and sexual development • Contact with radio-metals/radiation is contraindicated in pregnant women owing to risk of teratogenesis in developing embryo • Women receiving pelvic radiation during treatment have impaired vaginal lubrication, vaginal stenosis, and fibrosis
	Male	• Spermatogenesis is impaired by very low RT doses, with immature spermatogonia followed by spermatocytes and spermatids

9.8 MECHANISMS OF TOXICITY

The mechanisms of toxicity have the following five steps:

Step 1: Oxidative Stress: An imbalance between the production of free radicals and ability of the body to counteract or detoxify their detrimental effects through neutralization by antioxidants.

- Radioactive metals can directly or indirectly interact with the *in vivo* antioxidant defence system resulting in an imbalance of redox homeostasis with impairment of cellular components.
- Uranium depletes the antioxidant defence system by interacting with thiobarbituric acid reactive substances, reduced/oxidized glutathione, superoxide dismutase (SOD), and glutathione reductase.

Step 2: DNA Damage:

- Radioactive metals are chemically geno-toxic. Thus, these radicals can attack DNA and form DNA damage markers, e.g., thymine glycol and 8-oxoguanine.

- Predominant uranium accumulation in the cell nucleus induces DNA single- or double-strand breaks and further promotes phosphorylation of γ-H2AX and formation of micronuclei, suggesting that uranium can induce a rapid manifestation of geno-toxic stress.

Step 3: Protein Binding and Abnormal Function:

- Hexavalent uranyl (UO_2^{2+}) tends to bond by the carboxyl, phosphoryl, and amide groups in protein molecules.
- Proteins have an inherent binding affinity to UO_2^{2+} ions, leading to the formation of uranyl protein complexes and molecular impairments.
- Uranyl ions can interact with the proteins involved in the reaction with Ca^{2+} ions such as C-reactive protein and osteopontin.

Step 4: Inflammation Activation:

Radioactive metals activate the NF-κB pathway and up-regulate the expression of inflammatory genes TNF-α, iNOS, and COX-2.

Step 5: Cell Death:

- Radioactive metals like uranium can cause various molecular and cellular damages, which may eventually trigger cell death.

Uranium at low concentration induces apoptosis through the activation of caspase-3, caspase-8, and caspase-9 and leads to necrosis at higher concentrations. Radiation biology or radiobiology is a branch of medical science that deals with the study of the biological effects of ionizing radiation on living tissues. Radiobiology evolved after scientists' made observations about ionizing radiation on humans and other life forms. It was not until the discovery of DNA that the physical effects of radiation on living cells were linked to DNA damage. Wilhelm Roentgen, a German engineer and physicist, was the first to discover and characterize X-rays. During experiments with a Crooks tube, he found that photographic plates distant to the tube began to develop in spite of the distance and absence of visible light reaching them. Roentgen presented his findings in late 1895; by January 1896 the news was well-known and within months the probable use of the new X-rays was tested on cancer patients. It was soon noticed that extended exposure to X-rays led to the development of dermatitis and several doctors proposed using X-rays in the treatment of cancer. Living beings are intensely influenced by radiation-induced cellular damage, which not only threatens the health of an individual but leads to diseased tissue. There is a variety of responses to radiation, which is governed by the following parameters:

- Source of radiation
- Radiation dose or the amount of radiation energy received
- Duration of exposure
- Genetic and epigenetic configuration of the exposed individual

These parameters can range extensively, and humans may be exposed to low-dose radiation used for diagnostic purposes, e.g., CT scan, or high-dose radiation, such as that used for radiotherapy or that produced by nuclear disasters. On interaction with a cell, the ionizing radiation can:

1. Pass through the cell without harming the DNA
2. Harm the cell's DNA, but the DNA can repair itself
3. Stop DNA from duplicating properly
4. Harm or injure the DNA so badly that the cell dies (this phenomenon is known as apoptosis)

Though much advancement has been made in realizing the basic principles of IR-induced implications on individual constituents of biological systems, less is known about how localized IR effects on target molecules control the cellular network that contain these modified species, the interfaces between complexes, and the complete state of a biological system. The accessibility of high-end technologies and detection methods to observe single or global radiation-induced modification of biomolecules has increased alarmingly in recent years, revealing an extensive context of how radiation affects the cell's lifecycle. The following paragraphs discuss the impacts of IR on biological macromolecules.

9.8.1 Cell Sensitivity toward Radiation

Sensitivity of cells toward radiation varies and depends on their type and division mechanism. Some cells, such as blood and reproductive cells, divide more frequently than others, and hence they are more sensitive to radiation, e.g., embryos containing fast-dividing cells are very sensitive to radiation; therefore, it is recommended that pregnant women limit their exposure to radiation. Likewise, fast-growing tumour cells are also sensitive to radiation; hence, cancer therapy makes use of radiation to destroy cancer cells. Ionizing radiation can affect cells through direct and indirect action, causing DNA damage and mutations. This can particularly be harmful to cells which multiply very fast. Radiation therapy (RT) is one of the most effective treatments both for primary and secondary tumours; however, irradiation may also affect the adjacent organs close to the irradiated area and result in their dysfunctioning, e.g., cranial irradiation prompts cognitive decline and intellectual dysfunctioning, resulting in impaired learning and memory loss. Like various cellular stress factors, ionizing radiation brings about the following changes in cells:

- Generates reactive oxygen species (ROS) and reactive nitrogen species (RNS)
- Damages DNA strands by disturbing their sugar-phosphate backbone
- Brings about cellular toxicity, thus pushing cells toward:
  ✓ Apoptosis
  ✓ Necrosis
  ✓ Autophagy
  ✓ Senescence

- Reduction and oxidation systems manifest in the form of acute radiation syndrome which has several early- and late-stage side effects
- There are two mechanisms by which radiation eventually affect cells, as described below.

Direct Effects: These effects are produced when radiation directly interacts with the atoms of a DNA molecule or other cellular component required for cell survival. The probability of the interaction of radiation with a DNA molecule is very small as these components constitute a very small fragment of the cell.

Indirect Effects: Indirect effects are generally produced when radiation interacts with water molecules. Each cell mostly contains water. Ionizing radiation breaks the bonds holding the water molecule together, generating radicals such as hydroxyl OH, superoxide anion O^{2-}, and others, which add to the cell damage. The impact of radiation can be divided into high and low doses. The impact of a high dose can be noticed in a short time interval, producing acute effects. High-dose radiation has a tendency to kill cells and cause direct problems to organs. Skin burn, hair loss, infertility, and cataracts are some impacts of high-dose radiation. Radiation produces:

Acute Radiation Syndrome: This syndrome develops after acute whole-body exposure to low or moderate doses of radiation, which, based on the period of exposure, is further divided into two periods:

Prodromal Period: Symptoms in this phase develop within the first few minutes to hours after exposure to whole-body irradiation. The individual may develop:

- Anorexia
- Nausea
- Vomiting
- Weakness
- Diarrhoea
- Fatigue

Latent Period: In this phase the symptoms appear after the prodromal period. The extent of the latent period is also dose-related. It can be prolonged from hours or days at supra-lethal exposures to a few weeks at sub-lethal exposures.

Hematopoietic Syndrome: Whole-body radiation injures the hematopoietic stem cells of the bone marrow and spleen. It brings about a rapid fall in the number of circulating granulocytes, platelets, and erythrocytes, which are radio-resistant. On account of the high mitotic activity of cells and the presence of differentiated cells, the bone marrow is a very radiosensitive tissue. However, scarcity in the peripheral blood after irradiation reveals the radio sensitivity of their predecessors. Clinical signs of hematopoietic syndrome include:

- Anaemia
- Haemorrhage
- Infection
- Death usually occurs 10–30 days after irradiation

Gastrointestinal Syndrome: Whole-body radiation injures the promptly multiplying basal epithelial cells of the intestinal villi, resulting in loss of the epithelial layer of the intestinal mucosa. The symptoms include:

- Dehydration
- Diarrhoea
- Electrolyte loss
- Loss of weight
- Septicaemia
- Ulceration

Cardiovascular and CNS Syndrome: This is the irreparable clinical progression that runs from a few minutes to 48 hours before death. Cardiovascular/central nervous system sickness is the type of acute radiation sickness produced by high total body doses of radiation. Symptoms include:

- Nausea
- Vomiting
- Anxiety
- Confusion
- Loss of consciousness
- Tremors
- Convulsions
- Coma
- Death

Radiation Sickness: This includes the most detrimental acute, prolonged, or chronic effects produced by exposure to ionizing radiation. A noticeable effect consequent to radiation exposure becomes quite obvious after a single dose of several hundred rads. In general, large doses of radiation are of concern on account of their instant effect on the body (somatic), while low doses are of concern because of the possible late somatic and lasting genomic impacts. The radiation exposure effects on an individual increase.

Cancer: Normal cells undergo carcinogenesis and are converted into cancer cells. Radiation acts as a motivator as well as an organizer of multiplication. It can cause gene mutation and cancer. Thyroid cancer, leukaemia, and salivary gland cancer are some of the consequences of exposure to radiation.

9.8.2 Biological Outcomes of Radiation

The human body is composed of many organs, and each organ is made up of specialized cells. Ionizing radiation can affect the normal functioning of these cells. By ionization of atoms in the material, radiation damages human tissue. Ionizing radiation absorbed by human tissue has enough energy to eliminate electrons from the atoms that constitute molecules of the tissue. The bond is broken and the molecule falls apart as a result of the dislodging of the electron shared by two atoms forming a molecular bond. When ionizing radiation interacts with cells, it may or may not attack a critical part of the cell. Chromosomes are believed to be the most important part of the cell as these contains the genetic information and instructions essential for the cell to function and multiply. However, at the same time, there exists an effective repair mechanism which fixes cellular damage, including chromosome damage (Figure 9.6). The possible effects of radiation include:

Cells are unharmed by the dose: Ionization may form chemically active substances which at times alter the cell structure. These changes may be the same as those that occur naturally in the cell and may have no negative effect.

Cells are harmed, repair the damage, and function normally: Ionizing events produce substances which are not normally found in the cell, and may result in the breakdown of the cell structure and its components. If it is restricted, cells can repair the damage. Even chromosomal damage is generally repaired. Thousands of chromosome aberrations occur continuously in the human body and the body has efficient mechanisms to repair these changes.

Cells are impaired, repair the damage, and function atypically: If a damaged cell needs to perform a function before it has had time to repair itself, it will

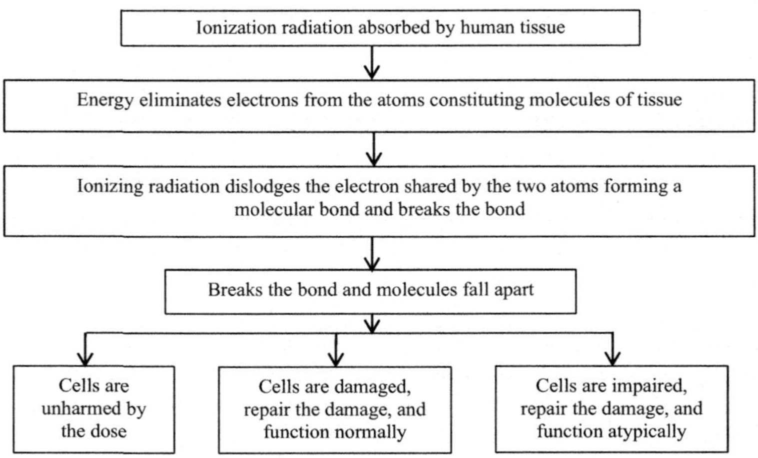

FIGURE 9.6 Mechanisms of action of IR.

either be incapable of the repair function or function imperfectly or partly. Consequently, the cell may not perform its normal functions or it may be damaging to other cells. The transformed cells may be incapable of reproducing themselves or they may reproduce at an uncontrolled rate. Such cells can cause cancer.

Cell death due to damage: If a cell is considerably harmed by radiation, or damaged in such a way that duplication is affected, the cell may die. Radiation damage to cells thus depends on the sensitivity of cells to radiation. All cells are not uniformly sensitive to radiation damage. Generally, cells which divide quickly and/or are comparatively non-specialized tend to exhibit effects at lower doses than those which are less quickly dividing and are more specialized, e.g., cells of the haemopoietic system are the most sensitive biological indicator of radiation exposure.

Variations in Cellular Radiation Sensitivity

Cell organelles have diverse sensitivities to radiation. Alpha particles have a short range in which selective irradiation of cellular components is possible. On the whole, the nucleus is the most sensitive organelle of the cell and is the target of ionizing radiation in radiotherapy, owing to the presence of DNA. The cell membrane and cytoplasm are comparatively resistant. It is possible to examine the effects of radiation on different parts of the cell cycle.

- Cells in the late G2 and M phases are most sensitive
- Cells in S phase, mainly in the late S phase, are more resistant
- Cells in G1 phase have normal sensitivity

Cells may be somatic or gametic. Somatic cells are those which form the body of an organism, and are diploid, i.e., they contain two pairs of chromosomes. Since they do not give rise to future organisms, irradiation of somatic cells may lead to lasting effects in the exposed person only. Some somatic cells are comparatively radioresistant. A haploid gametic cell involved in reproduction unites with its opposite type to form a diploid cell.

Gametic cells arise through meiosis, where the cell divides to form two daughter cells, which later divide and give rise to four haploid daughter cells. Gametic cells once irradiated suffer damage, and long-term effects are not likely to affect the person but may affect their children because of mutations in the genetic code of the gamete. Gametes tend to be considerably more radio-sensitive than most somatic cells.

9.8.3 Chromosome Damage

Following irradiation, the breaking of a double strand in DNA is the most harmful event in a cell, the reason being the likelihood of chromosomal modifications caused by the presence of exposed ends of DNA. Chromosome damage arises when a single chromosome is broken before duplication in the S phase. Chromatid damage results from impairment to one arm of the replicated chromosome, with no harm to the other. When double-strand breaks are induced in a population of chromosomes, the broken ends may re-join to their original positions with no apparent chromosomal imperfections being recognizable. A fragment of the chromosome may be lost if the number of double-strand breaks occur in a single chromosome. Broken ends may adhere to incorrect broken ends, resulting in lethal abnormalities.

When part of an arm of two separate chromosomes is separated, a dicentric chromosome is formed. Two chromosomes then attach to each other rather than to their detached arms. New chromosomes contain two centromeres, making it difficult for them to divide normally at mitosis. An anaphase bridge is formed when a duplicated chromosome loses both ends of a paired arm. Arms then join, and when the cell tries to divide during mitosis it is unable to detach the joined arms. A ring is formed when both ends are lost from the same chromosome. The chromosome then fastens its new ends together, resulting in the formation of a ring. It is likely for chromosomes to re-attach in a non-lethal way, e.g., the ends of two chromosomes may be exchanged. This may have no effect, or may lead to unusual gene expression if there is orientation of promoter regions with incorrect genes. Such translocations are seen in many malignancies, either causing deactivation of tumour suppressor genes or activation of oncogenes.

DNA Damage and Repair: In human cells, DNA is the main target of RT. DNA damage occurs due to the ionization of molecules as they interact with radiation. The direct effect of radiation is due to the primary radiation beam interacting with the DNA. The indirect effect depends on the secondary radiation produced by the primary radiation to deposit the damage. The ratio of direct to indirect effects varies with different radiation attributes. Alpha particles have a large direct effect, while photons are more dependent on the generation of secondary electrons and radical ions. Damage caused by ionizing radiation involves harm to the bases, which are repaired through base excision repair. Some base damage is mended through nucleotide deletion repair. Single-strand breaks are repairs through single-strand break repair. Double-strand breaks are repaired through homologous recombination or non-homologous end joining. DNA damage attributable to other mechanisms may take place, but mutations in the genes which code for their restoration proteins do not affect the sensitivity to ionizing radiation. Common sources of damage include:

- Normal metabolic processes
- Free radical generation during metabolism
- DNA replication
- Transcription errors
- Non-ionizing radiation
- Chemicals
- Heat

Cells that are more sensitive to radiation show increased disintegration of DNA, detectable with pulsed field gel electrophoresis or the comet assay. The micronucleus assay is capable of spotting DNA breaks by assessing the formation of micronuclei, which occur when DNA fragments are not aligned on the mitotic spindle. DNA repair can also be measured using these methods. By allowing cells to survive for some time after radiation exposure, and linking the disintegration of DNA with instantly lost cells, the extent of repair that takes place is measurable.

Classes of Damage: Damage imposed on cells is categorized into three categories:

Lethal Damage: Widespread or complex damage that is irreparable for the cell

Sub-lethal Damage: Sub-lethal damage is revealed when a single radiation dose is instead given as two split doses. The number of cells that die from the split dose is less than the single large dose

Potentially Lethal Damage: Damage that can be repaired under certain conditions, such as the presence of favourable environmental features or cell cycle arrest. Cells in S-phase are capable of repairing DNA damage. Cells in M-phase cannot repair the DNA damage as there are many other cellular processes in action. Tumour cells having defective cell cycle controls are unable to stop their development throughout the cycle

Mechanism of Radio-toxicity

Effectors of IR: Gamma radiation of cellular water rapidly generates:

- Reactive oxygen species (ROS)
- Hydroxyl radicals (•OH)
- Ionized water (H_2O+)
- Hydrogen radicals (H•)
- Hydrated electrons (eaq−)

Within one ps (10–12 s), secondary ROS products of IR are formed as:

- Superoxide (O_2•−)
- Hydrogen peroxide (H_2O_2)

Subsequent chemical cascades affect the intracellular stoichiometry of these reactive species and generate additional cell-damaging molecules, e.g., metal catalysis by intracellular ferrous and/or cuprous ions convert O_2•− and H_2O_2 to form additional amounts of •OH. In a separate, but critically significant process:

- O_2•$^-$ couples with endogenous nitric oxide (•NO) and forms peroxynitrite anion ($ONOO^-$)
- These species, along with peroxynitrous acid (ONOOH), nitrogen dioxide (NO_2•), dinitrogen trioxide (N_2O_3), and others, are referred to as reactive nitrogen species (RNS)
- The increased formation of RNS and additional ROS equivalents are mainly harmful to the cell, as the reaction products are more reactive with biomolecules than their precursors

Cellular macromolecules are reformed by direct ionization and through the reactivity of the high-energy species originating from water radiolysis (indirect effects of ionization), affecting an estimated 2000 primary ionization event. Direct radiation damage is initiated in the range of 10–14–10–12 s with the breaking of S–H, O–H, N–H, and C–H bonds. Widespread bio-molecular damage induced by radiolytic products of water initiates along with thiol depletion and further breaking of bonds (e.g., C–C and C–N). Cumulative effects of early, rapid biochemical processes are manifested in later stages of cellular damage, including:

- Slowing of mitosis
- Damage to protein signalling networks
- Membrane rupture

Endogenous Propagation of IR-induced ROS

The overall amount of ROS generated from primary ionization events is further propagated via the intracellular activation of endogenous ROS-producing systems such as:

- Nicotinamide adenine dinucleotide phosphate, (NADPH) oxidase (reduced form)
- Mitochondrial electron transport chain (ETC)
- IR exposure is associated with mitochondria-dependent ROS/RNS generation in tumour cells
- Increased ROS generation in mitochondria after low-dose IR significantly contributes to radio-sensitivity and cell survival

Temporal propagation of IR effects is also achieved through nitrosative stress mechanisms. In addition to directly modifying tyrosines, cysteines, and haems, •NO is the endogenous precursor to $ONOO^-$ and other RNS. Activation of ROS- and RNS-producing pathways by IR is very important, as it shows a targeted localized increase in these reactive species as a result of global IR-induced ROS production, selectively altering cellular signalling and metabolic pathways.

Cellular defence against IR-generated ROS

The capability of cells to survive an IR-based attack depends on a complex network of ROS-metabolizing molecules and detoxifying enzymes that comprise cellular oxidative stress defence. $O_2^{\bullet -}$, generated through water radiolysis, is dismutated to H_2O_2 and O_2 by the activity of superoxide dismutases (SODs). Enzymatic detoxification of H_2O_2 is achieved through the activity of:

- Catalase
- Peroxiredoxins (Prx)
- Glutathione peroxidases (GPxs)

The roles of these antioxidant enzymes are vital even under normal cellular conditions to keep ROS levels and pro-oxidant mechanisms in check. The process is further assisted by low-molecular-weight (MW) endogenous antioxidants such as:

- Ascorbate (vitamin C)
- Glutathione (GSH)
- Melatonin, lipoic acid
- Ubiquinone (coenzyme Q10)
- Vitamin E

Endogenous antioxidants target water radiolysis products, partially oxidized biomolecules, and peroxynitrite. Almost immediately after cellular exposure to IR, the low-MW antioxidant supply becomes compromised, leading to a rapid decrease in reduced GSH levels.

In an effort to combat oxidative eruption imparted by IR, cellular transcription factors, including nuclear factor and nuclear factor kappa-light-chain-enhancer of activated B cells (NF-κB), are activated, resulting in increased expression of ROS detoxifying enzymes, including catalase and SOD along with GPx, glutathione S-transferase (GST), and haem oxygenase-1. Both cumulative and acute IR readily disrupt the cellular redox balance, ultimately overwhelming the cellular antioxidant system, an event marked by enzyme inactivation, a low GSH/glutathione disulphide ratio, and a decreased pool of low-MW antioxidants. The result of such a redox imbalance is revealed in the efficient modifications of nucleic acids, lipids, proteins, and other biomolecules.

The interaction of IR with the cell nucleus for many years has been considered the primary mechanism responsible for the genotoxic effects of radiation fuelling the central dogma of radiation biology.

DNA Damage Response: Homologous recombination is the most precise form of double-strand break repair. The complementary sister chromatid is used as a template. Homologous recombination is triggered by the attraction of the ATM/ATR protein and MRE complex of proteins to the site of damage. ATM is involved in the signalling and activation of repair pathways, which causes a halt in the cell cycle as well as cumulative making of repair molecules. Repair is initiated by the MRE complex, which eliminates several bases from a single strand on each side of the break. This exposes a single strand on each side of the break, which is later coated by Replication Protein A, preventing aberrant binding of the single strands. Replacement of RPA by RAD 51 occurs under the supervision of the BRCA complex. This leads to invasion of the sister chromatid by the damaged strand, and DNA polymerase uses the undamaged strand to substitute the damaged region of each side of the original strand. The final step is cutting of crossover points, where non-homologous end joining is a well-understood path for repairing double-strand breaks. It can be used at any stage in the cell cycle, but is not as precise as it does not use a template for repair. Repairing base deletion and single-strand break pathways are related. If a base is changed by any means, it causes an aberration in the shape of the DNA helix. This can be spotted by glycosylases which eliminate the damaged base. The sugar/phosphate backbone of the affected base is also detached by APE1, leaving a single-strand break. Long repairing is more complex, involving the elimination of a section of the DNA around the single break and renovation of the region by polymerases. Repair of nucleotide deletion is carried out by an array of proteins. It is involved in the deletion of bulky

DNA lesions and cross-links, which are not distinctive of ionizing radiation. The damaged strand is identified, and incisions made downstream of the lesion. The entire section is detached and polymerases and ligases replace the strand. Incompatibility repair occurs during DNA replication, and confirms highly accurate transformation of DNA.

9.8.4 Mechanisms of Cell Death

In radiobiology "*cell death*" signifies that a cell is rendered unable to revive the tumour mass due to its damage through apoptosis, autophagy, or necrosis. It may also occur if the cell is made senescent or if it experiences mitotic misfortune. Apoptosis is the controlled method of cell death that occurs in response to several extra- or intracellular pointers. It includes hypoxia, excessive oncogene activation, or cytokines from neighbouring cells. Apoptosis is induced by natural killer cells through the induction of granzyme-B into the cell. Apoptosis is a visually distinct form of cell death. It starts when the cell separates from its neighbours, followed by condensation of its nucleus and cytoplasm. A rounded outgrowth develops on the cell membrane, and the nucleus splits into pieces. The outgrowth of cytoplasm or the apoptotic bodies may or may not contain nuclear fragments. These remains are consumed by macrophages to prevent accidental exposure of cytoplasmic and nuclear proteins to antigen-presenting cells. There is a lack of an acute inflammatory response following this mechanism of cell death, due to "hiding" of the intracellular components from the immune system.

Autophagy is another form of automated cell death, but this is used by the cell when it is deprived of nutrients or stirred by some growth hormones. Autophagy does not require caspases and is morphologically different from apoptosis. It is another method of tumour dominance because mutations in the component genes are viewed in some cancers and after treatment with anticancer agents, including radiotherapy, mainly when the apoptosis pathway is impaired.

Necrosis is generally considered an unrestrained form of cell death owing to loss of nutrients or other critical problems in the cellular environment. Degradation of lysosomes releases enzymes that attack and destroy the cell proteins. If a cell continues mitosis without proper alignment of chromosomes on the metaphase plate, the cell division may lead to aneuploidy in both daughter cells. This is known as mitotic catastrophe as the cells die due to a loss or gain of significant genetic material that allows mitosis to occur or due to failure of the cell to pass on genetic material once the catastrophe has happened.

In radiobiology, the relationship between the fraction of cells retaining their reproductive integrity and the absorbed dose of radiation are depicted in cell-survival curves. Normally, the surviving fraction is shown on a logarithmic scale, and is plotted on the y-axis against dose on the x-axis.

9.8.5 Impact of Radiation on Biological Molecules

Ionizing radiation (IR) is a universal natural phenomenon, which represents an effective therapeutic modality for the treatment of some tumours. Owing to its negative effects on organisms, IR has gained attention but also played an important role in the creation of life and evolution. Though the IR effect on important molecules can be destructive because of chemical modifications, some may perhaps be useful for the development of an organism during evolution. Organisms are exposed to natural IR sources including extra-terrestrial and terrestrial sources and also to artificial sources during radiotherapy. IR causes negative impacts on living organisms such as a direct interface with DNA or the generation of reactive oxygen species (ROS) which harm DNA, proteins, and lipids. Mutations are among the most serious effects that can bring about malignant transformations. Various impacts of IR on bio-macromolecules are discussed next.

Nucleic Acid: Damage to nucleic acid in response to cellular IR is extensively documented and is the classic example for comprehending the harmful effects of ionizing radiation, where direct and indirect effects of IR on cells can be differentiated. Direct damage is believed to be caused by the direct interaction of bio-macromolecules and IR particles, or secondary electrons in the case of roentgen or gamma radiation. In general, this completely damages the nucleic acids, as it displaces hydrogen bonds between complementary bases. The indirect effect is associated with water radiolysis and ROS generation. The incidence of DNA damage is increased when chemotherapeutics are used in combination with radiation. DNA-damaging events imposed by radiation include:

- Damaging alteration of bases and sugars
- Formation of cross-links
- Single- and double-strand breaks
- DNA assembling

Among water radiolysis products, OH is the most abundant and generally most damaging nucleic acid molecule. Radiation damage to deoxyribose is the main event underlying strand breakages, which occurs at a high rate and erratically along the DNA backbone in response to both direct OH attack and the activity of nucleic acid-binding enzymes. Both nucleobases and deoxyribose are the targets for OH-facilitated damage. For purine nucleobases, OH adds at C_4, C_5, and C_8, producing reactive adduct radicals that lead to a range of products, with the most common being the 8-hydroxypurines, particularly 8-oxod G, which serve as recognized hallmarks for oxidative DNA damage. Radiation-induced nucleo-base abrasions include oxidatively modified bases, but do not instantaneously result in strand breakage.

The biological consequences of IR on nucleic acids are alarming. It is distinguished by the fact that IR causes a wide variety of nucleic acid damages, namely nucleotide damage, and single-strand and double-strand breaks and IR is also able to induce a variety of products within nucleic acids such as 8-hydroxydeoxyguanosin. However, reparative mechanisms efficiently reduce altered nucleotides, which play a minor role in radiation-induced mutagenesis. A majority of single-strand breaks are repaired by ligases. In contrast, double-strand breaks caused by IR or chemical compounds are regarded as the most serious damage, which often induces mutations and brings about carcinogenesis because of the inability to get repaired correctly. Double-strand breaks may eventually lead to chromosomal aberrations, damage genes, and cause their malfunctioning and cell death. MRN complex and kinase ATM (Ataxia Telangiectasia Mutated) are stimulated in response to double-strand breaks induced by IR. ATM phosphorylates the double-stranded base, which results in activation and phosphorylation to adjacent substrates like H2AX on chromatin. Phosphorylation of H2AX leads it to switch to H2AX which interacts with MDC1 (mediator of DNA-damage checkpoint 1) and magnify the signals vital for other proteins involved in reparative mechanisms. These proteins bind to breaks by ubiquitin ligase RNF 8. The signal cascade leads to CHK2, p53, and cdc25 phosphorylation and stops the cell cycle in the G1/S or G2/M phase, which provides sufficient time for DNA repair. IR is capable of influencing nucleic acid which is not exposed directly to it. This radiation-induced bystander effect (RIBE) has three steps:

- Cells exposed to IR secretly transport molecules like TGF-β1 or interleukin-8, which signal cascades and further induce nucleic acids damage
- Propagation of RIBE via GJIC (gap-junctional intercellular communication)
- RIBE can be activated by oxidative metabolism

The immediate response to IR-induced ROS/RNS-mediated DNA damage is basically:

- Activation of the cell cycle checkpoint response
- An intricately controlled network involving sensor, transducer, and effector proteins responding to the DNA damage signal by initiating a cyto-protective response – the DNA damage response (DDR)

IR-induced protein adjustments and chromatin remodelling eventually meet to repair nucleic acid damage or signal to initiate the cell death pathways. Sensor and transducer proteins identify DNA damage sites, and then start and intensify a biochemical cascade. Cellular DNA damage after IR exposure mainly depends on the ability of the cellular proteins to prevent and repair the IR-induced DNA modification.

A DNA molecule has a twisted ladder-like structure. Long rails are made of sugar and phosphate molecules, which are the "backbone" of a DNA molecule. Each rung is a combination of four nucleotide bases. These are adenine ("A"), guanine ("G"), cytosine ("C"), and thymine ("T"). Ionizing radiation can affect DNA through either direct or indirect action. Ionizing radiation can interact directly with a DNA molecule's atoms, which stops cells from reproducing. Direct action of IR can also damage critical cellular systems and even cause cancer. Alpha particles, beta particles, and X-rays can directly affect a DNA molecule in one of three ways:

- Changing chemical structure of bases
- Breaking the sugar-phosphate backbone
- Breaking hydrogen bonds connecting the base pairs

Direct effects of ionizing radiation can:

- Change the chemistry of a nucleotide
- Break the sugar-phosphate backbone
- Break the hydrogen bonds between bases

Direct action can either result in DNA damage or DNA mutations. DNA lesions involve physical damage. This includes changes to the chemical structure of the DNA molecule. DNA mutations involve changes to the sequence of base pairs. If DNA damage is not repaired, it can lead to mutations. Mutations can stop genes from making the correct proteins and can be harmful to an organism. Ionizing radiation can affect important molecules other than DNA. For example, it can break the bonds holding water molecules together, which form free radicals, namely hydrogen (H^+) and hydroxyls (OH^-) ions. Free radicals are highly reactive and easily join with other ions inside cells. Hydroxyl ions (OH^-) can react with hydrogen atoms inside a DNA molecule to form hydrogen peroxide (H_2O_2). Over time, the damage from free radicals can build up, with free radical damage contributing to aging and diseases like cancer, Alzheimer's, and Parkinson's. DNA-damaging events imposed by IR alone include the deleterious alteration of bases and sugars, cross-link formation, single- and double-strand breaks (SSBs/DSBs), and DNA clustering. Of the water radiolysis products, •OH is the most abundant and is mainly destructive to nucleic acid molecules. Radiation damage to deoxyriboses is the primary event underlying strand breakages, which occur in a high frequency and randomly along the DNA backbone in response to both direct •OH attack and the activity of nucleic acid-binding enzymes.

Double-strand breaks, in particular, initiate through the coordinated reactivity of two •OH radicals at adjoining ribose sites, ultimately leading to strand breaks through subsequent radical pathways. Both nucleobases and deoxyribose are targets for •OH-mediated damage. Radiation-induced nucleobase lesions include oxidatively modified bases as well as abasic sites, but do not instantly result in strand breakage.

Lipids: The lipid layer within cell membranes is another biomolecule target of radiation-generated ROS. The lipid component of cell membranes is usually estimated to be ~5 nm in thickness with significant exposure to the aqueous cellular environment. Though IR is capable of directly damaging lipids, lipid bilayer mimetics have showed that indirect damage induced by water radiolysis is a larger contributor toward overall lipid modification by radiation.

Radiation induces lipid peroxidation, mainly peroxidation of polyunsaturated fatty acids (PUFAs), which increases membrane penetrability, disturbance of ion gradients, other trans-membrane processes, and altered activity of membrane-linked proteins. Studies of radiation targeted at the cell membrane have shown the induction of apoptosis at 5–10 Gy through increased ceramide levels. This is observed even in cells without a nucleus, showing another pathway for cellular IR damage independent of nucleic acid damage.

Unsaturated fatty acids are unreactive toward molecular oxygen, but freely oxidized in a number of radical-mediated processes. Peroxidation of polyunsaturated fatty acids (PUFAs), such as linoleic and arachidonic acids, generates diene hydroperoxides, isomerization of *cis*-alkenes, and/or degradation to form small-molecule reactive carbonyls such as malondialdehyde (MDA), acrolein, and 4-hydroxy-2-nonenal (HNE). Alkene isomerization is started by the removal of hydrogen from a PUFA bis-allylic site by OH, thiyl radical (RS), or other radicals, producing a carbon-centred PUFA radical. Subsequent capture by oxygen results in a *cis-trans* isomerization and forms a peroxyl radical in a diffusion-controlled reaction.

On the other hand, *cis-trans* isomerization occurs via the direct addition of RS to a PUFA alkene. Discharge of the RS radical occurs together with isomerization, a process documented even in the presence of oxygen. Peroxyl radicals are released by the formation of lipid hydroperoxides (LOOH) and PUFA fragmentation to MDA, HNE, and acrolein. Reactive aldehyde products of PUFA peroxidation persist intracellularly for about 2 minutes, offering ample time for the modification of proteins, nucleic acids, and other biomolecules. A major factor influencing the fate of LOOH and the degree of its impacts on cells is the level of lipid peroxidation, linked directly to the cellular redox status. Moderate levels of LOOH activate the oxidative stress response, which, when exceeded, leads to:

- Apoptosis
- High-level and widespread LOOH
- Damage to the cell membrane and intracellular content
- Triggering of membrane lysis and necrosis

Proteins: In addition to nucleic acids and lipids, proteins are highly abundant cellular biomolecules and highly targeted by IR, bringing about changes in their expression and activity as well as oxidative or reductive post-translational modifications (PTMs). Single-cell exposure to micro beam-directed alpha particles in cytoplasm mutagenic with severity of outcome depends on the intracellular GSH and radicals. Importantly, these mutations are shown to occur without much, if any, cell death, demonstrating the potential for transmuting cell signalling along with cell persistence. PTMs critically affect protein structure and function through:

- Conformational changes
- Modulation of enzymatic activity
- Degradation and cellular trafficking

More than 35 types of oxidative protein modifications are currently known, including:

- Direct amino-acid oxidation (e.g., for cys and met)
- Oxidative cleavage of the protein backbone and/or amino acid side chains
- Carbonylation
- Addition of lipid oxidation products

Besides RT, protein oxidation has been associated with the advancement of many disease states, including diabetes, inflammation, sepsis, Alzheimer's disease, multiple sclerosis, Parkinson's disease, and many cancers.

9.8.6 Radiation Protection

Radiation protection, also known as radiological protection, is defined by the International Atomic Energy Agency (IAEA) as:

> "...protection of people from harmful effects of exposure to ionizing radiation, and the means for achieving this..."

Exposure can be from a source of radiation external to the human body or due to internal irradiation triggered by the ingestion of radioactive contaminated foodstuff. IR is extensively used in industry and medicine, and can present a momentous health hazard by causing microscopic damage to living tissue. There are two main groups of ionizing radiation health effects.

At high exposures, it can cause "tissue effects," also called "deterministic effects" due to the certainty of happening, usually specified by the unit gray and resulting in acute radiation syndrome. For low-level exposures there can be statistically increased risks of radiation-induced cancer, called "stochastic effects," owing to the uncertainty of them happening, and conventionally indicated by the unit sievert.

Fundamental to radiation protection is prevention or dose reduction using the simple protective measures of time, distance, and shielding. The duration of exposure must be limited, distance from the radiation source must be maximized, and the source should be shielded wherever possible.

To measure the personal dose uptake in occupational or emergency exposure for external radiation, personal dosimeters are used, and for internal dose due to the ingestion of radioactive contamination, bioassay techniques are useful.

Principles: Based on the assessment of a large body of scientific studies accessible to associate risk to received dose levels, the ICRP recommends, develops, and maintains the International System of Radiological Protection. The system's health objectives are "to manage and control exposures to ionizing radiation so that deterministic effects are averted, and the risks of stochastic effects are lessened to the extent sensibly attainable." **The** ICRP recognizes planned, emergency, and existing exposure situations, as follows:

Planned exposure: Radiological protection can be planned in advance, before exposures occur, and when the magnitude or extent of exposure can be realistically foreseen.

Emergency exposure: Unanticipated circumstances that need urgent protecting actions.

Existing exposure: These can be from naturally occurring radioactive materials which exist in the environment.

Mechanism of Radio-Protection: Individuals can shield themselves against IR by using compounds that avoid indirect action, and repair direct or indirect damage to cells after IR exposure. Radio-protective agents repress reactive-compound formation (free-radical scavengers), decontaminate radiation-induced species, target equilibrium of vital biomolecules, and boost the repair and recovery progressions. Radio-protectors can also act as immunomodulators, i.e., they encourage the proliferation of hematopoietic and immune-poietic stem cells. Radio-protective agents can be classified as chemical radio-protectors/prophylactic preparations, mitigators, and therapeutic preparations. Prophylactic formulations, also known as classical radio-protectors are used before radiation exposure to prevent tissue damage. This class of radio-protectors includes compounds with thiol (sulphhydryl) groups and/or compounds with antioxidant properties.

Mitigators are useful during or after IR exposure before the manifestation of radiation symptoms, with the aim of minimizing toxicity, and avoiding or lessening the adverse effects of radiation on cells/tissues. This class of compounds includes chelating agents (chelators) and blocking agents, such as KI, which shields the thyroid gland from radioactive isotope iodine-131, by checking its amassing in the thyroid. Therapeutic preparations are applied after radiation exposure, for curing, treating, and healing acute radiation syndrome and the delayed effects of radiation exposure, such as fibrosis, vascular damage, and damage to organs.

Types of Radio-Protectors

(a) Synthetic Radio-Protectors

- Amino acid cysteine, containing SH group
- WR agents, structurally differ in the length of amino alkyl groups, by the Me-group presence/absence on the terminus, and/or by the presence of an –OH group on the alkyl chain. The most prominent compound in this class is amifostine

(b) Non-Thiol Radioprotectors

- Besides compounds with SH groups, a number of compounds of the classes amino sulphides, thio-urea derivatives, thiosulfates, thiophosphates, diethyldithio carbamates, and thiazoles were formed and scrutinized, including some biogenic amines and their derivatives
- Radioprotectors are mainly synthesized in the form of esters with phosphoric acid to achieve better solubility in water
- Nitroxides are the most promising, low-molecular-weight compounds, which represent stable organic free radicals. These act as superoxide dismutase mimics, defending cells against oxidative stress, inhibiting lipid peroxidation, and preventing disruption to DNA helix

(c) Radio-Protectors of Natural Origin

Compounds of natural origin are less toxic and do not show serious side effects. The best-known radio-protectors are β-glucans and polysaccharide ginsan, which are known to have multiple immuno-modulatory effects but also confer strong radioprotection. This group of compounds stimulates endogenous production of cytokine (interleukins IL-1 and IL-6), and increases the number of bone marrow cells, circulating neutrophils, lymphocytes, and platelets in irradiated mice. A hormone, 5-androstenediol, produced by the human adrenal cortex, has been examined as a potential radio-protector which stimulates myelopoiesis, and increases the number of circulating neutrophils and platelets, but not erythrocytes. It has an effective radio-protective effect and low toxicity. There are problems with its application owing to low oral efficiency and local inflammatory response at the site of injection. The application of oxymetholone (a steroid with an androgenic/anabolic effect) offers a good protective effect, as it can be orally administered

(d) **Phyto-Radio-protectors**
- Many plants and herbs, as well as the complexes isolated from plants, are very promising radio-protectors
- Unlike synthetic compounds, herbal products are non-toxic, without detrimental effects on human health, economical, and are normally administered orally

External Dose Uptake: Radiation exposure can be managed by a combination of the following factors:

- **Time**: Decreasing exposure time proportionally reduces the effective dose.
- **Distance**: Amplifying the distance decreases the dose because of the inverse square law. Distance can be as simple as handling a source with forceps rather than fingers.
- **Shielding**: Sources of radiation can be protected with solid or liquid materials, which absorb the energy of radiation. The term "biological shield" is used for absorbing material placed around a nuclear reactor to decrease the radiation emitted to a level safe for humans.

Internal Dose Uptake: The internal dose, due to the inhalation or ingestion of radioactive substances, can result in stochastic or deterministic effects, depending on the amount of radioactive material ingested and other bio-kinetic factors. The risk from a low-level internal source is categorized by the dose quantity, which has the same risk as the same amount of external effective dose. The intake of radioactive material can occur through four pathways:

- Inhalation of airborne contaminants such as radon gas and radioactive particles
- Ingestion of radioactive contamination in food or liquids
- Absorption of vapours such as tritium oxide through the skin
- Injection of medical radioisotopes such as technetium-99m

Occupational hazards from airborne radioactive particles in nuclear and radiochemical applications can significantly be decreased by extensive use of glove boxes to contain such material. A large-scale glove box is used to contain airborne radioactive particles. To protect against breathing in radioactive particles in ambient air, respirators with particulate filters are worn. To scrutinize the concentration of radioactive particles in ambient air, radioactive particulate monitoring instruments are used for measuring the concentration or presence of airborne materials. Special laboratory radiometric assay methods are used to determine the concentration of ingested radioactive materials in food and drink.

A radiation dosimeter is an important personal dose-measuring instrument worn by the person being monitored and is used to estimate the external radiation dose deposited in the individual wearing the device. They are used for gamma, X-ray, beta, and other strongly penetrating radiation, but not for feebly penetrating radiation such as alpha particles. Workers exposed to radiation, such as radiographers, nuclear power plant workers, doctors giving radiotherapy, and those in laboratories using radionuclides must wear dosimeters to record their occupational exposure. Dosimeters can be worn to obtain a whole-body dose and there are special types that can be worn on fingers or clipped to headgear, to measure the localized body irradiation for specific activities. Common types of wearable dosimeters for ionizing radiation include:

- Electronic personal dosimeter
- Film badge dosimeter
- Quartz fibre dosimeter
- Thermo-luminescent dosimeter

Almost any material can act as a shield from gamma or X-rays if used in sufficient amounts. Different types of ionizing radiation interact in diverse ways with shielding materials. The effectiveness of shielding is dependent on the discontinuing power of radiation particles, which varies with the type and energy of radiation and the shielding material used. Different shielding techniques are thus used depending on the application and the type and energy of the radiation. Shielding reduces the strength of radiation depending on the thickness. The effectiveness of a material as a biological shield is related to its cross-section for scattering and absorption, and to a first approximation is relative to the total mass of material per unit area interrupted along the line of sight between the radiation source and the region to be protected. Hence, the shielding strength or thickness is usually measured in units of g/cm^2.

9.9 DIAGNOSIS

The detection and diagnosis of radioactive metals in the body of an exposed person are important before treatment and management of radioactive metal toxicity.

Analytical methods used for the diagnosis of radioactive metals from urine, blood, or bone samples are listed in Table 9.10.

9.10 TREATMENT AND MANAGEMENT OF RADIOACTIVE METAL POISONING

- Accumulation of radioactive metals in body may be mitigated by decreasing its absorption, distribution, and deposition, and increasing its elimination with the help of chelating agents.
- Chelating agents are chemical compounds that react with metal ions to form a stable, water-soluble complex.
- The formation of soluble chelates may enhance the mobilization of uranium deposited in tissue and expedite its transportation to and elimination from the renal system.
- Chelating agents have a ring-like centre which forms at least two bonds with metal ions, allowing it to be excreted.
- The removal of plutonium by chelating agents is of great importance because plutonium is extremely dangerous and induces cancer in humans owing to its radiation toxicity.

TABLE 9.10
Analytical Methods for Determining Radioactive Metals in Biological Samples

Radioactive Metal	Sample matrix	Analytical method
Uranium	Urine	Fluorometric (total uranium)
		Inductively coupled plasma – mass spectrometry
		Isotope dilution-MS (isotopic quantification)
		Laser-induced fluorometry (total uranium)
		Neutron activation analysis (isotopic quantification)
		Spectro-photometric (total uranium)
		α-Counting (total uranium)
		α-Spectrometry (total uranium)
	Blood	Inductively coupled plasma – mass spectrometry (isotope quantification)
	Soft tissue	α-Spectrometry (total uranium)
		Isotope dilution-MS (isotopic quantification)
	Bone/bone ash	α-Spectrometry (isotopic quantification)
Iodine	Urine	Inductively coupled plasma – mass spectrometry
Thorium	Bone	γ-Ray spectrometry

- Chelating agents such as ethane-1-hydroxy-1,1-bisphosphonate (EHBP), 3,4,3-LIHOPO, and CBMIDA have the ability to remove plutonium and uranium.
- Removing the patient from the workplace or source of overexposure can minimize the impact.
- Ensuring proper ventilation at the workplace is important.
- The use of a radiation suit or nuclear full suit at the workplace can help in avoiding exposure.
- External decontamination of patients to reduce further overexposure if they have residue from the exposure on them can be beneficial.
- Routine analysis of urine samples after large accidental overexposures is required.

9.11 TOXICOLOGICAL EFFECTS OF RADIOACTIVE METALS ON THE ENVIRONMENT

The debate on the exploration and utilization of radioactive metals centres around safety concerns, especially on what it foretells to humans, plants, animals, and the environment on exposure to radioactive metals. Proponents as well as opponents are sharply divided on the socioeconomic as well as environmental sustainability and suitability of nuclear programs. Ironically, both sides of this debate agree that nuclear programs create significant safety risks to users as well as to the natural environment. Radioactive metals are highly toxic to the environment and induce severe health impacts in the various living entities of an ecosystem. Radioactive metals exist naturally in almost all soils, rocks, and water sources across the world. They can move through the environment in rain, wind, and through other natural processes. People have been using radioactive metals such as uranium for building military shielding, weapons, planes, and helicopters. These radioactive metals are used to fuel nuclear weapons and nuclear power plants. People can come into contact with toxic metals by spending time at abandoned mines, living in a home that was built with material from a mine or mill site, drinking unregulated water, or not following recommended water-sourcing practices.

Everyone comes in contact with a low level of radioactive metals because they are present in the environment. However, coming into contact with higher levels of radioactive metals can be harmful to health. The toxicological effects of radioactive metals on the environment, including plants and humans, are discussed in Table 9.11.

TABLE 9.11
Toxicological Effects of Radioactive Metals on Humans, Plants, and the Environment

Radioactive Metal	Toxicological Effects		
	Human	Plants	Environment
Uranium	Autoimmune diseases Bone cancer Decrease in oocytes Oedema Fatty liver Focal necrosis Foetal malformation Glomerular injury Kidney damage Lung cancer Neurotoxicity Pulmonary toxicity Reproductive problems Sperm defects	Cell damage Highly toxic to plants Oxidative stress	Groundwater contamination Increased level of background radiation environmental pollution Radioactive dust Soil contamination Waterborne toxins
Thorium	Bone marrow cancer Colon damage Damage to liver, spleen, and pancreas Damage to lymph nodes Foetal malformation Lung cancer	Increase in content of photosynthetic pigments	Damage to aquatic ecosystem Soil Contamination
Radon	Alzheimer's disease DNA damage Enlargement of brain Lung cancer Multiple sclerosis Parkinson's disease Placenta damage	–	Indoor air pollution

TABLE 9.11 (Continued)
Toxicological Effects of Radioactive Metals on Humans, Plants, and the Environment

Radioactive Metal	Toxicological Effects		
	Human	**Plants**	**Environment**
Polonium	Foetal deformities Liver and kidney damage Miscarriage Mutation Ovaries infertility	Bioaccumulation in leaves	Freshwater and groundwater contamination
Plutonium	Damage to gastrointestinal tract Leukaemia Lung cancer Lung tissue damage	Bioaccumulation in plants	Ground water contamination
Iodine-131	Nausea Neck swelling Thyroid gland disorders	Cellular damage	Radioactive dust
Radium	Anaemia Cataracts Loss of appetite Lung and bone cancer Reduced bone growth	Reduced seed germination Long-term reduction in life span Reduction in viability	Bio-magnification

Biomagnification in the Terrestrial and Aquatic Food Chains: Uranium mining is an ecologically destructive and unsustainable process which needs to be either stopped or regulated by an international organization. Uranium mining results in higher levels of radium in water close to uranium mines. Plants absorb radium from contaminated soil. Primary consumers such as herbivores consume these plants and accumulate radium within their body. Finally, radium gets accumulated in top predators and has severe impacts. Radioactive metals also contaminate the water bodies and may concentrate in aquatic organisms and bio-magnify.

Uranium Mill Tailing: Uranium mill tailings are generally disposed of near surface impoundments close to mines, which poses serious environmental and health risks in the form of radon emission, windblown dust dispersal, and leaching of contaminants including HMs and arsenic into water.

Soil Infertility: Improper disposal of radioactive waste can severely contaminate soil and result in soil pollution, rendering it highly toxic and infertile.

Groundwater Contamination: Plutonium may enter surface water from accidental releases and disposal of radioactive wastes. Soil also becomes contaminated with plutonium through fallout during nuclear weapons testing. Plutonium slowly moves downwards in the soil and contaminates the groundwater.

Toxic Waste Dumping: Dumping of radioactive toxic waste produces destructive effects on the environment. There is a need for proper disposal techniques for dumping of radioactive wastes.

Radioactive Metal Pollution: Nuclear power plants are the main sources of radioactive pollution. U-238, Th-232, Ra-222, and Rb-87 are major radionuclides that exist in various environmental matrices such as rocks, soil, sediments, air, and water. On the other hand, the main sources of anthropogenic radionuclides like Cs-137, Pu-239, Am-241, Sr-90, and Y-91 are nuclear weapons and nuclear power plants. Contamination caused by radioactive pollutants in the environment is a crucial issue that affects the quality of public health and the environment.

BIBLIOGRAPHY

Baldwin, A. S. (2001) Control of oncogenesis and cancer therapy resistance by the transcription factor N F kappa-B. *Journal of Clinical Investigation* 1073: 241–246.

Bentzen S. M. (2006) Preventing or reducing late side effects of radiation therapy: radiobiology meets molecular pathology. *Nature Reviews Cancer* 6: 702–713.

Berlett, B. S. and Stadtman, E. R. (1997) Protein oxidation in aging, disease, and oxidative stress. *Journal of Biological Chemistry* 272(33): 20313–20316.

Chaturvedi, A. and Jai, A. (2019) Effect of ionizing radiation on human health. *International Journal of Plant and Environment* 53: 200–205.

Coia L. R., Myerson R. J., and Tepper J. E. (1995) Late effects of radiation therapy on gastrointestinal tract. *International Journal of Radiation Oncology Biology Physics* 31: 1213–1236.

Cucinotta, F. A. and Durante, M. (2006) Cancer risk from exposure to galactic cosmic rays: Implications for space exploration by human beings. *Lancet Oncology* 75: 431–435.

Dartnell, L. R. (2011) Ionizing radiation and life. *Astrobiology* 11(6): 551–582.

Emami B., et al. (1991) Tolerance of normal tissue to therapeutic irradiation. *International Journal of Radiation Oncology Biology Physics* 21(1): 109–122.

Gregory, C. (2013) *Radiochemistry and Nuclear Chemistry*, 4th edition, pp. 88–99. Amsterdam, Netherland: Elsevier/North Holland.

Hagen, U. (1994) Mechanisms of induction and repair of DNA double-strand breaks by ionizing radiation: Some contradictions. *Radiation and Environmental Biophysics* 33(1): 45–61.

Han, W. and Yu, K. N. (2009) Response of cells to ionizing radiation. *Advances in Biomedical Sciences and Engineering* 6: 204–262.

Hsiao, C. P., Daly, B., and Saligan, L. N. (2016) Etiology and management of radiotherapy-induced fatigue. *Expert Review of Quality of Life in Cancer Care* 1(4): 323–328.

Hymes, S. R., Strom, E. A., and Fife, C. (2006) Radiation dermatitis: Clinical presentation, pathophysiology, and treatment 2006. *Journal of the American Academy of Dermatology* 54(1): 28–46.

Keith, S., et al. (2013) *Toxicological Profile for Uranium*. Georgia: Agency for Toxic Substances and Disease Registry (US).

Khanna, K. K. and Jackson, S. P. (2001) DNA double strand breaks: Signalling, repair and the cancer connection. *Nature Genetics* 27(3): 247–254.

Kuntic, V., et al. (2013) Radioprotectors – the evergreen topic. *Chemistry and Biodiversity* 10: 1791–1803.

Lannunziata, M. F. and Burkart, W. (2007) *Beta Radiation in Radioactivity*, pp. 119–186. Amsterdam: Elsevier Science.

Libermann, T.A. and Zerbini, L.F. (2006) Targeting transcription factors for cancer gene therapy. *Current Gene Therapy* 61: 17–33.

Monje, M. L., Mizumatsu, S., Fike, J. R., and Palmer, T. D. (2002) Irradiation induces neural precursor-cell dysfunction. *Nature Medicine* 8: 955–962.

Murphy, P.H. (1987) NCRP Report 91: Recommendations on limits for exposure to ionizing radiation. *Journal of Nuclear Medicine* 28(12): 1929.

Nambiar, D., Rajamani, P., and Singh, R. P. (2011) Effects of phytochemicals on ionization radiation-mediated carcinogenesis and cancer therapy. *Mutation Research* 728(3): 139–157.

Pillay, V. V. (2013) *Modern Medical Toxicology*. New Delhi: Jaypee Brothers Medical Publishers.

Preston, D. L., et al. (2007) Solid cancer incidence in atomic bomb survivors: 1958–1998. *Radiation Research* 168(1): 1–64.

Reisz, J. A., Bansal, N., Qian, J., Zhao, W., and Furdui, C. M. (2014) Effects of ionizing radiation on biological molecules – mechanisms of damage and emerging methods of detection. *Antioxidants & Redox Signaling* 21(2): 260–292.

Sachs, R. K. (1992) DNA damage caused by ionizing radiation. *Mathematical Biosciences* 1122: 271–303.

Stalter, R. and Howarth, D. (2012) *Gamma Radiation*, pp. 43–54. London, UK: Intech Open.

Tchounwou, P. B., Yedjou, C. G., Patlolla, A. K., and Sutton, D. J. (2012) Heavy metal toxicity and environment. *Experientia Supplementum* 101: 133–164.

Walker, C. H., Sibly, R. M., Hopkin, S. P., and Peakall, D. B. (2004) *Principles of Ecotoxicology*, p. 386. London: Taylor & Francis Group.

Wang, K. and Tepper, J. E. (2021) Radiation therapy-associated toxicity: Etiology, management, and prevention. *A Cancer Journal for Clinicians* 71(5): 437–454.

Wrixon, A., Barraclough, I., and Clark, M. J. (2004) *Radiation, People and Environment*. International Atomic Energy Agency.

10 Pesticide Toxicity

The more we pour the big machines, the fuel, the pesticides, the herbicides, the fertilizer and chemicals into farming, the more we knock out the mechanism that made it all work in the first place.

David R. Brower

After going through this chapter, the reader will learn about the pesticides, the classification of pesticides, effects of pesticide toxicity, measurement of toxicity, the mechanisms of action and impacts of pesticides on human health and the environment, as well as impacts on non-target organisms, diagnosing and treating pesticide poisoning, and managing pesticides.

FACT FILE

- Pesticides are used to protect crops against pests, insects, weeds, and fungi
- Pesticides can be possibly toxic to humans and can have both acute and chronic health effects, depending on the amount and manner of exposure
- Some of the older and cheaper pesticides can remain in soil and water for years
- People who face the greatest health risks from exposure to pesticides are those who come into occupational contact with them
- There are more than 1000 pesticides in use around the world and each pesticide has different properties and toxicological effects
- Some pesticides have been banned by nations which signed the 2001 Stockholm Convention
- Organophosphate pesticides disrupt enzymes of the nervous system in insects and humans and can cause paralysis and death
- Organophosphate pesticides were used as nerve agents by the Germans, Italians, and the Japanese during the Second World War
- Baking soda and canola oil are included in making a bio-pesticide, a type of pesticide derived from certain plants, animals, minerals, or bacteria
- Molluscicides contain desiccants that extract moisture from the living tissue of slugs and snails

DOI: 10.1201/9781003343721-10

- Ovicides are unique pesticides that kill the eggs laid by mites and insects
- Insect growth regulators are a type of pesticide that disturbs the life stages from pupa to adult so that the insect is not capable of maturing normally
- Hydramethylnon is an effective, slow-acting pesticide used to inhibit the cell metabolism of ants and termites
- There are almost 300 products belonging to seven chemical classes used for controlling bed bugs
- Pyrethroids and pyrethrins are the most commonly used insecticides for controlling and eradicating bed bugs
- Larvicides and adulticides are used to control mosquito larvae and adult mosquitoes
- Malathion, an organophosphate pesticide, is the most common insecticide used in mosquito control programmes

10.1 INTRODUCTION

A pesticide is a substance used for killing a pest. It may be a mixture of substances used for inhibiting, abolishing, preventing, or moderating any pest. Usually, a pesticide is a chemical or biological agent such as a virus, bacterium, antimicrobial, or disinfectant that discourages, disables, and destroys pests. The Food and Agriculture Organization (FAO) defines a pesticide as:

> any substance or mixture of substances intended for preventing, destroying, or controlling any pest, including vectors of human or animal disease, unwanted species of plants or animals, causing harm during or otherwise interfering with the production, processing, storage, transport, or marketing of food, agricultural commodities, wood and wood products or animal feedstuffs, or substances that may be administered to animals for the control of insects, arachnids, or other pests in or on their bodies.

Pesticide is a common term that describes numerous groups of insecticides, herbicides, fungicides, rodenticides, wood preservatives, and household disinfectants used either to kill or protect against pests. Pesticides vary in their physical, chemical, and other properties from one class to other. Thus, it is important to classify them. There may be naturally available or synthetic pesticides/manmade chemicals. Drum (1980) classified pesticides based on:

- Mode of entry
- Pesticide function and the pest organism they kill
- Chemical constitution of pesticide

A pesticide, synonymous with a plant protection product, is usually used to remove or control a variety of pests that can harm crops and livestock and reduce farm output. The most frequently used pesticides are:

- Fungicides, used to control fungi and mildew
- Herbicides, used to kill weeds

- Insecticides, used to kill insects
- Rodenticides, used to kill rodents

The use of pesticides has taken place since the ancient civilizations. The Ancient Sumerians used sulphur to protect their crops from insects, while medieval farmers used chemicals including arsenic and lead on common crops. The Chinese used combinations of arsenic and mercury to control body lice and other pests; while the Greeks and Romans used oil, ash, sulphur, and other materials to protect livestock and crops from various pests.

In the nineteenth century, scientists focused on natural techniques including the roots of tropical vegetables and chrysanthemums for managing pests. After the discovery of DDT in 1939, it was considered extremely effective and it was increasingly used as an insecticide. However, owing to its biological effects and human safety, its use was banned.

10.2 CLASSIFICATION

10.2.1 Based on Mode of Entry

The manner a pesticide comes into contact with or enters the target is defined as the mode of entry. These include:

Systemic pesticides: These are absorbed by plants or animals and transfer to the tissues, e.g., the systemic herbicide moves through the plant and reaches the leaves, stems, or roots. They can kill weeds with partial spray coverage. They can efficiently penetrate plant tissues and move through the vascular system and destroy particular pests. The movement of pesticides may be unidirectional or multidirectional. Likewise, some pesticides are locally systemic and cover a short distance from the point of contact. Examples of systemic pesticides are 2,4-D and glyphosate.

Non-systemic pesticides: These are also known as contact pesticides as they act on target pests after coming into contact with them. After coming into physical contact, these pesticides enter the body of pests through their epidermis and cause death. These pesticides do not essentially enter the plant tissues and thus are not transported through the plant vascular system. Examples are paraquat and diquat dibromide.

Stomach toxicants: These toxicants enter the pest's body via the mouth and digestive system and result in death. These poisons are assimilated during the feeding of pests, when they consume the insecticide applied on leaves and other plant parts. These toxicants may also enter the body, mouth and digestive tract, where they are absorbed into the insect's body, thus destroying the midgut. This is more suitable mainly in as a vector control, e.g., Malathion.

Fumigants: These pesticides act on or kill the target pests by producing vapour. They are in the form of poisonous gases and enter the bodies of pests through the tracheal system/spiracles and cause death. They are used to remove stored product pests from fruits, vegetables, and grains, and also in controlling soil pests.

Repellents: These do not kill but are repulsive and keep pests away from treated areas/commodities. They also inhibit pests' ability to trace a crop.

10.2.2 Based on Pesticide Function and Pest Organism they Kill

Based on pesticide function and the pest organism they kill; these pesticides are classified in Table 10.1.

10.2.3 Based on Chemistry

On the basis of their chemistry, these pesticides are classified in Table 10.2.

10.2.4 Based on their Nature

Based on their nature, these pesticides can be considered as:

- **Biodegradable**: Those which can be broken down by microbes and other living beings into non-toxic compounds
- **Persistent**: Those which may take months or years to break down

10.2.5 Based on Chemical Composition

The most common method of categorizing pesticides is based on their chemical composition and ingredients. This type of grouping gives an idea about the efficiency and physico-chemical characteristics of pesticides, which determine the mode of application and precautions that need to be taken during their application. Based on the chemical composition, pesticides are classified as organochlorines, organophosphorus, carbamates, pyrethrin, pyrethroids, etc. Generally, pesticides are organic chemicals both of synthetic and plant origin (see Figure 10.1). However, some inorganic compounds are also used as pesticides. The insecticides are the main pesticides that can further be categorized into many sub-classes (Table 10.1):

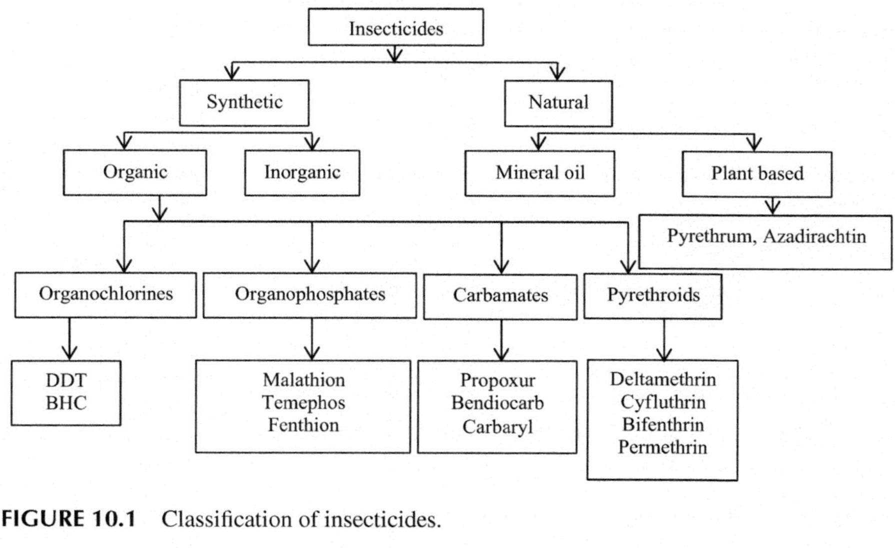

FIGURE 10.1 Classification of insecticides.

TABLE 10.1
Classification of Pesticides Based on Their Functions and the Organisms they Kill

Acaricides	Used to kill or disrupt the growth or development of mites and ticks	DDT Dicofol Carbofuran Flufenoxuron
Algicide	Used to kill or prevent algae	Copper sulphate Diuron Isoproturon Oxyfluorfen
Anti-feedants	Chemicals which stop an insect or other pest from feeding	Chlordimeforn Fentin Azadirachtin
Avicides	Chemicals used to kill birds	Fenthion Strychnine
Bactericides	Isolated from or produced by a microorganism or produced artificially to kill or inhibit bacteria in plants or soil	Copper hydroxide Kasugamycin Streptomycin Tetracycline
Bird repellents	Chemicals which repel birds	Copper oxychloride Diazinon Methiocarb Thiram Ziram
Chemosterillant	Chemicals that render an insect infertile and consequently stops it from reproducing. They act by: (a) Inhibiting production of egg or sperm (b) Causing death of the sperm or eggs (c) Bringing about lethal mutation on the sperm or eggs	
Desiccants	Act on plants by drying their tissues	Boric acid
Fungicides	Chemicals used to prevent, cure, or eradicate fungi	Cymoxanil Bordeaux mixture Copper oxychloride Flusilazole
Herbicide softeners	Chemical that protects crops from injury by herbicides	Benoxacor Cloquintocet Cyometrinil Cyprosulfamide
Herbicides	Substances used to kill plants or to inhibit their growth or development	Alachlor 2,4-D Sulfosulfuron Atrazine Isoproturon
Insect attractant	A chemical that lures pests to a trap thus removing them from crops, animals or stored products	Gossyplure Gyplure Muscalure
Insect Growth regulator	Substance that works by disrupting the growth or development of an insect	Diflubenzuron Buprofezin

TABLE 10.1 (Continued)
Classification of Pesticides Based on Their Functions and the Organisms they Kill

Insect repellents	Chemical that deters insects from landing on a human or an animal	Citronella oil Permethrin
Insecticides	Used to kill insects, or to disrupt their growth or development	Azadirachtin Pyrethrins Carbofuran Carbosulfan Methomyl Buprofezin DDT
Lampricides	Target larvae of lampreys which are jawless fish-like vertebrates in rivers	Trifluromethyl nitrophenol
Larvicides	Inhibit growth of larvae	Methoprene
Mammal repellents	A chemical that deters mammals from approaching or feeding on crops or stored products	Ammonia Vinegar Tiger urine Fox urine, Bobcat urine
Mating disrupters	Chemicals that interfere with the way that male and female insects locate each other using airborne chemicals (pheromones), thereby preventing them from reproducing	Alarm pheromones Sex pheromones
Moth balls	Stop damage to cloths by moth larvae or moulds	Dichlorobenzene
Molluscicides	Substances used to kill slugs and snails	Copper sulphate Metaldehyde Thiacloprid Thiodicarb
Nematicides	Chemicals which are used to control nematodes	Carbofuran Carbosulfan Methyl bromide
Plant growth regulators	Substances that alter the expected growth, flowering, or reproduction rate of plants	2,4-D Gibberellic acid Ethephon Metoxuron
Piscicides	Act against fish	Rotenone
Rodenticides	Substances used to kill rats and related animals	Strychnine Bromadiolone Warfarin Zinc phosphide Aluminium phosphide
Silvicides	Act against woody vegetation	Tebuthiuron
Synergists	Chemicals that enhance the toxicity of a pesticide to a pest, but that are not themselves toxic to the pest	Piperonyl butoxide
Termiticides	Kill termites	Fipronil
Virucide	An agent having the capacity to destroy or inactivate viruses	Ribavirin

TABLE 10.2
Classification of Pesticides Based on Their Chemistry

Insecticides	Antibiotics Carbamates Neonicotinoids Organo halogen Organophosphorus Pyrethroids
Fungicides	Acylamino acid fungicides Amide fungicides Anilide fungicides Antibiotic fungicides Aromatic fungicides Benzimidazole fungicides Carbamate fungicides Copper fungicides Morpholine fungicides
Herbicides	Acid herbicides Anilide herbicides Ether herbicides Halogenated aliphatic herbicides Phenylurea herbicides Quaternary ammonium herbicides Sulfonylurea herbicides Urea herbicides
Rodenticides	Coumarin Rodenticides Inorganic Rodenticides

Organochlorine: These were commonly used earlier, but because of their health and ecological impacts, they are no longer used. They are also known as chlorinated hydrocarbons with five or more chlorine atoms and were used in agriculture and public health for controlling a wide range of insects, and have a long-term residual impact on the environment. They disrupt the nervous systems of insects leading to convulsions and paralysis followed by death. The most common examples include DDT, endosulfan, aldrin, dieldrin, etc.

Organophosphates: Most organophosphates affect the nervous system by disturbing the enzyme that controls a neurotransmitter. Organophosphate pesticides control a wide range of pests owing to their multiple functions and are characterized by stomach poisons, contact poisons, and fumigant poisons. These pesticides are decomposable, and cause the least ecological pollution. They are more fatal to vertebrates and invertebrates as cholinesterase inhibitors leading to a permanent overlay of acetylcholine neurotransmitter across a synapse. Some of the most widely used organophosphorus insecticides include parathion, malathion, diaznon, and glyphosate.

Carbamates: Carbamate pesticides also disturb the nervous system by disturbing an enzyme that regulates the neurotransmitter; however, the enzyme effects are generally reversible. Carbamates are the derivatives of carbamic acid. At times, they are also used as stomach and contact poisons and fumigants. They easily degrade in the environment with low pollution. These insecticides include carbaryl, carbofuran, propoxur, and aminocarb.

Synthetic pyrethroids: These are a synthetic version of pyrethrin, and are developed in such a way as to maximize their permanence in the environment. They are comparatively more stable, with longer residual effects than natural pyrethrins. The main active components are pyrethrin I and pyrethrin II. These are lethal to insects and fish but also somewhat toxic to mammals and birds.

10.2.6 Based on Mode of Action

Based on the mode of action, pesticides are categorized as:

Physical pesticides: These classes of pesticides kill insects by exerting a physical effect, e.g., activated clay

Protoplasmic pesticides: These pesticides are responsible for precipitation of protein, e.g., arsenicals

Respiratory pesticides: These pesticides inactivate the respiratory enzymes, e.g., hydrogen cyanide

Nerve pesticides: These pesticides obstruct impulse conduction, e.g., Malathion

Chitin inhibitors: These chemicals inhibit chitin synthesis in pests, e.g., Diflubenzuron

10.2.7 Based on Sources of Origin

Based on the sources of origin, pesticides may be classified into chemical pesticides and bio-pesticides. Bio-pesticides are derived from natural sources, e.g. animals, plants, microbes (bacteria, viruses, fungi, and nematodes), and minerals. The main advantage of using biological pesticides is host specificity, as these act on the target pest, whereas chemical pesticides usually affect a wide group of non-target organisms. Bio-pesticides are generally eco-friendly as they are less poisonous, decay easily, and are required in small quantities. They are classified as:

- **Microbial Pesticides**: The active ingredient in these pesticides can be a bacterium, fungus, or protozoan. These pesticides kill insects either by toxins released by microbial organisms, or by infecting the organisms. Most microbial pesticides are more selective than biochemical pesticides.
- **Plant Protectants**: These pesticides are naturally produced by plants. A gene vital for production of the pesticide is introduced into the plant through genetic engineering.
- **Biochemical Pesticides**: These pesticides include natural materials having non-toxic mechanisms to control pests, e.g., insect sex pheromones and aromatic plant extracts.

10.2.8 Based on the Range of Targets They Kill

In this classification, the pesticides are categorized as broad-spectrum pesticides and selective pesticides. Broad-spectrum pesticides are designed to kill a variety of pests and other non-target organisms. They are non-selective and are often fatal to reptiles, fish, pets, and birds, e.g. chlorpyrifos and chlordane. Selective pesticides, in contrast, kill only a specific or group of pests, leaving other organisms unaffected or only slight affected, e.g., 2,4-D.

10.2.9 Based on Type of Pesticide Formulation

Pesticide preparations consist of active ingredients (AI) and inert ingredients. Active ingredients control target pests, whereas the inert ingredients are the materials added to AIs to make the pesticide safer, more effective, and easier to mix and apply. The inert ingredient may be water, petroleum solvent, wetting agents, spreaders, stickers, etc. They are also easier and more suitable for use. One group of pesticides is mixed with another group of pesticides in such a manner that the efficacy of one pesticide is increased and it provides better protection than a single pesticide compound. Pesticide formulations can be solids, liquids, or gases. Some preparations are ready to use while others need further dilution with water, a petroleum-based solvent, or air before application. The most commonly used formulations are:

Baits: These have active ingredients mixed with a pest food or attractant. The baits are used to control domestic pests such as ants, mice, rats, cockroaches, and flies. They can also be used to control outdoors pests like birds, ants, and snails, and agricultural pests such as crickets and grasshoppers.

Dust: These formulations are intensely crushed combinations of an active ingredient and a carrier material. They are planned for direct application without mixing. They are applied to burrows or bait stations to control fleas.

Granules: In these, the active ingredient is mixed with different inert clays to form particles of different sizes ranging from 20 to 80 mesh. They are prepared for direct application and require specific dispersal equipment. They can be applied from the air or on ground, and are particularly useful in treating mosquito larvae.

Liquids: These are comprised of concentrated oil solutions of technical-grade pesticides mixed with an emulsifier that allows the suspension of small oil droplets in water to form an emulsion.

Powders: These consist of active pesticide ingredients mixed with other ingredients to help in mixing and dispersion. They are of two types: wettable and soluble powder. Wettable powders can mix with a liquid, usually water, for use through spray equipment. In soluble powders, the active ingredient and all other ingredients are soluble in water.

Ultra-low-volume liquid: These are designed to be used as they are or in dilute form with small quantities of specific solvents. These are used in farming, forestry, ornamental, and mosquito control programmes.

10.2.10 Based on Toxicity

Depending on the health risk linked with the pesticides and their toxic behaviour, the World Health Organization (WHO) has classified pesticides into four categories from lowest to highest toxicity as:

- Extremely toxic
- Highly toxic
- Moderately toxic
- Slightly toxic

10.3 TOXICITY OF PESTICIDES

The toxicity of a pesticide is its capability to bring about injury to a living system, whether a human body, or parts of the body, a pond, a forest, or the organisms living in these. The toxicity of a pesticide depends on.

Dose: This is the amount of a pesticide that a surface, plant, or animal is exposed to.

Time: This is the duration for which the exposure occurs. Depending on the duration of exposure, the toxicity may be acute or chronic. Acute toxicity denotes how toxic a pesticide is for a human, animal, or plant after a single short-term exposure. A pesticide with a high acute toxicity is fatal even when a very small amount is absorbed. Chronic toxicity is a prolonged toxic effect after exposure to a pesticide.

Routes of Entry: There are three ways a pesticide enters a body, as described below.

Dermal Route: Wet, dry, or gaseous pesticides can be absorbed through the skin. Oil or paste forms allow more absorption than water-based pesticides. The eyes, eardrums, scalp, and groin areas absorb pesticides more rapidly than other body areas. Once pesticides are absorbed through the skin, they enter the bloodstream and are carried throughout the body.

Inhalation Route: Dust, spray mist, or fumes can enter the lungs during inhalation. This can occur during mixing, fumigating, or spraying. The larger particles stay on the surface of the throat and nasal passages but smaller particles enter the lungs directly and later the bloodstream.

Oral Route: Pesticides can enter through the oral route. This occurs when the hands are not washed properly before eating, or consuming accidently or intentionally. Ingested pesticides can be absorbed anywhere along the digestive tract, but primarily in the small intestine. Once absorbed, they circulate all through the body in the bloodstream. The toxicity of a pesticide depends on:

- **Physical wellbeing**: People with poor health are more vulnerable
- **Age**: The youngest and oldest tend to be most sensitive
- **Gender**: Females are more susceptible
- **Body size**: The effect of a dose is linked with the body weight

10.3.1 Effects of Pesticide Toxicity

Local or Systemic: Local effects occur at the point of contact with a material, e.g., skin inflammation, irritation of the mucous membrane of the lungs. Systemic effects occur away from the point of contact. Both effects can occur with some pesticides.

Immediate or Delayed: Immediate effects are those which are felt shortly after exposure, e.g. sneezing in response to the inhalation of pesticides. Delayed effects occur after some time, e.g. tumours in chronically exposed people after 20–30 years of exposure. Both effects can occur with some materials.

Reversible or Irreversible: Reversible effects are not long-lasting and can be transformed or cured, e.g. skin rash, nausea, eye irritation, dizziness, etc. Irreversible effects are permanent and cannot be changed once they have appeared, e.g. injury to the nervous system, birth defects, mutations, cancer, etc.

Additive, Antagonistic, or Synergistic: An additive effect is one in which the joint effect of two pesticides is equal to the sum of the effects of each. An antagonistic effect occurs when the toxic effect of a mixture of pesticides is less than what would be expected from the individual toxicities. A synergistic effect occurs when the combined toxic effects of two pesticides are much greater, or worse, than the sum of the effects of each by itself. Exposure to pesticides may produce the following effects:

- **Carcinogenic**: Produces cancer in living tissues
- **Immunosuppression**: Delaying of the natural response of the immune system in shielding the body
- **Mutagenic**: Permanent effect on genetic material which can be inherited
- **Neurotoxicity**: Poisoning of the nervous system, including the brain
- **Reproductive**: Effect on the reproductive system
- **Teratogenic**: Effect on unborn offspring causing birth defects

10.3.2 Measurement of Toxicity

Animals such as rats, rabbits, mice, guinea pigs, etc. are used for testing pesticide toxicity. Due to some differences between the ways in which human and animal bodies work, toxicity studies suggest estimation and comparison of the toxic effects of pesticides. In acute toxicity measures, experimental doses are given orally, as well as through the eyes, skin, and in the air that the test animals inhale. The animals are then studied for changes.

Lethal Dose Fifty (LD_{50}): The amount of a pesticide that can kill half of the animals in a test is the LD_{50}. The lower the LD_{50} value of a pesticide the more poisonous it is. It is measured in units, such as milligrams per kilogram (mg/kg).

Lethal Concentration Fifty (LC_{50}): This is a measure of the acute inhalation toxicity of a pesticide. The concentration of a pesticide that causes half of the animals to die is called the LC_{50}. It is measured in milligrams per litre (mg/L) or ppm and sometimes in milligrams per cubic metre (mg/m^3). No standard measure exists for studying the chronic toxicity. Often, the duration of the experiment is in days, months, or years, and the amount of each dose is specified.

10.3.3 Mechanisms of Action

Analysis of the mechanism of toxic action is useful to recognize how the human system functions and to understand the similarities and differences between humans and the pests. It helps in preventing the development of pesticide resistance. Insecticides usually target:

1. Nervous system
2. Growth and development
3. Energy production of insects

Mechanism of Action on the Nervous System: The nervous system has two components, the peripheral nervous system (PNS) to accept and transfer external and outgoing signals to the muscles and other organs, efficiently guiding them on how to respond, and the central nervous system (CNS) to interpret the signals and coordinate the body's responses and movements. A neuron connects with other neurons and muscle fibres through a synapse. Incoming stimuli are transformed by the neuron to an electrical charge (called ions) which moves down the length of the neuron through:

(a) Sodium channels
(b) Potassium channels
(c) Calcium channels
(d) Chloride channels

After reaching the end of the neuron, an electrical charge stimulates the neurotransmitter, a chemical messenger which crosses the synapse and binds to a receptor on the receiving end of the subsequent neuron.

Binding to the receptor converts the signal back to an electrical charge in the second neuron, the signal is conducted along the length of the neuron and the neurotransmitter is absorbed back into its starting neuron, and the nerve cell is in a resting stage until a subsequent signal is received. This process recurs until the signal reaches the CNS to be understood. Impulses from the CNS to the peripheral nervous system remain until the signal extends to the relevant muscles or organs. Both humans and insects have different neurotransmitters that work at different sites all through the nervous system. Among them, acetylcholine (ACh) and gamma-aminobutyric acid (GABA) are important targets of some insecticides. ACh can either excite or inhibit its target neurons – depending on the specific neuron and receptors at the site. In contrast, GABA is an inhibitory neurotransmitter – when GABA is the neurotransmitter stimulated at a synapse, the nerve impulse ends.

Cholinesterase Inhibition: Organophosphate and carbamate insecticides attach to the enzyme that is usually responsible for breaking down ACh after it has passed its communication across the synapse. The cholinesterase is not accessible to help break down the ACh, and the neurotransmitter continues to cause the neuron to "fire" or send its electrical charge. This results in overstimulation of the nervous system and the insect dies. Cholinesterase inhibition by carbamates is slightly reversible. Organophosphate poisoning is irreversible and the insecticide does not release the attached cholinesterase.

Acetylcholine Receptor Stimulation: Neonicotinoid insecticides mimic the action of a neurotransmitter, acetylcholine (ACh). Cholinesterase remains unaffected by these insecticides, but the nerve is constantly accelerated by the neonicotinoid results in overstimulation of the nervous system which leads to poisoning and death.

Chloride Channel Regulation: Avermectins attach to the chloride channel and stimulate it causing an inhibitory effect, which in excess causes the death of insects. The cyclodiene group of organochlorine insecticides disturbs the chloride channel by stopping the GABA receptor and the GABA receptor does not close the chloride channel, which over-stimulates the nervous system. The sodium channel modulators pyrethrins and pyrethroids act on tiny channels through which sodium is pumped which causes the excitation of neurons. They prevent the sodium channels from closing, resulting in constant transmission of nerve impulses, tremors, and finally death.

Mechanism of Action on Growth and Development: Insects regularly shed their skin during their growth and development to enter the next phase of life. Insecticides affect an insect's growth and development either through interfering with hormones or blocking the production of a structural component of the exoskeleton.

Chitin Synthesis Inhibitors (CSIs): These inhibitors block the production of chitin. An insect poisoned with a CSI can neither moult and attain adult stage nor reproduce but dies. However, the CSIs are not regarded as toxic to humans.

Insect Growth Regulators (IGRs): IGRs attack an insect's endocrine system, which produces the hormones (juvenile hormone) required for growth and development. Insects poisoned with IGRs cannot moult or reproduce, and ultimately die. Humans neither produce nor utilize these hormones, which an insect uses in moulting. Hence, IGRs are less toxic to humans.

Energy Production: Food consumed by organisms helps in producing energy, which is stored in the form of ATP and is used for thinking, moving, growth, and synthesis of chemicals and structures required by the body. Certain insecticides prevent or interrupt energy production, and as a result the insect stops eating and moving, and dies. Some insecticides obstruct feeding. On the basis of the mode of action, they can also be categorized as contact and stomach insecticides, systemic or translocated, and with or without fumigant action. The contact pesticide acts after coming into contact with an insect which chews the plant material, e.g., the endosulfan, malathion, etc. A systemic pesticide moves within a plant from the point of application to the point where the insect will contact or ingest it.

10.3.4 Impact of Pesticides on Human Health

Pesticides mixed with food, air, or water enter the human body through the oral, dermal, or inhalation routes. The spraying of pesticides releases them into soil and groundwater and from there they are carried away by air or water from one place to another. The toxicity of chemicals, and duration and extent of exposure govern the degree of harmful impact on human health. The toxicity of chemicals is governed by the nature of the toxicant, route of exposure (oral, dermal and inhalation), dose, and organism, and its toxicity can be either acute or chronic. The toxicity of a pesticide is usually expressed in terms of lethal dose 50 per cent (LD_{50}) or lethal concentration 50 per cent (LC_{50}). LD_{50} is the single exposure dose of the pesticide per unit weight of the organism required to kill 50 per cent of the test population, where the population is genetically homogeneous. It is expressed in milligrams per kilogram body weight. LC_{50} is the concentration of pesticide in an external medium (usually air or water surrounding experimental animals), which causes 50 per cent mortality of the test population, where the population is genetically homogeneous. It is expressed in parts per million (ppm).

An acute effect occurs from a single exposure by any route, such as dermal (skin), inhalation (lungs), oral (mouth), and the eyes. Acute toxicity is measured by analysing the dermal toxicity, inhalation toxicity, and oral toxicity of test animals. Furthermore, eye and skin irritation are also observed. Acute illness usually appears a short time after contact or exposure to the pesticide. Pesticide drift from agricultural fields, exposure to pesticides during the application, and deliberate or unintended poisoning usually lead to acute illness in humans. Symptoms of pesticide poisoning include:

- Body aches
- Cramps
- Dizziness
- Headaches
- Impaired vision
- Nausea
- Panic attacks
- Poor concentration
- Skin rashes

In severe cases, coma and death can occur due to pesticide poisoning. Every year, approximately 3 million cases of acute pesticide poisoning are reported globally. Chronic effects occur from small doses recurring over a period of time. Suspected chronic effects from exposure to certain pesticides include:

- Birth defects
- Blood-related disorders
- Effects on reproduction
- Endocrine disruption
- Genetic disorders

- Nerve disorders
- Poisoning in a foetus
- Production of benign or malignant tumours

It is more difficult to determine the chronic pesticide toxicity than acute toxicity through laboratory analysis. Sustained and frequent exposure to sub-lethal quantities of pesticides for a longer time causes chronic illness in humans.

The general symptoms of pesticide poisoning are immediately noticed but appear at a later stage. Symptoms of pesticide poisoning are discussed in Table 10.3. Usually, farmers are at a higher risk though there is a substantial risk of the general population being affected by consuming contaminated food and water or pesticides in the air drifting from fields. Numerous studies have established a relationship between exposure to pesticides and the occurrence of human chronic illnesses affecting the nervous, reproductive, renal, cardiovascular, and respiratory systems. The effects of pesticide chemicals on the human body are listed in Table 10.4.

Some common chronic ailments due to long-term exposure to pesticides include:

- Alzheimer's disease
- Birth defects
- Brain cancer
- Cancer
- Cardio-vascular diseases
- Diabetes
- Hormonal imbalances
- Infertility
- Lymphocytic leukaemia
- Neuro-degenerative diseases

TABLE 10.3
Symptoms of Pesticide Poisoning

Mild	Moderate	Serious
Irritation of nose, throat, eyes, or skin	Abdominal cramps	Chemical burns on skin
Diarrhoea	Blurring of vision	Death
Dizziness	Coughing	Extra phlegm or mucus in airways
Headache	Excessive perspiration	Inability to breathe
Insomnia	Unnecessary salivation	Increased rate of breathing
Loss of appetite	Feeling of constriction in throat and chest	Loss of reflexes
Nausea	Mental confusion	Small or pinpoint pupils
Nervousness	Muscular incoordination	Unconsciousness
Restlessness	Intense weakness	
Sweating	Fast pulse	
Thirst	Trembling	
Weakness or fatigue	Vomiting	

TABLE 10.4
Effects of Pesticide Chemicals on the Human Body

Chemical Family	Type of Pesticide	Effect on Human Body	Symptoms of Internal Exposure	Symptoms of External Exposure	Symptoms of Chronic Exposure
Acetamides	Herbicides	Irritants	–	Moderately irritating to skin and eyes	–
Acetanilides	Herbicides	Irritants	–	Mild irritants	–
Alumino Fluoride Salt	Insecticide	Irritant	–	Slight irritant of eyes, nose, and skin	–
Arsenical Pesticides	Rodenticides, insecticides, acaricides, marine anti- fouling com- pounds, desiccants, herbicides, fungicides	Toxic to liver,kidney, brain, bone marrow, and nervous system	Headache, burning stomach pain, vomiting, diarrhoea, dizziness. Garlic odour on breath and faeces	Swelling of mouth and throat, irritating to nose, throat, and eyes	Accumulates in body. Chronic headaches, dizziness, stomach-aches, salivation, low fever, garlic breath
Bensoic and Benzilic Derivatives	Insecticides, herbicides	Irritants	–	Irritating to skin and respiratory tract	–
Benzonitriles	Fungicides, herbicides	Irritants	Moderately irritating to lungs	Moderately irritating to skin	–
Carbamates	Insecticides, acaricides	Reversible changes in acetylcholinesterase enzyme of tissues	Headache, dizziness, weakness, shaking, nausea, stomach cramps, diarrhoea, sweating	Minimal rashes but readily absorbed through the skin	Loss of appetite, weakness, weight loss, and general feeling of sickness
Carbanilate	Herbicide	Irritant	–	Irritant and skin sensitizer	–
Carbon disulphide	Fumigant	Injury to nervous system	Headache, dizziness, nausea, and disorientation	Irritates eyes, nose, and throat. Blisters and redness on skin	Pain, tingling and weakness of arms and legs; loss of mental functions

(Continued)

TABLE 10.4 (Continued)
Effects of Pesticide Chemicals on the Human Body

Chemical Family	Type of Pesticide	Effect on Human Body	Symptoms of Internal Exposure	Symptoms of External Exposure	Symptoms of Chronic Exposure
Chloro- phenoxy Pesticides	Herbicides	Irritant to lung, stomach, and intestinal linings.			
Injure liver, kidney, and nervous system	Prompt vomiting, burning sensation in stomach, diarrhoea, muscle twitching	Moderately irritating to eyes, skin, and lungs	Do not remain in body; passed out within hours or days		
Copper Salts and Organic Complexes	Fungicides	Injure intestinal lining, brain, liver, kidneys, and blood	Prompt vomiting; burning pain in chest; diarrhoea, headache, sweating	Irritate skin and eyes; damage mucous membranes	–
Coumarins, Indandiones, and Other Anti-coagulants	–	Prevent blood from clotting	Reaction if low accidental dose ingested	–	–
Cycloheximide	Fungicide	Irritates stomach and intestine; injury to kidneys, brain lining, and nervous system	Excitement, tremors, salivation, diarrhoea	Minimal	
Dicarboximides	Fungicides	Irritants	–	Skin irritant	–
Dinitroaniline Compounds	Herbicides	Irritants	–	Slightly to moderately irritating to skin, eyes, nose, and throat	–
Dithio-carbamates	Herbicides, fungicides	Do not inhibit cholinesterase; mild irritants	–	Mild irritant to skin, eyes, nose, and throat	–

Endothall	Herbicides, algicides	Damages heart, blood vessels, nervous system, and intestinal lining	Convulsions, shock, lack of coordination	Irritating to eyes, skin, and mucous membranes	–
Halocarbons	Fumigants	Injures heart muscles; also injures lungs, brain, liver and kidneys	Shock, drowsiness, shaking weakness	Irritates eyes, nose, throat; blisters and redness on the skin	Liver damage, weight loss, and jaundice
Hydrogen Cyanide, Acrylonitrile, and Sodium Cyanide	Fumigants, rodenticides	Injury to brain and heart tissues	Headache, nausea, constriction of throat, dizziness, nervousness. Sudden coma	Irritant	–
Metal Phosphides	Fumigants	Injure lungs, liver, kidneys, heart, and nervous system	Nausea and vomiting followed by weakness, shaking, and dizziness	Irritants	–
Methyl Bromide, Ethylene Oxide, and Propylene Oxide	Fumigants	Serious injury to lungs; injure nervous system	Coughing of frothy fluid; severe shortness of breath, drowsiness, shaking, weakness	Irritate eyes, nose, and throat	Lack of coordination
Nicotine Sulphate	Insecticide	Injures nervous system	Nausea, headache, diarrhoea, dizziness, shaking, abdominal pain, lack of coordination, sweating, salivation	Minimal but readily absorbed through the skin	–
Organo- phosphates	Insecticides, acaricides	Inhibit acetyl-cholinesterase (an enzyme) in the tissues	Headache, dizziness, weakness, shaking, nausea, stomach cramps, diarrhoea, sweating	Minimal rashes but readily absorbed through the skin	Loss of appetite, weakness, weight loss, and general feeling of sickness

(Continued)

TABLE 10.4 (Continued)
Effects of Pesticide Chemicals on the Human Body

Chemical Family	Type of Pesticide	Effect on Human Body	Symptoms of Internal Exposure	Symptoms of External Exposure	Symptoms of Chronic Exposure
Organochlorines (Chlorinated Hydrocarbons)	Insecticides, acaricides (HCB is a fungicide)	Disrupt function of nervous system, mainly the brain	Headache, dizziness, weakness, shaking, nausea, excitability, disorientation	Minimal rashes but readily absorbed through the skin	Some build up in the fat tissues. May cause nervousness, weakness, and shaking
Oxadiazola	Herbicide	Irritant	–	Irritating to skin and eyes	–
Paraquat and Diquat	Herbicides	Injure skin, nails, cornea, liver, kidney, linings of stomach and intestine, and respiratory system	Burning pain, nausea, vomiting, and diarrhoea	Irritate and injure skin and nails	–
Pentachlorophenol	Herbicides, defoliants, molluscicides, germicides, fungicides, and wood preservatives	Toxic to liver, kidneys, and nervous system	Headache, weakness, nausea, excessive sweating	Highly irritating to skin, eyes, nose, and throat	Weight loss, weakness, anaemia
Phenylmercuric Salts	Fungicides	Injure nervous system and kidneys	Delirium, muscle weakness, lack of coordination	Minimal	Weakness and lack of coordination in arms and legs; difficulty in talking and swallowing
Phosphine	Fumigants	Injures lungs, liver, kidneys, and nervous system	Weakness, shaking, vomiting, cough, difficulty in breathing, intense thirst	Irritates eyes, nose, and throat	Pain in eyes and nose; nosebleeds; abdominal pain

Phosphonomethyl Glycine	Herbicide	Irritant	Irritates lungs	Irritates eyes	–
Picolinic Acid	Herbicide	Irritant	Irritates lungs	Irritating to skin, eyes, nose, and throat	–
Pyrethrins and Pyrethroids	–	Very low human toxicity	Slight toxic reaction	–	–
Pyridazinone	Growth retardant	Irritant	–	Slightly irritating	–
Sodium Chlorate	Herbicide, defoliant	Injury to intestines, nervous system, and kidneys	Swelling of mouth and throat; pain in oesophagus, stomach, and intestinal lining	Irritant	–
Sodium Fluoroacetate	Rodenticide	Extremely toxic. Affects heart tissue and brain	Stomach pain, vomiting, hallucinations, nervousness	Minimal	–
Strychnine and Crimidine	Rodenticides	Act directly on cells in the brain and spinal cord to cause convulsions	Blue skin colour; violent convulsions	Minimal	–
Sulphur Dioxide, Formaldehyde, Chloropicrin, and Acrolein	Fumigants	Strong irritants of lungs and throat	Headache, dizziness, nausea, wheezing, cough	Severe irritation of eyes, nose, and throat. Blisters on skin	–
Sulphuryl Fluoride	Fumigant	Injures lungs and kidneys	Muscle twitching, convulsions	Irritant	Injury to kidneys and lungs
Thiadiazin(ol)	Herbicide, fungicide	Irritant	Vomiting, diarrhoea, shaking, and weakness caused by bentazon herbicides	Moderately irritating to skin, eyes, nose, and throat	–

(Continued)

TABLE 10.4 (Continued)
Effects of Pesticide Chemicals on the Human Body

Chemical Family	Type of Pesticide	Effect on Human Body	Symptoms of Internal Exposure	Symptoms of External Exposure	Symptoms of Chronic Exposure
Thiocarbamates and Dithiocarbamates	Fungicides	Low human toxicity	Nausea, vomiting, diarrhoea, weakness, and nasal stuffiness	Irritating to skin, eyes, nose, and throat	–
Triazines	Herbicides	Irritant	–	Mildly irritating to skin, eyes, nose, and throat	–
Uracils	Herbicides	Irritant	Irritating to lungs	Irritating to skin, eyes, nose, and throat	–
Urea	Herbicide, insecticide	Irritant	–	Moderately irritating to skin, eyes, nose, and throat	–
Yellow phosphorus	Rodenticide	Highly toxic; blood system injury; injury to liver, nervous system, heart, and kidneys	Breath has garlic odour; faeces may glow and smoke from phosphorus fumes; vomiting and diarrhoea; burning pain in throat, stomach, and intestines	Dermal exposure usually low; irritates nose and throat	–
Zinc Phosphide	Rodenticide	Highly toxic; severe intestinal irritation; severe injury to liver, kidneys, nervous system, and heart	Intense nausea, stomach pain, excitement, chills, cough	Minimal	–

Note: Table modified based on a similar table in "Applying Pesticides Correctly", USDA, USEPA.

- Parkinson's disease
- Prostate cancer
- Renal cell carcinoma
- Reproductive diseases
- Respiratory diseases (asthma, chronic obstructive pulmonary disease)

10.4 IMPACTS OF PESTICIDES ON THE ENVIRONMENT

The extensive use and disposal of pesticides by farmers, institutions, and others are the probable sources of pesticides in the environment. It is virtually impossible to fully control the area of a pesticide's effect. Even when applied in a limited area, it disperses in the air, is absorbed in the soil, or gets dissolved in water and ultimately reaches a much larger area. Once released into the environment, pesticides may have different fates. When sprayed on crops, they may find their way through the air and finally end up in other parts of the environment, e.g., soil or water. Directly applied pesticides may be washed off and spreads to adjacent water bodies through surface runoff or may percolate through the soil to lower layers of soil and groundwater. The effects of pesticides on the environmental system include deviances in normal functioning of the ecosystem and loss of biodiversity. At times, the use of pesticides may cause long-term residual effects including fatal results, For example organochlorine pesticides result in the contamination of groundwater, surface water, food products, air, and soil.

10.4.1 Impacts on Non-Target Organisms

Most insecticides applied to kill pests may also adversely impact non-target organisms, e.g., earthworms, natural predators, and pollinators. The use of pesticides can decrease the population of earthworms. For example, carbamate insecticides are toxic to earthworms and organophosphates reduce earthworm populations. Natural predators such as parasitoids and other predators essential for controlling pests are most vulnerable to insecticides and are severely affected. The destruction of natural predators can aggravate pest problems. If the natural enemies are lacking, extra insecticides are required to control the target pest. Moreover, pesticides also affect the behaviour of predators and their lifecycle, including growth, development, and reproduction. Pollinators, e.g., bees, fruit flies, beetles, and birds used as bio-indicators of ecosystem processes are affected by ecological stress caused by the application of pesticides and habitat modifications. The use of pesticides causes direct loss of insect pollinators and indirect loss of crops owing to the lack of an adequate population of pollinators.

Biodiversity is often regarded as a measure of the health of biological systems. The greater the number of organisms, the healthier is the environment. An environment contains diverse life forms, all of which are dependent and these may range from microbes to insects, e.g., ants, beetles, and wasps to birds to large animals and predators. Such a system has the capability to maintain its equilibrium so that no one species becomes dominant. At times, pests may also be useful to biological systems as they consume and control other pests. Thus, eradicating even a single species by use of a pesticide can cause important changes and may result in others changes. Sometimes, a pesticide may remove a species essential to the functioning of the entire community, or it may encourage the dominance of undesired species, or it may just reduce the number and variety of species in a community. This may disturb the dynamics of the food webs by breaking the prevailing dietary connections between species.

Non-target pesticides from agriculture and other sources may gather in the soil, which becomes aggravated by indiscriminate and repeated use of pesticides. Soil properties and soil microflora are also affected by pesticides which may undergo degradation, transport, and adsorption/desorption processes. Degraded pesticides interact with the soil and with native microbes, thus modifying its microbial diversity, biochemical reactions, and enzymatic activity. Any change in the microbial diversity and soil biomass finally disturbs the soil ecosystem and affects soil fertility. Pesticide application may also inhibit or kill certain microorganisms and outnumber others. They may adversely affect the biochemical reactions of soil, including nitrogen fixation, nitrification, and ammonification by activating/deactivating soil microbes and/or enzymes. Pesticides impact the mineralization of soil organic matter, which is an important soil property that regulates soil quality and productivity.

Pesticide residues in water pose a serious threat to biological communities, including humans. There are diverse methods by which pesticides can get into water such as accidental leakage, industrial sewage, surface run-off, transport from pesticide-treated soils, and washing of spray equipment flows into water bodies. Pesticides usually travel from fields to various water bodies by runoff or in drainage induced by rain or irrigation. Likewise, the presence of pesticides in the air can be caused by numerous aspects including spray drift, volatilization from treated surfaces, and the aerial application of pesticides. The extent of release depends on the size of droplets and wind speed. The rate of volatilization depends on the time after pesticide treatment, the surface on which the pesticide settles, ambient temperature, humidity, wind speed, and the vapour pressure of the ingredients.

10.5 DIAGNOSING AND TREATING PESTICIDE POISONING

Since pesticide-related complications have signs and symptoms similar to general medical conditions, a systematic and complete environmental and occupational history is required for properly diagnosing pesticide poisoning. Some extra screening questions about the patient's working atmosphere and home environment can indicate whether there was a pesticide poisoning, e.g., if one uses carbamate and organophosphate pesticides, it is important to seek a cholinesterase test.

Cholinesterase is a vital enzyme of the nervous system, and carbamate and organophosphate pesticides injure or kill humans by inhibiting the action of cholinesterase. The particular treatment for acute pesticide poisoning depends on the pesticide or group of pesticides accountable for the poisoning. However, there are basic management methods suitable for most acute poisonings, including skin decontamination, airway protection, gastrointestinal decontamination, and seizure treatment. Decontamination of the skin is done while other life-saving measures are also being enacted. After removing the clothes, the patient/affected area is wiped with soap and water to remove chemicals from the skin and hair. The patient is incubated and oxygen is administered, if required. In more severe cases, mechanical pulmonary ventilation may be required.

Seizures are typically managed with lorazepam, phenytoin, and diazepam. Gastric lavage is not suggested in managing pesticide poisoning. An orogastric tube is inserted and the stomach is flushed with saline to remove the poison. If the patient is neurologically affected, a cuffed endotracheal tube is inserted beforehand for airway protection. However, lavage is contraindicated in the case of hydrocarbon ingestion. Activated charcoal is sometimes given as it has been shown to be effective for some pesticides. Accidental poisonings can be avoided by:

- Proper labelling of storage containers
- Reducing exposure by protecting body parts, such as the scrotal region, underarms, face, scalp, and hands
- Following safety protocols to reduce exposure
- Use of personal protective equipment (respirator, chemical-resistant gloves, goggles, and protective clothing)
- Washing hands after work
- Changing clothes between work shifts
- Having first aid training and protocols at the workplace

10.6 MANAGING PESTICIDES

Pesticide management includes the control of production, distribution, sale, transportation, quality, and use of pesticides to:

(i) Allow its use only after evaluating its efficiency and safety
(ii) Control pests
(iii) Create awareness among consumers about safe and sensible use of pesticides
(iv) Ensure the availability of quality pesticides
(v) Minimize pollution of farming commodities by pesticide residues
(vi) Take essential measures to continue, limit, or prohibit the use of pesticides on the basis of reconsideration with a view to avoid its risk to humans, flora, fauna, or the environment

Pesticide management is carried out in the overall framework of the plant production and protection division of the FAO. It is designed to work together with member countries and other international organizations as a partner to introduce sustainable and environmentally sound agricultural practices that reduce health and environmental risks associated with the use of pesticides.

The International Code of Conduct on the distribution and use of pesticides is a global guidance document on pesticide management for all public and private entities engaged in or linked with the distribution and use of pesticides. This document focuses on:

(a) Risk reduction
(b) Protection of human and environmental health
(c) Support for sustainable agricultural development by using pesticides in an effective manner and applying Integrated Pest Management (IPM) strategies

The governments of the respective nations are accountable for controlling the accessibility, delivery, and use of pesticides in their countries and must confirm the allocation of adequate resources for this mandate. The pesticide industry must follow the conditions of this code as a standard for the manufacturing, distribution, and promotion of pesticides, especially in those nations lacking suitable legislation and advisory services. The pesticide industry and dealers must observe the following practices in pesticide management:

- Adhere closely to the provisions of the FAO guidelines
- Be capable of offering effective technical support on the disposal of pesticides and used pesticide containers; maintain an active interest in labelling, directions for use, packaging, formulation, or product availability. Preference must be given to pesticides that require low-cost personal protective and application equipment and techniques suitable to the conditions under which the pesticides are to be used
- Pay special attention to the choice of pesticide formulations to reduce the risk to users and reduce harmful effects on the environment
- Provide, with each package of pesticide, information and instructions sufficient to enable effective use and moderate hazards during management
- Supply only pesticides of suitable quality, packaged and labelled as applicable for each specific market

National and international organizations, governments, and the pesticide industry must take organized action to publicize educational materials of all types to pesticide users, farmers, farmer organizations, agricultural workers, unions, and other interested parties. Similarly, users must look for and comprehend educational materials before applying pesticides and must follow proper techniques. Combined efforts must be made by governments to evolve and support the use of IPM. Likewise, lending institutions, donor agencies, and governments must support the development of national IPM policies and amended IPM concepts and practices. Governments, with the support of relevant international and regional organizations, should:

- Encourage and promote research on the development of alternatives posing fewer risks
- Encourage the use of biological control agents and techniques
- Encourage and promote non-chemical pesticides
- Focus on desirable, target-specific pesticides that degrade into harmless constituent parts or metabolites after use and have low risk to humans and the environment

It must be ensured that each pesticide/pesticide product is effectively tested by standard procedures and test methods to completely assess its efficacy, behaviour, fate, hazard, and risk regarding various anticipated conditions. For reducing health and environmental risks, governments should:

- Implement a pesticide registration and control system
- Periodically review the pesticides marketed in the respective countries
- Undertake health surveillance programmes for those who are occupationally exposed to pesticides, and examine and document poisoning cases
- Provide guidance and instructions to health workers, physicians, and hospital staff on pesticide poisoning
- Establish national or regional poisoning information and control centres to provide immediate guidance on first aid and treatments
- Provide extension and advisory services about practical IPM strategies and methods
- Implement programmes to monitor pesticide residues in food
- Make every rational effort to reduce risks posed by pesticides by:
  - ✓ Making less toxic formulations
  - ✓ Familiarizing products in ready-to-use packages
  - ✓ Developing application methods and equipment that reduce exposure to pesticides
  - ✓ Using returnable and refillable containers
  - ✓ Using containers that are not attractive to or easily opened by children
  - ✓ Protecting biodiversity and decreasing the adverse effects of pesticides on the environment (water, soil, and air) and on non-target organisms

For ensuring the regulatory and technical requirements governments must:

- Conduct risk evaluations and make risk management decisions based on available information
- Detect and control the illegal trade in pesticides
- Introduce necessary legislation for the regulation of pesticides

Governments should take the required controlling actions to enforce disciplinary measures that efficiently discourage faulty practices.

BIBLIOGRAPHY

Chakraborty, S., et al. (2009) Chronic exposures to cholinesterase inhibiting pesticides adversely affect respiratory health of agricultural workers in India. *Journal of Occupational Health* 51: 488–497.

Cocco, P., et al. (2013) Lymphoma risk and occupational exposure to pesticides: Results of the Epilymph study. *Occupational and Environmental Medicine* 70: 91–98.

Dawson, A. H., et al. (2010) Acute human lethal toxicity of agricultural pesticides: A prospective cohort study. *PLoS Medicine* 7: 1–10.

Eddleston, M., et al. (2002) Pesticide poisoning in the developing world – A minimum pesticide list. *The Lancet* 360: 1163–1167.
Elbaz, A., et al. (2009) Professional exposure to pesticides and Parkinson disease. *Annals of Neurology* 66: 494–504.
FAO/WHO. (2000) *Pesticide residues in food-1999 evaluations Part II- toxicological*, Joint FAO/WHO Meeting on Pesticide Residues. Geneva: World Health Organization.
Greenlee, A. R., Arbuckle, T. E., and Chyou, P. H. (2003) Risk factors for female infertility in an agricultural region. *Epidemiology* 14: 429–436.
Heck, J. E., et al. (2010) Occupation and renal cell cancer in Central and Eastern Europe. *Occupational and Environmental Medicine* 67: 47–53.
Hoppin, J. A., et al. (2009) Pesticide use and adult-onset asthma among male farmers in the agricultural health study. *European Respiratory Journal* 34: 1296–1303.
Hussain, S., et al. (2009) Impact of pesticides on soil microbial diversity, enzymes, and biochemical reactions. *Advances in Agronomy* 102: 159–200.
Kevan, P. G. (1999) Pollinators as bio-indicators of the state of the environments: Species, activity and diversity. *Agriculture, Ecosystems and Environment* 74: 373–393.
Lee, W. J., et al. (2005) Agricultural pesticide uses and risk of glioma in Nebraska, United States. *Occupational and Environmental Medicine* 62: 786–792.
Mesnage, R., Clair, E., de Vendômois, J. S., and Seralini, G.E. (2010) Two cases of birth defects overlapping Stratton Parker syndrome after multiple pesticide exposure. *Occupational and Environmental Medicine* 67: 359.
MunozLeoz, B., RuizRomera, E., Antiguedad, and I., Garbisu, C. (2011) Tebuconazole application decreases soil microbial biomass and activity. *Soil Biology and Biochemistry* 43: 2176–2183.
Petrelli, G. and Mantovani, A. (2002) Environmental risk factors and male fertility and reproduction. *Contraception* 65: 297–300.
Son, H. K., et al. (2010) Strong associations between low dose organochlorine pesticides and type 2 diabetes in Korea. *Environment International* 36: 410–414.
WHO. (2009) *The WHO Recommended Classification of Pesticides by Hazard and Guidelines to Classification*. Geneva, Switzerland: World Health Organization.
Winchester, P. D., Huskins, J., and Ying, J. (2009) Agrichemicals in surface water and birth defects in the United States. *Acta Paediatrica* 98: 664–669.
Xavier, R., Rekha, K., and Bairy, K. L. (2004) Health perspective of pesticide exposure and dietary management. *Malaysian Journal of Nutrition* 10: 39–51.
Yadav, I. C., et al. (2015) Current status of persistent organic pesticides residues in air, water, and soil, and their possible effect on neighbouring countries: A comprehensive review of India. *Science of the Total Environment* 511: 123–137.

11 Recent Advances in the Remediation of Heavy Metals

The Truly Healthy environment is not merely safe but stimulating.

William H. Stewart

After going through this chapter, the reader will learn about: the types of physico-chemical remediation methods, adsorption, ultrafiltration membrane technology, sediment remediation technologies, soil washing, electro-dialysis, filtration, flocculation, flotation, electro-kinetic remediation, the ion-exchange method, ball milling, chemical precipitation, hydroxide precipitation, sulphide precipitation, alternative methods of chemical precipitation, and the role of nanotechnology in HM removal.

FACT FILE

- Heavy metal (HM) pollution has emerged as one of the most important environmental problems causing ecological imbalance worldwide
- Increased risk of human exposure to HMs has serious health implications
- Seepage of HMs destroys crops and land
- A number of physicochemical and biological methods have been evolved for the remediation of HMs from the environment
- Remediation of HMs needs proper attention to protect soil quality, maintain the ecosystem, and take account of human health
- Most of the physicochemical methods are less eco-friendly and less cost-effective
- Recently, nanoparticles have emerged as effective substitutes for existing treatment methods, both in resource conservation and environmental remediation of anthropogenic compounds
- Nanotechnology has developed as a universal solution for the remediation of HMs
- Biological synthesis of nanoparticles has emerged as promising for creating novel materials that are eco-friendly, cost-effective, and stable, with wider applications

DOI: 10.1201/9781003343721-11

- *In situ* and *ex situ* remediation techniques have been developed to rectify HM-contaminated sites, including landfilling, soil flushing, soil washing, stabilization, solidification, phytoremediation, and bioremediation

11.1 INTRODUCTION

Toxic substances in the environment produce adverse effects at all levels of biological organization, from the molecular level to the community level, and eventually to the ecosystem level. It is not surprising that eco-toxicologists facing this complexity are struggling to address the consequences of toxic substances released at the ecosystem level, which include the recent research on remediation of HMs from the environment. On account of their tendency to form complexes, high reactivity, and increased biochemical activity, HMs remain persistent in the environment. HMs are transferred through aqueous media and accumulate in land and water bodies, which makes them unsafe to all life forms and the environment. Hence, to prevent their harmful consequences, it is important to eliminate toxic metals from the environment. Conventional techniques such as membrane filtration, ion exchange, chemical precipitation, etc. have been in use for removing HMs from the environment, however these techniques have some drawbacks.

Owing to their toxicity, bio-accumulative nature, and persistence in the environment, HMs are well-known environmental pollutants. The main problem linked with HMs is the difficulty in biodegradation and bioaccumulation in tissues. HMs are capable of biomagnification. Remediation is defined as the removal of pollutants or contaminants from the environment to safeguard human health and reinstate the environment. Depending on the physicochemical properties of metals, various physical, chemical, and biological methods have been evolved for removing HMs from soil and water. Some of the physico-chemical methods employed for the removal of HMs are discussed in this chapter. Additionally, biological systems are employed for the removal of HMs, which include bioremediation techniques, namely *in situ* and *ex situ* bioremediation.

11.2 TYPES OF PHYSICO-CHEMICAL REMEDIATION METHODS

11.2.1 Adsorption

A process wherein a substance (adsorbate or sorbate) gets accumulated on a solid surface (adsorbent or sorbent) is defined as adsorption. The adsorbate can be in a gas or liquid phase. The propelling force for adsorption is unsaturated force at a solid surface which can form bonds with adsorbate. It is a surface phenomenon and the attractive force on the surface is restricted to a one molecule deep distance. The extent of adsorption depends on the natures of both the adsorbing agent and the substance adsorbed. The greater the surface area of the adsorbing agent, the greater will be the adsorption. Adsorption finds application in industry. Principles regulating adsorption include:

- Adsorption is a reversible process, which decreases with increasing temperature
- The process occurs rapidly and equilibrium is achieved within one hour
- Adsorption is proportional to the surface area and it varies with the nature of the surface of the adsorbent and substances to be adsorbed
- Adsorption occurs best with dilute solutions
- Narrow pores on the surface of the adsorbing agent are more effective than globular openings
- Heat is released in adsorption
- Molecules adsorbed on the surface are oriented and arranged in a set manner

Due to its low cost, easy availability, and eco-friendly nature, adsorption is widely used for removing HMs from wastewater. Both commercially developed adsorbents and bio-adsorbents with high removal capacity are used for removing HMs from wastewater. The adsorption process is reversible in some cases so that the adsorbent can be reused, adding another advantage to this technique. Adsorption is a surface phenomenon wherein a solution containing the adsorbate gets adsorbed on the surface of an adsorbent. Adsorption is of two types:

- Physio-sorption (where the adsorbate binds to the adsorbent due to van der Waals forces)
- Chemisorption (which occurs due to chemical reactions between the adsorbate and adsorbent)

Physio-sorption is reversible, weak, and generally endothermic, while chemisorption is irreversible, selective, and exothermic.

Types of Adsorbents: On the basis of their origin, adsorbents are generally classified as natural or synthetic;

- Natural adsorbents include activated carbon, clays, minerals, charcoal, ores, and zeolites
- Synthetic adsorbents are prepared from industrial wastes, agricultural wastes, waste sludge, etc.

For HM remediation, adsorption is believed to be a more effective and economical technique as compared to other wastewater treatment technologies. In most cases, the adsorbent can be restored back and can be used further. Moreover, it is easy to use and does not produce any toxic pollutants. The criteria used for selection of adsorbents include:

- Cost-effectiveness
- High surface area
- Porosity
- Distribution of functional groups
- Polarity

Carbon and its by-products are the most widely used adsorbents owing to their great adsorption efficiency. Their structural characteristics provide them with a large surface area with easy chemical modifications which makes them suitable for a wide spectrum of pollutants.

Activated Carbon Adsorbents*:* Owing to its high efficiency, porosity, and high surface area, activated carbon (AC) is one of the most widely used adsorbents and is commercially produced from the carbonization of coal and wood, hence it is costly and its use is restricted. Activated carbon is mainly produced by pyrolysis of carbonaceous material at temperatures lower than 1000°C.

Biochar: This is the burnt solid material acquired after carbonization of biomass. Biochars are less carbonized than AC; hence, more carbon, hydrogen, and oxygen remain in their structure. Biochar has greater potential for remediation of HMs from wastewater than other conventional and low-cost adsorbents. Due to its mesoporous structure, it possesses a high surface area and the presence of different functional groups and low ash content it is an outstanding adsorbent. Feedstocks such as rice husk, corn husk, tea waste, and digested sludge have been used for the removal of HMs from wastewater.

Zeolites: These are alumina-silicates with a crystalline structure that are found naturally or manufactured. They are among the best adsorbents for HM removal as they consist of hydrated alumina-silicate minerals comprised of interlinked alumina and silica. They have considerable ion-exchange capacities, hydrophilic properties, and high specific surface area, which make them remarkably suitable adsorbents for remediation of HMs. Zeolites can also be modified for better adsorption capacity as compared to unmodified ones. NaX zeolite is one of the most widely used nano-sized zeolites for removal of HMs from wastewater.

Clay Minerals: Bentonite, the clay mineral, possesses the highest cation exchange capacity, is regeneratable, and is approximately 20 times more economical than activated carbon. When compared to zeolites, clay minerals possess less ability for the removal of HMs. On account of their advantages, for instance, brilliant physical, chemical, and surface properties, they are still in use.

Nanostructured Materials: In recent years, carbon nanotubes, fullerenes, and graphene have gained importance in the adsorption of HMs from effluents as they have remarkable mechanical and chemical properties, such as:

- Electrical conductivity
- Exchange capacity
- High surface area
- Numerous intermolecular interactions
- Strength
- Thermal stability

Carbon Nanotubes, Fullerenes, and Graphene: Carbon nanotubes (CNTs) are long carbon cylindrical-shaped structures with an unbroken hexagonal graphite sheet. They have excellent potential for the removal of HMs from wastewater. CNTs are excellent adsorbents due to their mechanical and surface properties, electrical and semiconductor properties, high specific surface area, and the presence of mesopores, which increase their adsorption efficiency. The presence of functional groups containing elements, for example, oxygen, nitrogen, and sulphur, directly and indirectly affect the adsorption mechanisms and improve the adsorption of HMs.

Fullerenes have a closed-cage structure comprised of pentagonal and hexagonal carbon rings. Their adsorption efficiency is due to their surface morphology and the presence of mesopores, which imparts higher ion affinity and specific surface area for remediation of HM ions from water and wastewater.

Graphene is a 2-D hexagonal lattice of carbon atoms, which have structural, chemical, and mechanical properties that help in wastewater treatment. It is a potential adsorbent for safe drinking water with disinfection control and cost-effective water detoxification. It has a high surface area, active functional groups, and sites on its surface which increase its adsorption capacity. Graphene can be activated by oxidation to increase the functional groups, which improves the adsorption capacity for removing HMs. It has high dispersity in water, greater surface area, and more functional groups on the surface as compared to other nanomaterials. It has been developed as an advanced nano-electrocatalyst for the construction of electrochemical sensors. Cr^{6+}, As^{5+}, and Pb^{2+} ions are successfully removed by adsorbent combinations with graphene. Graphene oxide nanotubes coated with *in situ* bismuth film absorb HMs such as lead, cadmium, silver, copper, and zinc ions with high adsorption capacity.

Low-Cost Adsorbents: Although activated carbons, carbon nanotubes, fullerenes and nano-composites are the most extensively used adsorbents, their use is limited due to their high cost and low regeneration ability. Therefore, non-conventional materials that are easily available and cost-effective are preferred such as agricultural and industrial waste.

Agriculture Waste: Agricultural wastes comprise of lignin, cellulose, hydrocarbons, sugars, water, and starch, along with other functional groups that enhance the adsorption capacity. These wastes include rice husk, wheat shells, egg shells, coconut husk, palm fruit, bagasse, groundnut shell, fruit peels, biochar, etc., and can be used directly or in modified form.

Industrial Waste: Industrial activities produce large amounts of waste which is generally sent to landfill sites for disposal. These wastes have a good adsorption capacity and can be effective in waste treatment. Waste products like fly ash, red mud, and slag are effectively used due to their capacity for removal of HMs from wastewater.

The rate of HM removal increases with increasing initial concentration, temperature, adsorbent dose, contact time, and stirring speed.

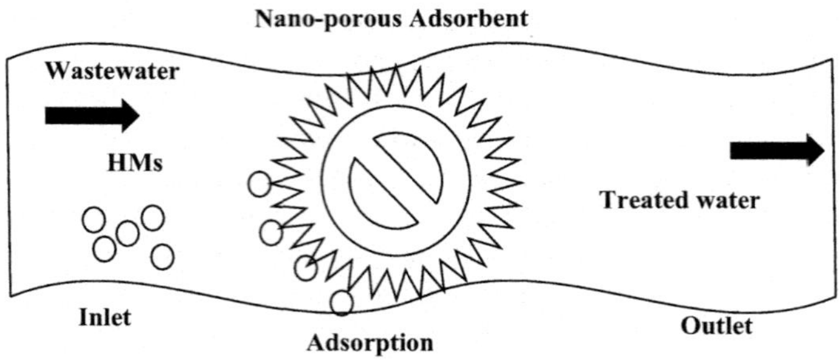

FIGURE 11.1 Adsorption of metal ions onto the adsorbent's surface.

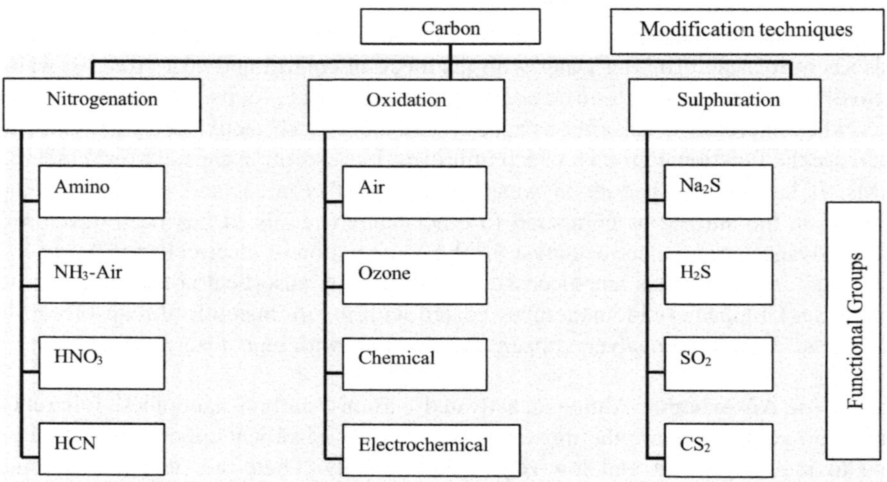

FIGURE 11.2 Functionalization for improving adsorption.

Adsorption-Based Separation: Generally, HM ions can be adsorbed onto the adsorbent's surface, as shown in Figure 11.1.

(a) **HM Ions Adsorption Process**: Metal ions from wastewater stick to the surface of nano-porous adsorbents with high surface area owing to their permeability.
(b) **Modification Techniques** (nitrogenation, oxidation, and sulphuration) are used to functionalize carbon with different functional groups (see Figure 11.2).

Functionalization improves adsorption capacity and stability.

Chitosan-Based Adsorbents: Chitosan (CS) is a natural adsorptive polymer that has an affinity towards pollutants in wastewaters because it possesses amino ($-NH_2$) and hydroxyl (–OH) groups. Despite its unique features, it has low mechanical strength, poor stability, low porosity, low surface area, resistance to mass transfer, and high crystallinity. Therefore, structural and chemical modifications have been made to overcome these shortcomings. Cross-linking chemical modification imparts strength to CS by connecting polymer chains and the functional groups. Grafting is another chemical modification technique that involves the covalent bonding of functional groups (like amine and hydroxyl) on the backbone of CS, leading to increased adsorption capacity. Combining CS with other adsorbent materials enhance CS's adsorption capacity, mechanical strength, and thermal stability. The ion-imprinting technique has been adopted to prepare adsorbents.

Mineral Adsorbents: Mineral adsorbents such as zeolite, silica, and clay are considered good for water purification with low operating costs, for instance, clay possesses:

- Extraordinary cation exchange capacity (CEC)
- Cation exchange selectivity
- Surface hydrophilicity or wettability
- High swelling/expanding capacity
- Surface electronegativity

Moreover, acid washing, thermal treatment, and pillar bearing can enlarge the pore size, pore volume, and specific surface area, leading to a remarkable increase in the adsorption efficiency. Using natural minerals could be cost-effective. However, the removal efficiency might decrease after a few cycles. Therefore, different modification methods, such as calcination and impregnation, have been proposed to improve the exclusion efficacy of such adsorbents. However, these alterations incur additional costs to the technique and release new chemical agents into the environment.

Magnetic Adsorbents: Magnetic adsorbents are specific material matrices that host iron particles (usually magnetic nanoparticles, such as Fe_3O_4). The base material could be carbon, CS, polymers, starch, or biomass. The adsorption process is governed by the magnetic field, surface charge, and redox activity characteristics. They revealed low cost, easy synthesis, amazing surface charge, and reusability. The mechanism and kinetics of the sorption process depend on the surface morphology and adsorbent magnetic behaviour, besides pH, irradiation time, adsorbent concentration, wastewater temperature, and initial dosage of contaminants. The presence of iron particles in the adsorbent is effective in metal ion removal from effluent. Magnetic adsorbent particles adsorb metal ions which are sequentially accumulated; thus, the wastewater is treated (see Figure 11.3).

Biosorbents: The presence of numerous functional groups (carboxyl, amino, hydroxyl, phosphate, thiol, etc.) on the surface expedites the biosorption process. Usually, the interaction between pollutants and the surface of the biosorbent can occur through:

- Aggregation
- Complexation/coordination

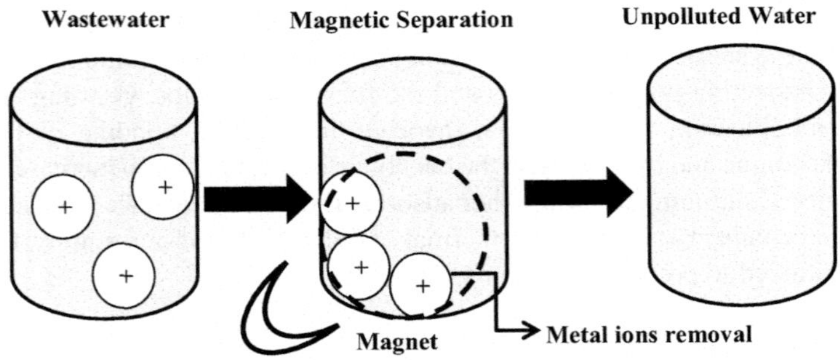

FIGURE 11.3 Magnetic adsorption.

- Electrostatic interaction
- Ion exchange
- Microprecipitation
- Oxidation
- Reduction

The solution pH affects the biosorbent surface charge density and ionization of functional groups located on the biosorbent surface. When the pH is low, cations are stable and can be attached to the biosorbent surface. Conversely, at higher pH values, the solubility of metal cations decreases with the likelihood of a precipitation phenomenon.

11.2.2 Ultrafiltration Membrane Technology

Innovative processes for remediation of HMs from wastewater involve equipment for decreasing toxicity to meet technology-based treatment principles. In recent times, many approaches have been investigated for evolving cheaper and more effective technologies, both to decrease the amount of wastewater and to improve the quality of treated effluent. Most removal methods for metals involve sedimentation, filtration, or chemical treatment. Since the solubility of different metals varies with the acidity and alkalinity of the wastewater, chemical physical technology usually includes pH adjustment systems to precipitate the metals, subsequently treating extra chemicals, flocculants, and coagulants to make a sludge particle large and heavy enough to settle. Traditional separation methods are very labour-intensive as machinists must perform periodic analytical tests and control operating parameters to uphold consistent effluent quality that meets the rigid seepage limits. Recent studies show that the latest cross-flow ultrafiltration (UF) membrane technology offers a cost-effective option for separating HMs from industrial wastewater.

Metal solubility is affected by the pH of the solution. A UF membrane has a submicroscopic pore size of < 0.01 micron and thus retains the metal precipitate. In ultrafiltration, hydrostatic pressure forces a liquid against a semipermeable membrane. In this process, the suspended solids and solutes of high molecular weight are retained, but water and low-molecular-weight solutes pass through the membrane. Cross-flow

UF is a pressure-driven filtration process in which the process liquid runs parallel to the membrane surface. Under a pressure of 10 pounds per square inch (psi) to 100 psi, the filtrate passes through the membrane and leaves as clear permeate. Rejected species are recollected for disposal or reprocessing. The membrane's performance is determined by the infiltrate flux and the rejection of constituent metals.

Besides pore size, pore construction is also important for the performance of a membrane. Conventional filters sometimes shape pores that allow accumulation of particles at bottlenecks and cracks within the cross-section of the filter. The UF membrane pores are uneven and shaped like reversed cones, with smaller diameters on the feed side and larger diameters on the infiltrate side. Since any particle that passes through a pore continues unrestricted without collecting within the membrane, UF membrane pores do not block. Cleaning of these filters is therefore trouble-free and economical, and routine washing allows repeated use over long periods of time. With proper maintenance, UF membranes work for years without replacement.

Advancements in polymer science enable membrane scientists to utilize engineering plastics to improve the structure–property relationships of membrane products. The most commonly used membranes are poly-vinylidene difluoride (PVDF), polysulphone (PS), and polyacrylonitrile (PAN). The membrane life expectancy depends on the setting of the process and cleaning frequency. Membranes can be packed into tubular, spiral, and hollow-fibre configurations, with the difference being the features of the flow channel.

Ultrafiltration membranes used in wastewater treatment have proved to be very successful and offer economic benefits over other filtration and gravimetric methods. Membrane systems are simple in design and operation, which is particularly important in wastewater treatment. The entire system requires less cleaning. UF membranes incur an initial higher investment than MF technology. Thus, ultrafiltration is an efficient and economical replacement for traditional and microfiltration technology for removing HMs from wastewater. Due to the sub-microscopic pore size and unique pore construction, UF membranes yield consistent wastewater quality that runs into stringent discharge and recycling regulations.

11.2.3 Sediment Remediation Technologies

Sediment remediation involves the use of physical, chemical, or biological treatment technologies to minimize the contaminant concentrations in sediment. Remediation techniques undertake separating, destroying, or transforming of the contaminants from the sediment to less toxic forms, or even stabilizing the contaminants so that they are fixed to the solids and are resilient to damage by ecological pathways.

These techniques may be selected in line with sediment textures of coarse sand, very coarse sand, and gravel, permeability, and particle size diameter. Usually, the site of the original pollution has the maximum concentration of pollutant and on moving away from the source the concentration decreases. The main objective of most sediment remedial alternatives is to isolate the solids from the water fraction of the sediment (i.e., dewater) to the maximum possible extent. This is mainly done to reduce the disposal cost for the solids and is a prerequisite of treatment technologies, such as removal and non-removal technologies. Sediments may be dewatered through various methods to a solids content greater than 50 per cent. Depending on the process used, there may be little or no volume reduction, because water is substituted by air in the voids between the sediment solids. One of the most important factors to be considered during a sediment remediation process is how the sediments are removed, to abate the release of pollutants. Some of sediment remedial methods (except non-removal) involve the digging of sediment from the bottom of a watercourse and moving the dug sediment to a location where the material is kept in a holding area, moved to pre-treatment units, and then transferred to the treatment units. Some of the most common strategies include:

- ***In situ* Containment**: In this remediation choice, a layer of clean segregating material is placed in CDFs (confined disposal facilities), CADs (contained aquatic disposal), and landfills to contain and stabilize the polluted sediment. A variety of capping materials and cap placement techniques are available for this purpose
- ***In situ* Treatment**: This treatment technology remediates polluted sediments based on the biological properties of the pollutants
- ***Ex situ* Containment**: This land disposal technology is used to contain or minimize the impact of pollutants released around the dredging operation
- ***Ex situ* Treatment and Extraction**: This refers to a remediation technique that is used off-location, based on the physico-chemical properties of the contaminants.

The cost of each of these remediation technologies may be significant. In some cases, the physical properties of the polluted sediments (such as particle size and solids/water composition) may involve the use of one or more pre-treatment technologies prior to the processing of the sediment through a treatment unit. The concerns about which remediation technique to use include the costs of equipment, materials, reagents, and labour required to treat the sediments. *In situ* treatment can produce negative impacts on the water column by releasing residue from the treatment. In contrast, some *in situ* treatments may be less cost-effective than *ex situ* treatments.

11.2.4 Soil Washing

Soil contamination by HMs or metalloids is one of the most serious environmental problems which have emerged as a result of rapid industrialization and urbanization. Generally, anthropogenic actives such as mining, domestic waste discharge, agricultural production, and industrial activities have contributed to the contamination of soil, which through the food chain has adverse health implications. Soil washing is one of the *ex situ* traditional remediation methods employed for the removal of HMs from polluted soils. It involves dissolving of metals/metalloids with extractants followed by reducing them to small volumes through particle size separation (see Figure 11.4). Soil washing has proved to be effective for soil with higher permeability. Successful soil washing usually depends on the concentration of soluble trace metals, while most metals are less mobile owing to their dispersal and speciation in soils.

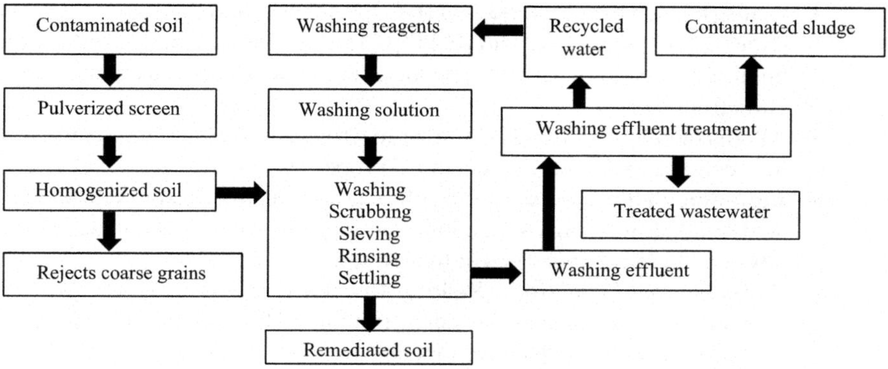

FIGURE 11.4 Soil washing mechanism.

The low solubility and accessibility of trace metals at times affect the washing efficiency. Optimization of washing variables such as extractant dosage, contact time, and washing temperature can efficiently improve the remediation. Conversely, reducing soil pH and the application of assisted amendments can release metals from soil–solid components through ion exchange, dissolution, and metal sorption/desorption processes. Additionally, other improved technologies such as electro-kinetic remediation and ultrasonication are feasible to couple with soil washing. Soil washing is considered to be an effective measure that can permanently remove HMs from soil with high remediation efficiency despite the concern about losing some nutrients and causing secondary pollution. It employs physical processes to separate the most polluted soil particles, which in turn go through chemical extraction with specific washing reagents to transfer HMs from soil to solution. The remediation efficacy of soil washing is closely linked to the extracting reagents.

Types and Characteristics of Washing Reagents

Washing reagents used for remediating heavy-metal-polluted soils are divided into four categories, namely inorganic reagents, chelating reagents, surfactants, and compound chemical reagents.

Inorganic Reagents (water, inorganic acids, alkali, inorganic compounds, and reducing or oxidizing [redox] reagents): The mechanism of soil washing with inorganic reagents is done mainly to destroy the functional groups of the soil surface and enable complex formation with HMs through washing, acidolysis, and complexation or ion exchange, thus exchanging and desorbing the HMs from the soil to the leachates.

- **Water**: Water, being economical, flushes away soluble contaminants. Deionized water can remove over 70 per cent of hexavalent chromium. However, HMs are sparsely soluble and mainly in the sorbed state; therefore, washing soils with water alone cannot remove a significant amount of HMs

- **Acids** (include hydrochloric acid [HCl], sulphuric acid [H_2SO_4], nitric acid [HNO_3], and phosphoric acid [H_3PO_4]): Owing to ion exchange and dissolution of soil components/discrete metal compounds, acids usually remove HMs from soil
- **Alkalis** (sodium hydroxide): Similar to strong acids, alkalis also cause serious damage to the physico-chemical properties of soil
- **Inorganic Salts**: The use of salt solutions can be an effective substitute to acid or alkali washing as inorganic salt reagents cause less damage to the physico-chemical properties of treated soils
- **Reducing or Oxidizing (Redox) Reagents**: These reagents provide another option to enhance the solubilization of HMs because the chemical oxidation/reduction can convert metals to more soluble or less toxic forms by changing their valency. Based on the removal mechanism, the use of reducing or oxidizing reagents is generally restricted to the HMs, which are varied in solubility and toxicity under different valences

Chelating Reagents: These reagents are able to form stable water-soluble complexes with a range of metal ions through chelating and desorption, consequently creating a suitable environment for soil washing. Numerous chelating reagents have been tested for soil washing, and their roles are becoming increasingly important. There are generally two kinds of chelating reagents that are commonly used in study and practice, namely aminopolycarboxylic acids, also known as synthetic chelating reagents, and natural low-molecular-weight organic acids (NLMOAs). The aminopolycarboxylic acid group includes diethylenetriamine five acetic acid (DTPA), ethylene glycol double tetraacetic acid (EGTA), ethylenediamine tetraacetic acid (EDTA), etc. EDTA and its homologues form stable complexes with many HMs, and have consequently become one of the most regularly studied chelating reagents. EDTA is recognized as the most effective synthetic chelating reagent because it has a strong chelating ability for cationic HMs; it can treat a broad range of soil types; and it is recoverable and reusable owing to its low biodegradability.

Low-molecular-weight natural organic acids include acetic, citric, oxalic, tartaric, formic, malic, lactic, succinic, and fumaric acids, which are natural products of root exudates and microbial secretions in addition to plant and animal residue decomposition in soils. The natural organic acids are regarded as biodegradable and eco-friendly chelating reagents and are capable of promoting the dissolution and transformation of HMs from soils by forming complexes. The washing mechanism of using NLMOAs includes:

(1) Desorption of soil minerals by organic acids
(2) Displacement of hydrogen ions
(3) Formation of positively charged metal complexes and exchange ions with soil organic acids
(4) After adsorption of organic acids on the soil surface, the functional groups complex with HMs to form ternary complexes
(5) Coordination between organic acids and HMs, and absorption of the complexes to the soil

Surfactants: These are surface-active compounds capable of considerably changing the interface state of the solution system. Since surfactants are capable of supporting solubilization, dispersal, and desorption of hydrophobic compounds and HMs owing to their hydrophobic and hydrophilic portions, they are regarded as favourable washing solutions to remediate heavy-metal-contaminated soils. Generally, surfactants are classified into four categories:

Amphoteric (presents both positive and negative charges at an intermediate pH)
Anionic (negatively charged): These are the most productive and widely used surfactants, which are classified into sulphonate and sulphate salts according to their structure of hydrophilic groups
Cationic (positively charged): These mainly include nitrogen-containing organic amine derivatives which can dissociate cations in water
Non-Ionic (uncharged): Owing to their good compatibility, these can be used in combination with other surfactants because the main hydrophilic group of non-ionic surfactants is an undissociated ether group in an aqueous solution

Heavy-Metal Removal Mechanism using Surfactant

- Surfactant is first adsorbed on the soil surface and interacts with the organic molecules and heavy-metal ions
- It causes heavy-metal ions to leave the soil particle surface via ionic exchange and surfactant-associated complexation
- Later, a complex micelle form prevents the adsorption of heavy-metal ions to the soil

Chemical Surfactants: The commonly used chemical surfactants include:

- Cetyl trimethyl ammonium bromide (CTAB)
- Polyoxyethylene laurel ether
- Sodium dodecyl benzene sulphonate (SDBS)
- Sodium dodecyl sulphate (SDS)
- Tween, Tritonx

Bio-surfactants: These are surface-active compounds that are mostly produced during the movement of bacteria and fungi available in bio-surfactants, including:

- Cyclodextrin
- Rhamnolipids
- Saponin
- Sophorolipids
- Surfactin
- Tannic acid

Owing to different microorganisms, raw matter, and process conditions, biosurfactants have diverse structures and chemical compositions. They are helpful for improving the interaction between microorganisms and pollutants in soil and have a less harmful impact on soil structure and microbial growth. Bio-surfactants are used to a lesser extent because their cost is usually relatively high, there is a risk of secondary pollution, and several bio-surfactants are required for remediation of multiple-metal-contaminated soils because of the diverse affinities of bio-surfactants to different HMs.

Although a large number of washing reagents are used to treat heavy-metal-contaminated soil, it is important that selection of the most suitable reagents depends on the governing factors. Thus, it is important to understand the factors affecting the choice of reagents for washing remediation. There are various factors that affect the application of washing remediation, such as the buffering capacity of soils, cation exchange capacity, geochemistry of the soil, organic matter content, soil texture, etc.

11.2.5 Electro-Dialysis

Electro-dialysis is a membrane-based technology that can be used for treating acidic effluents containing metal species. Its continuous operation capacity, scalability, and easy operation can solve most of the drawbacks of current technologies. Electro-dialysis has emerged as a promising technology with huge growth potential to concentrate, separate, and selectively recover metals from acidic effluents. Electro-dialysis has been used in recent years for the treatment of metal-contaminated water. Electro-dialysis is a membrane-based technique in which the ions are passed through a semi-permeable barrier under the influence of electric potential (see Figure 11.5).

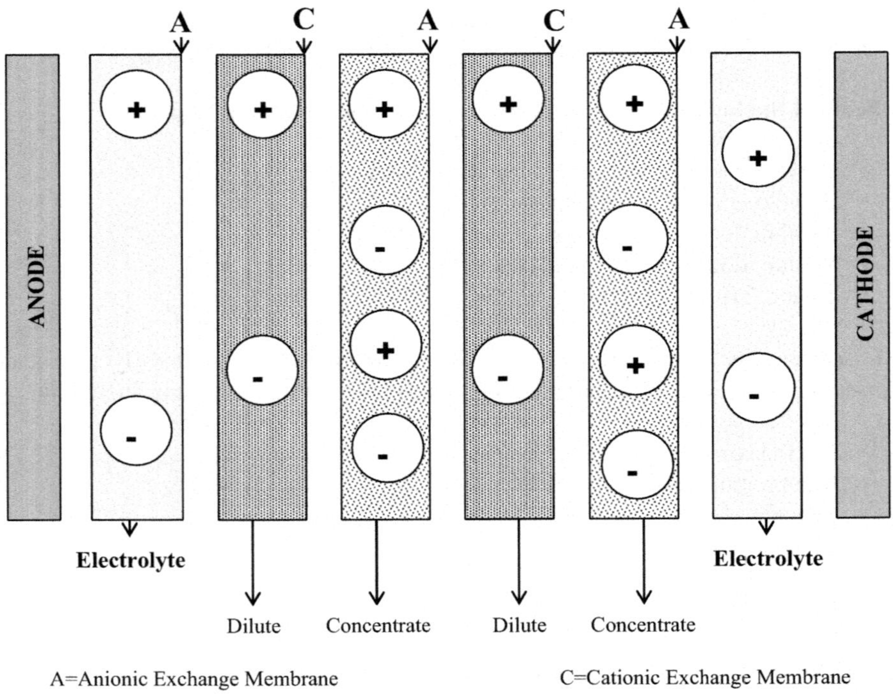

FIGURE 11.5 Electro-dialysis set-up.

Ion-exchange membrane and electric potential are the key features of electro-dialysis. It is a voltage-driven process containing positively and negatively charged membranes. It involves the deionization of solvent, and membranes are attached in series pattern. Membranes are 0.2–0.5 mm in thickness and impervious to water. Alternating anion-selective membranes (divinyl benzene) and cation-selective membranes (polystyrene sulphonic acid) are placed between the anode and cathode. These membranes act as a thin selective barrier that makes it possible to demineralize or concentrate acids and bases from saline solutions. The presence of an electric field increases the speed of removal of electrolyte from the solution. Inert metals like platinum-coated titanium (in the cathode) and stainless steel (in the anode) are used in electrodes. The performances of different electro-dialysis membranes aim to reduce the metal content from a zinc-rich acidic effluent. Electro-dialysis is used for the removal and recovery of metals like copper, iron, cadmium, chromium, lead, nickel, etc. During electro-dialysis, the solution to be desalinated is held in one compartment and when the current flows, anions move through the anion exchange membrane towards the anode into the adjacent compartment, while cations move in the opposite direction into an adjacent compartment on the other side.

Thus, the solution becomes depleted of ions in one compartment and the solution in the adjacent compartment becomes enriched in ions. Cadmium can be removed from sludge by using nitric and citric acid employing a single electro-dialysis cell. Lead, a toxic metal that has been regularly discharged into the environment, can be removed using electro-dialysis. Temperature, initial concentration, and voltage also play important roles in its complete removal under optimized conditions.

Chromium is a carcinogenic and non-biodegradable metal that is widely used in the electroplating industry. Electro-dialysis is used for the removal of direct concentrations of Cr (VI). Mercury behaves differently from other HMs due to its easy reduction to elemental mercury.

The addition of chloride as a chelating agent can increase the electro-dialysis removal rate of mercury. The poisonous effect of mercury complexes remains in the solution, which can be a drawback in mercury decontamination using electro-dialysis.

Electro-dialysis is an emerging technology for metal removal with resource recovery potential. Its continuous operation capacity and simple scale-up possibilities make it suitable for metal acidic effluent treatment in large-scale applications. Low energy consumption, high conversion ratio, low space, and less material requirement are the key features of this technique. It is an eco-friendly technology that does not involve chemicals, unless a pre-treatment is required. However, application of the classical electro-dialysis technology is still challenged by drawbacks including high costs, membrane fouling, concentrate streams, and specific treatment requirements of each industrial effluent.

11.2.6 Filtration

Filtration is a simple separation technique wherein solids are separated from a fluid by passing the solution through a porous membrane or filter. In this process, a filter permits the liquid to pass through it and solid particles become separated (see Figures 11.6 and 11.7). In wastewater treatment, various types of porous and thin-film

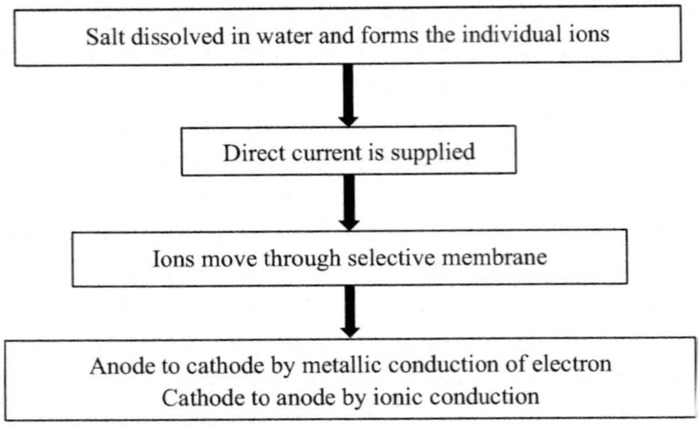

FIGURE 11.6 Steps involved in electro-dialysis.

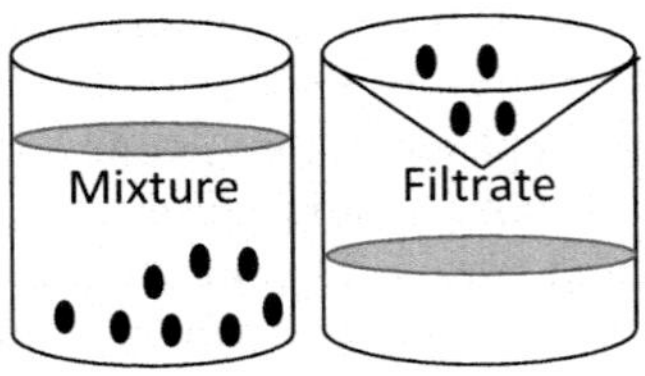

FIGURE 11.7 Filtration.

composite membranes are used. There are different types of driving forces regulating membrane processes, such as:

- Low pressure (microfiltration, ultrafiltration, and distillation)
- High pressure (nano-filtration, reverse osmosis)
- Osmotic pressure (forward osmosis, electro-dialysis)

Material used as filter media include:

- Woven material (wool, metal, or synthetic fibres)
- Perforated sheet metal (stainless steel plates having pores which act as channels)
- Pre-fabricated porous solid units (sintered glass, earthenware, and porous plastic bottles)

The main factors applied to the application of the membrane process are:

- Pore size
- Pore distribution
- Degree of hydrophilicity
- Surface charge
- Solution flow
- Presence of functional groups

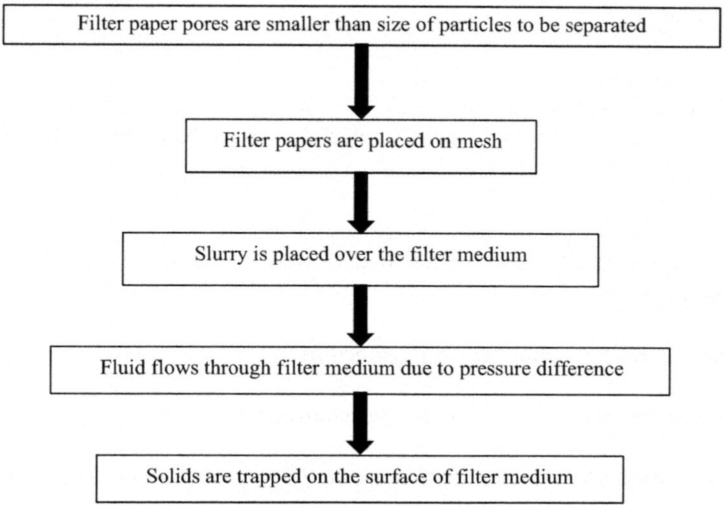

FIGURE 11.8 Steps involved in filtration.

The performance of the membrane is generally determined by its flux rate and selectivity. Cellulose acetate, polyamide, and polysulphide, being porous membranes, are used in this process where solution flow under pressure and thinning the membrane releases water at atmospheric pressure. Filtration is the process in which components of solution are separated based on their size during transfer through porous material (see Figure 11.8). The porous filter material acts like a sieve and solid particles are retained on the surface of the membrane of the filter material. The flow of solids is resisted by the filter medium, while the liquid is allowed to pass through the filter paper. As the filtration proceeds, retention of the solids on the filter media continues to increase, which acts as a secondary filtering medium. Filtrate is separated from the solution and solid residue is entrapped on the surface of the filter media. Filtration is used to clear HM ions such as cadmium, copper, aluminium, and ferrous ions from wastewater.

Types of Filtration Methods

General Filtration

(a) Basic types of filtration
(b) Gravity is used to pull down the liquid

Vacuum Filtration

(a) A vacuum is used to suck the fluid through the filter
(b) Slightly faster

Cold Filtration

(a) Formation of small crystals
(b) Used for initially dissolved solids
(c) The container with the solution is placed in an ice bath prior to filtration

Hot Filtration

(a) The solution, filter, and funnel are heated to minimize crystal formation
(b) Stemless funnels are useful

Membrane separation technology is governed by:

- Steric hindrance mechanisms or size exclusion
- Charge–charge repulsion
- Adsorptive ability toward contaminants

Membrane filtration has received considerable recognition for the treatment of inorganic effluents such as HMs, and it can also remove organic compounds and suspended solids. This technology has gathered caught the attention and has become a favourable approach to HM removal due to its cost-effectiveness and high efficiency.

The ultrafiltration technique is used at a low trans-membrane operating pressure. Ultrafiltration membrane pores are larger than HM ions. Therefore, additives become attached to metal ions and enlarge the size of the metal ions. Of the various types, micellar enhanced ultrafiltration (MEUF) and polymer enhanced ultrafiltration (PEUF) are mostly used for the removal of HMs. MEUF involves the combination of ultrafiltration and surfactant. MEUF has high flux and high selectivity, leading to low-energy consumption, high removal efficiency, and less space demand. MEUF is the most suitable for wastewater, whereas HMs are in low concentrations.

PEUF involves the integration of ultrafiltration and binding polymers. The functional groups of the bonding polymers could be sulphonate, phosphonic, carboxylate, or amine, and they are attached via chelating or ionic bonds. PEUF are also known as polymer-supported, complexation, polymer-assisted, size-enhanced, and complexation-enhanced ultrafiltrations. While allowing water and uncomplexed components to infiltrate the membrane pores, the PEUF process blocks and extracts polymer-bonded metal ions.

Nano-filtration (NF) is used to concentrate constituents whose molecular weight is greater than 1000 Da and to remove solutes with size of 0.0005–0.007 μm. The operating range of nano-filtration is between ultrafiltration and reverse osmosis. The nano-filtration membranes are composed of polymer composites of multiple-layer thin films of negatively charged chemical groups. Anti-fouling nano-filtration membranes are synthesized through phase inversion and used to extract Fe^{3+}, Al^{3+}, Co^{2+}, Cd^{2+}, Cu^{2+}, and humic acid from wastewater.

Microfiltration uses a microporous membrane to remove micron-sized particles, bacteria, viruses, protozoa, contaminants, and pollutants from solutions. Microfiltration is a low-pressure-driven membrane process, whose membrane pores are in the range

TABLE 11.1
Differences in Various Types of Filtration

Attribute	Microfiltration	Ultrafiltration	Nano-filtration	Reverse Osmosis
Membrane	Porous, isotropic	Porous, asymmetric	Finely porous, asymmetric	Non-porous, asymmetric
Pore size	50 nm–1 μm	5–20 nm	1–5 nm	–
Transfer mechanism	Sieving and adsorptive	Sieving and preferential adsorption	Sieving, electrostatic hydration, diffusive	Diffusive
Governing law	Darcy's law	Darcy's law	Fick's law	Fick's law

of 0.1–10 μm. The microfiltration membranes are made up of silica, ceramics, zirconia, alumina, PVC, polysulphide, PTFE, polypropylene, PVDF, polyamides, polycarbonate, cellulose acetate, cellulose esters, or composite materials. Reverse osmosis (RO) is a pressure-driven separation process that employs a semi-permeable membrane with a pore size of 0.5–1.5 nm and allows only smaller molecules to pass through it. The reverse osmosis process is compact and has a high rejection efficiency. However, membrane fouling and degrading are the major drawbacks of reverse osmosis systems. Metal ions like Cu^{2+}, Ni^{2+}, Cr (VI), and As (V) have been successfully removed by RO and nano-filtration. Differences in different types of filtrations are listed in Table 11.1.

11.2.7 Flocculation

Flocculation is a process that involves contact and adhesion through which the dispersed particles form large clusters. Flocculation simply means aggregation of particles. The coagulant is added to the solution to facilitate the bonding or aggregation of the particles in a liquid. The key features of flocculation are:

a) Use of a high-molecular-weight polymeric flocculant
b) Flocculant facilitates bridging of gaps between flocs
c) van der Waals attraction forces reduce the energy barrier
d) Aggregation, binding, and strengthening of flocs
e) Sedimentation occurs

Flocculation is based on the property of contact and adhesion through which the dispersed particles form large clusters which upon the aggregation is sedimented using chemicals. Flocculation is an efficient, cost-effective, convenient, and widely used method employed for the removal of HMs. The typical HMs removed by this method include Cu^{2+}, Pb^{2+}, Ni^{2+}, As^{2+}, Se^{2+}, Cr^{2+}, Sb^{3+}, and Ag^{2+}. Flocculation involves following five step steps (Figure 11.9). The flocculation technique bonds the particles forming large agglomerates with the help of a flocculant. In flocculation, one or more flocculants are added to HM-contaminated wastewater, and this method depends on

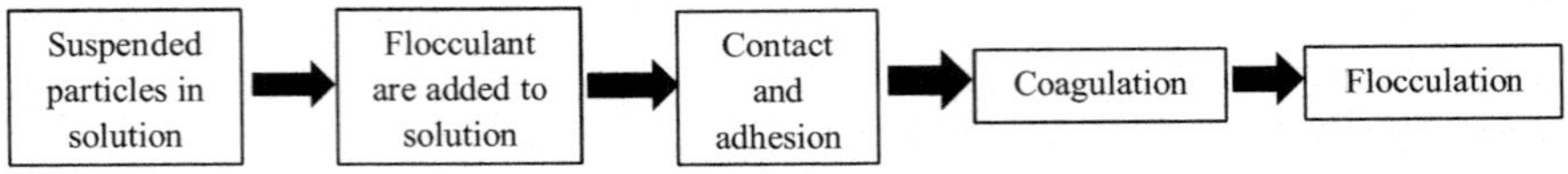

FIGURE 11.9 Steps involved in flocculation.

TABLE 11.2
List of Flocculants

Alum	Ferrous sulphate	Aluminium sulphate	Sodium silicate	Sodium aluminate	Calcium oxide
Calcium hydroxide	Ferric chloride	Guar gum	Isinglass	Gelatine	Polyacrylamide

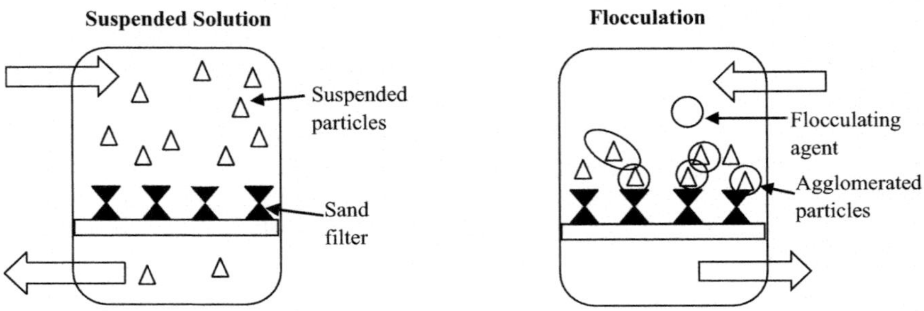

FIGURE 11.10 Mechanism of flocculation.

the nature of flocculation agents. Flocculants are natural or chemical substances that are added to a solution to accelerate the rate of flocculation and to strengthen the formation of flocs. Flocculants are generally man-made and non-biodegradable. Some common flocculants used for initiating flocculation process are listed in Table 11.2.

In contrast to general pollutants that can be removed through oxidative decomposition, HMs are non-degradable. Appropriate flocculants can be used to remove dissolved HM ions and compounds on the surface of suspended solids or colloidal particles through flocculation (see Figure 11.10). Chitosan is a common natural organic polymer material with a wide range of sources, nontoxicity, and environmental protection. Chitosan can be used as a flocculant and absorbent in HM removal because its macromolecular structures possess surface groups. Chitosan forms stable complexes with dissolved HM ions because of its active and adjacent hydroxyl and amino groups. However, given that chitosan has poor water solubility and can only function in a weakly acidic environment, its flocculation and chelation ability is significantly reduced. Modified chitosan polymer flocculants are used with specific methods, such as grafting, to:

- Introduce specific reactive groups into their chain units
- Provide an efficient performance of HM chelation

- Capture and improve the hydrophilic properties of chitosan and its ability to chelate and capture HM

The directional modification of functional defects has made advances in research on chitosan for the removal of HM. Modified chitosan polymer flocculants exhibit excellent performance in terms of chelation, flocculation, and sedimentation in the treatment of HM-containing wastewater. In contrast to the use of traditional synthetic high-molecular-weight flocculants, the use of chitosan can significantly reduce the cost of synthetic source materials because chitosan is an easily accessible natural polymer material; however, its use is restricted by a helical molecular chain structure.

Flocculation and chelation of HMs are achieved through the use of two bio-flocculants i.e., starch-graft-P [acrylamide-acrylic acid] (SAA) and starch-graft-P [acrylamide-2-acacrylamido-2-methyl-1-propanesulphonic acid] (SAS). These bio-flocculants are used to remove Cu (II), Zn (II), and Ni (II) from simulated wastewater and Ni (II) from actual electroplating wastewater. SAA and SAS have high selectivity for Cu (II) in the simulated water sample with mixed HM. The 3E model triangular evaluation model, i.e., environmental impact, economic benefit, and energy consumption, is established to evaluate the flocculation experiments for the removal of HM ions using SAA and SAS. Flocculation is a simple, cost-effective method that is affected by flocculant dilution, molecular weight, pH, polymer type, ionic strength, etc.

11.2.8 Flotation

Flotation is an extractive process in which particles are separated from a mixture by selective extraction from a surface. Being a physio-chemical method of separation, it is based on the difference in the surface chemical properties of the solids. The method for removing HM is a three-step process. The first step is true flotation (selective attachment to air bubbles), followed by entrainment (which occurs in water which passes through froth), and the final step is aggregation (physical entrapment between the particles in froth). It is a sorting technology which selectively improves various materials by using differences in the physico-chemical properties of the surface of minerals to separate them. There are two types of flotation:

- Direct flotation (mineral is attached to the froth and gangue remains in the tailing)
- Reverse flotation (gangue is attached to the froth and mineral remains in the tailing)

The flotation method has found applications in waste residue and wastewater treatment, metallurgy, the chemical industry, and other material separation and recovery processes (see Figure 11.12). Chemicals used for flotation include:

- **Collector**: Organic compound which increases hydrophobic properties of minerals, for example, sodium oleate
- **Frother**: Heteropolar surface-active reagent, for example, pine oil, cresol

- **Depressant**: Inorganic compounds, for example, cyanide
- **pH modifier**: Affects pH, for example, lime

Froth flotation is based on the wettability differences of mineral and gangue particles (see Figure 11.11). Particles are classified as hydrophobic (water-hating) and hydrophilic (water-loving). Hydrophobic minerals can only attach to air bubbles. Once they reach the surface, the particle–bubble contact can sustain while forming a stable froth. Metallic ore is ground to a fine powder in the presence of water, and slurry is produced. The collector, surfactant chemical, is added to the slurry to make the desired mineral slurry hydrophobic. This slurry or pulp is placed in a water bath containing a frother, which is aerated to create bubbles (Figure 11.2). The desired mineral escapes water by getting attached to the air bubbles, which rise to the surface and produce froth. This contaminated froth is removed and the concentrated mineral is refined.

In the process of mineral flotation, particle size is one of the influencing physical factors. HM-enriched material is crushed, ground to the correct particle size, and then subjected to flotation based on the properties of the material.

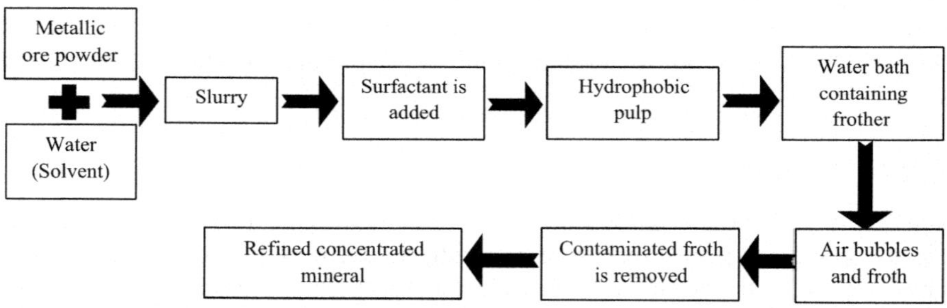

FIGURE 11.11 Steps involved in flotation.

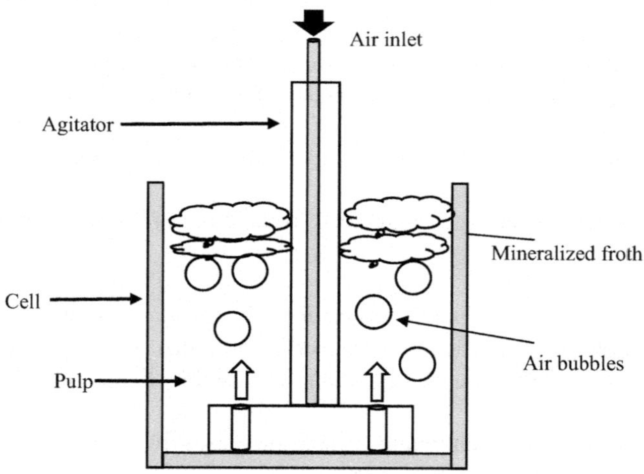

FIGURE 11.12 Flotation set-up.

Fly ash produced from the incineration of municipal solid waste is generally in the form of an off-white powder. The particles are fine, having a large specific surface area along with large gaps between the particles. Fly ash contains a large number of HMs such as Pb, Zn, and Cu. The HMs in fly ash are mainly treated by means of chemical agents, which can cause secondary pollution. Malachite is a common copper oxide mineral having poor floatability. Therefore, high-performance collectors are used for its flotation along with improved separation.

The direct flotation technique is extensively used for flotation of malachite with the application of phosphorus-based surfactants and is extremely selective for the separation of malachite from calcite or quartz. The separation of organic and inorganic elements in mineral processing and metallurgy are also done with the help of flotation. The high separation and mineral recovery efficiency are some of the advantages of flotation, along with disadvantages such as high reagent consumption, strict grinding size, high cost, and environmental pollution.

11.2.9 Electro-Kinetic Remediation

Electro-kinetic remediation is the best and most efficient green technology which is used to remove HMs from soil using a low-voltage direct current. This separation technique is suitable for low-permeability soils and removes radio-nucleotides, and organic or inorganic contaminants. A wide range of HM contaminants are removed by the electro-kinetic method. This method has four main four components:

1. Electrode compartment made up of graphite and platinum
2. Electrolytic solution reservoir
3. Direct current, solar cell, and pulse current are used for power supply
4. Soil cell

The electro-kinetic method is based on the principle that acid (H^+) is generated at the anode and base (OH^-) is generated at the cathode (see Figure 11.13). Hydrogen is more mobile than hydroxide; and an acid front can potentially travel toward the cathode. Consequently, a pH gradient is generated in the solution. The electrode polarity can be transformed during operation, efficiently transforming the cathode to the anode and vice versa, which can address pH changes and possibly help more effective distribution and transportation of contaminants.

Fly ash generated from municipal solid waste incineration (MSWI) has a high concentration of HMs such as copper, lead, cadmium, and zinc which are hazardous to the environment. A number of techniques, such as physical separation, isolation, flushing, washing, toxicity reduction, solidification/immobilization, and phytoremediation, are

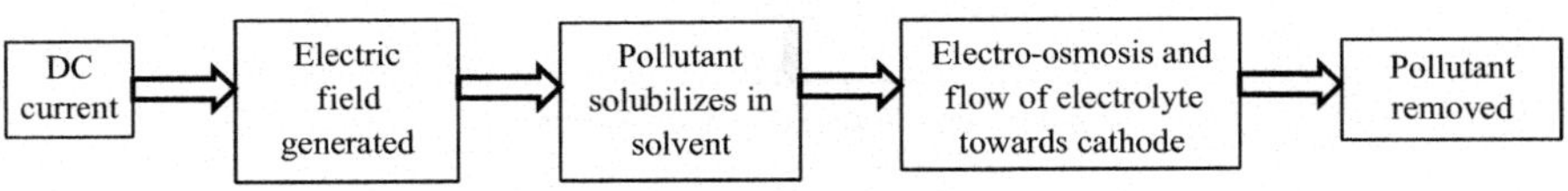

FIGURE 11.13 Mechanism of electro-kinetic remediation.

used to treat HM pollution. However, a number of limitations are linked with these techniques such as partial removal of HMs, high consumption of reagents and energy, low selectivity, and the generation of secondary wastes. These problems arise due to the high buffering capacity of MSWI fly ash.

Electro-kinetic remediation technology is the best solution to address HM pollution (see Figure 11.14). It can be used for both *in situ* and *ex situ* remediation and has achieved excellent effects in detoxifying contaminated soil, mud, and water. Therefore, electro-kinetic remediation technology has significant potential for the treatment of MSWI fly ash. Generally, the electro-kinetic remediation process is carried out in an electrolytic cell/chamber which consists of a sample area, a voltage source, and two electrodes, i.e., the cathode and anode. These electrodes are connected to a voltage source and placed in the sample area to generate an electric field that helps in the migration of HMs. The remediation mechanism is completed in four steps which are acidification, desorption, migration (electro-migration, electro-osmosis, and electrophoresis processes), and precipitation of metal ions. MSWI fly ash has numerous oxides, such as SO_3, SiO_2, CaO_2, and Al_2O_3, and several kinds of metallic oxide which react with water and generate carbonate and hydroxide compounds. These compounds are accountable for the high pH of MSWI fly ash. The pH can be regulated through the addition of acid in the fly ash sample before being subjected to electro-kinetic

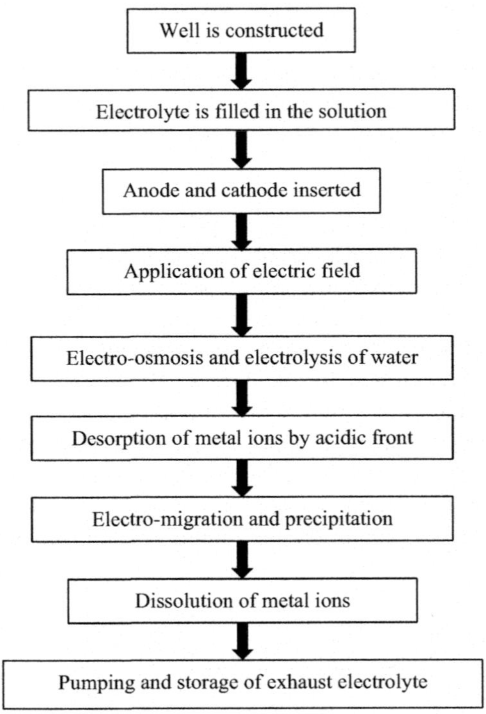

FIGURE 11.14 Steps involved in electro-kinetic remediation.

remediation. MSWI fly ash is pre-treated with ammonium citrate to remove HMs like Cu, Zn, Pb, Cd, and Cr from MSWI fly ash via electro-dialytic remediation. Another method to control pH during electro-kinetic remediation is by using an ion-exchange membrane which is placed between the electrolytic solution of the cathode region and the sample area. This membrane is selective in nature and permits only the passage of cations. In this way, the OH^- ions cannot cross the membrane and the cathode region has high pH. Metal cations cross this membrane through migration under the action of the electric field and get precipitated after reaction with OH^- ions near the cathode region. The electro-kinetic method involves three main processes:

Electro-osmosis (movement of water molecules towards the cathode)
Electro-migration (transport of metal ions towards the opposite charged electrode)
Electrophoresis (transport of charged colloidal particles)

HMs exist as complex, precipitated, and dissolved forms in MSWI fly ash. The removal mechanism depends on different morphologies of HMs in electro-kinetic remediation. HMs are categorized into four states:

- An acid extractable state released in an acidic environment
- A reducible state form complex with iron and manganese oxide which is released in reductive conditions
- Oxidizable states are bounded with organic matter and sulphide compound can be released in an oxidizing environment
- A residual state existing as a mineral-bounded form is released on weathering. The former three states can easily release their respective conditions when compared to the fourth state which is highly immobile

An increase in the acid extractible extraction fraction is due to the release and transformation of reducible and oxidizable states into acidic states. Heat generation, cracking of the electrode, pH value variation, and production of chlorine gas are the major disadvantages of the electro-kinetic method. Acidic pre-treatment has a strong effect on these two states, which are simply discharged during electro-kinetic remediation. The removal rate can be increased by extending the reaction time or by utilizing this method along with other techniques.

11.2.10 Ball Milling

Ball milling is an economic and facile technique widely used to transform larger substances into smaller ones through a grinding process (see Figure 11.15). Particles used for grinding in the ball milling method have different natures and a wide diversity of physical, mechanical, and chemical characteristics. It is generally a top-down approach of nanoparticle synthesis which is used to mechanically break down larger substances into smaller ones. Due to its cost-effectiveness and environmentally friendly nature, it finds a wide range of applications in many industries. The ball milling method follows the principle of impact and attrition according to which an impact occurs between fast-moving balls and the powder material, which cats to

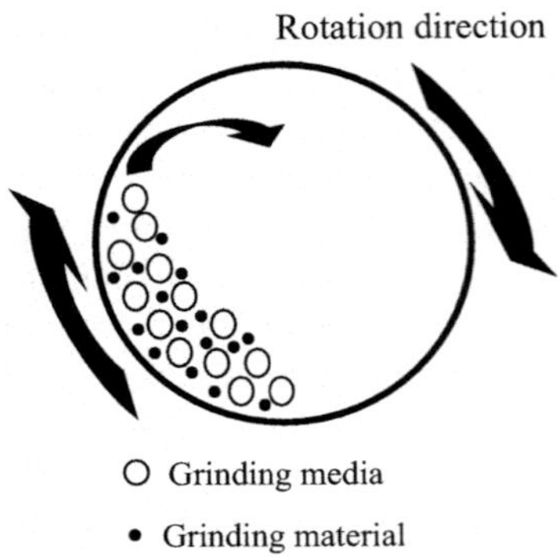

FIGURE 11.15 Schematic representation of ball milling.

break down the particles, while the attrition mode comes into effect when the hollow cylinder rotates on its longitudinal axis which allows obtaining of the desired reduction in the size of particles.

The structural design of the ball mill consists of a hollow cylinder shell mounted on a metallic frame which rotates around its longitudinal axis. The hollow cylinder or shell is partially filled with balls which form the grinding medium of the ball mill. Different kinds of grinding media used in ball mills include steel and other metal balls; metal cylindrical bodies called cylpebs; and ceramic balls with regular or high density. However, the inner wall of the shell is generally coated with an abrasion-resistant material, such as manganese steel or rubber (causes less wear to the products). This process occurs at a low speed and the balls roll over each other in a hollow cylinder. The working methodology of a ball mill consists of:

- A feed material meant for break down that is introduced in the cylinder, optimally at 60 per cent of the volume of the cylinder
- A fixed number of balls are inserted in the cylinder and the mill starts rotating
- The speed of the rotation is very important to obtain the desired size of particles. At low speed, the balls will roll over each other and only the attrition principle will apply
- At high speed, the centrifugal force tends to throw the balls to the walls of cylinder which means no grinding takes place. Here only compression by the balls against the wall occurs which is not sufficient for reducing size
- At optimum or critical speed, the results are significant due to the fact that the balls are picked up by the walls and carried to the top and finally fall to the bottom. This creates the impact and, due to the attrition principle, size reduction becomes more efficient

The major parameters of a ball mill include the size and number of balls; nature of balls; and the rotation speed. The grinding process in ball mills can be carried out either in dry or wet conditions depending on the wetting agents to be added at the start. The basic properties of the milling bodies include the mass and size of the

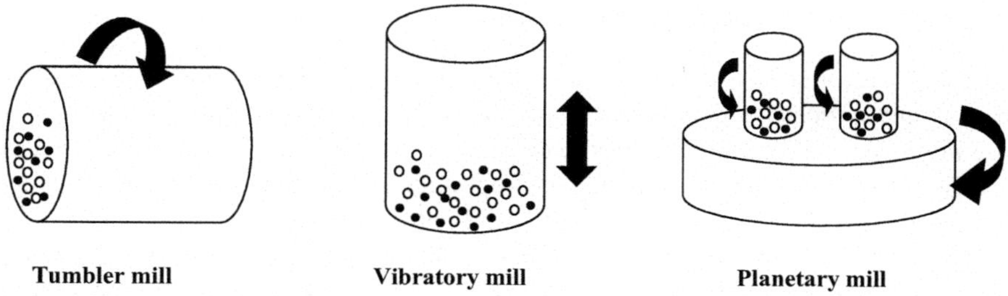

FIGURE 11.16 Different kinds of ball mill instruments.

particles, their wear rate, the influence on the particle breakage rate, and energy efficiency of the grinding process.

Types of Ball Mills: Ball mills are mainly of three types, i.e., tumbler mills, vibratory mills, and planetary mills (see Figure 11.16). A tumbler mill consists of a cylindrical shell partially filled with steel balls rotating on its longitudinal axis.

Due to the proportionality of dependence of efficiency on the diameter of the mill, the greater the diameter, the greater will be the height of the fall and consequently the higher will be the energy transmitted to the balls. In the case of vibratory mills, the sample material and grinding medium in the vessel are shaken back and forth at high vibration frequencies. A planetary mill consists of vessels placed on a rotating disk which rotate around their own axes. The size of the vessel acts as an essential parameter because a high distance increases the kinetic energy which ultimately enhances the impact of the process.

Ball milling is a non-combustion technology which follows simple grinding operations of particles, with no solvent, gas, or temperature requirements. The economical and efficient output of the ball mill method indicates its promising applications in the decontamination of soil pollutants. Ball milling is known for treating different kinds of contaminated soils, such as polychlorinated biphenyls (PCBs), organic bromide, and HMs like lead, zinc, and chromium. As ball milling processes have a low grinding cost in a closed system, they are increasingly used for the detoxification of persistent organic pollutants and HMs in contaminated soil. Ball milling was initially used for the effective removal of HMs, such as, Pb, Zn, and Cr. A high amount of energy is provided to the contaminated soil with innumerable collisions or shear to enable the mechano-chemical reactions. This supply of energy with respect to the particular soil or pollutant further promotes the physico-chemical transformations which may lead to the successful degradation or immobilization of various pollutants.

Ball milling has numerous functions and applications, and is not only used to treat various classes of pure waste chemicals or solid waste but also acts as a significant remediation of contaminated soil. In addition, the ball milling method is a relatively prominent energy-saving soil remediation technique when compared to the other conventional technologies. In brief, ball milling can be simply applied, and is efficient, economical, and has green soil remediation effects.

11.3 CHEMICAL PRECIPITATION

Chemical precipitation is the most common technology for the removal of HMs, phosphorus, and hardness in contaminated or waste water. This technique is widely used to remove dissolved ionic HMs from contaminated solutions. The chemical reaction is carried out between the soluble metal compounds and the precipitating agent which converts the ionic HMs into an insoluble solid particle form.

The solid form thus obtained is known as the precipitate, while the remaining liquid residue is called the supernatant. The effectiveness of a chemical precipitation process is dependent on several factors which include:

- The type and concentration of ionic metals present in the solution
- The precipitant used
- The reaction conditions (especially the pH of the solution)
- The presence of other constituents that may inhibit the precipitation reaction

Chemical precipitation is the most common method for removing dissolved HMs from wastewater. pH is one of the major parameters that regulates chemical precipitation. Initially, the precipitation agents (ligands) tend to increase the pH of the wastewater with the help of an alkaline agent which consequently reduces the concentration of metals in the solution causing them to precipitate.

The process of altering the ionic equilibrium of contaminated water to produce insoluble precipitates can further be removed easily by a sedimentation process. The chemical precipitation process is accomplished by following a solids separation operation which includes coagulation and/or sedimentation, or filtration to remove the obtained precipitates (see Figure 11.17). It is the least expensive technology, and is always combined with chemical coagulation for HM wastewater treatment. The entire process can be preceded by chemical reduction to change the characteristics of the metal ions that can be precipitated. The conventional chemical precipitation processes include hydroxide precipitation and sulphide precipitation; however, alternative methods are used for the precipitation process.

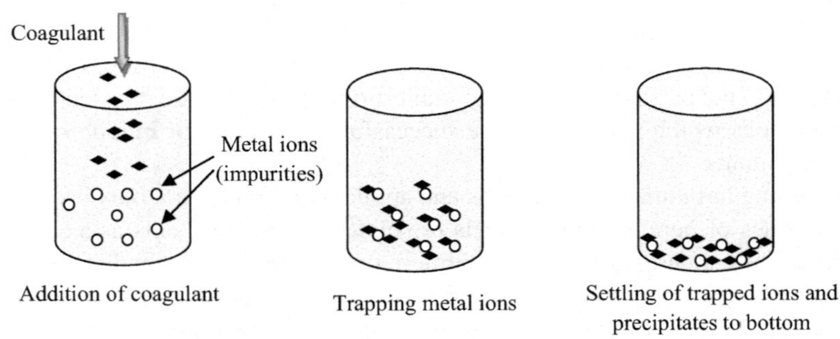

FIGURE 11.17 Schematic representation of the basic chemical precipitation process.

11.3.1 Hydroxide Precipitation

Relative simplicity, low cost, and ease of pH control make the hydroxide precipitation technology one of the most widely used precipitation techniques for decontamination of wastewater. In the process of hydroxide precipitation, an alkaline agent is used to increase the pH of the solution resulting in decreased metal ion solubility, and thus their precipitation from the solvent. To do this, lime or sodium hydroxide is considered best for carrying out this type of precipitation process. The mechanism of HM removal by precipitation of metal hydroxides is shown in Equation (11.1).

$$M^{2+} + 2\,(OH^-) \rightleftharpoons M\,(OH)_2 \qquad (11.1)$$

where M^{2+} and OH^- represent the dissolved metal ions and the precipitant, respectively, while $M(OH)_2$ is the insoluble metal hydroxide. In hydroxide precipitation, different kinds of coagulants are used for the precipitation of HMs from the solution, such as calcium carbonate, calcium hydroxide (slaked lime), calcium oxide (quicklime), sodium carbonate (soda ash), sodium hydroxide, and ammonium hydroxide. The process of metal hydroxide precipitation has certain advantages, such as the technique is relatively simple, implementation costs are low, and pH is easy to control. The supplementation of coagulants such as alum, iron salts, and organic polymers can relatively enhance the detoxification of HMs from wastewater. Besides these, the process of chemical precipitation has some limitations also. Firstly, hydroxide precipitation can cause a dewatering and disposal problem due to the deposition of large volumes of relatively low-density sludge. Some metal hydroxides are amphoteric in nature, which creates a problem for some of the metals to fall back into the solution due to variations in the ideal pH of the elements.

11.3.2 Sulphide Precipitation

Metal sulphide precipitation is also an effective method used for the detection and removal of HM ions. Metal sulphide precipitation can be achieved using a variety of precipitants such as sodium sulphide (Na_2S), sodium hydrosulphide (NaHS), calcium sulphide (CaS), barium sulphide (BaS), iron sulphide (FeS), ammonium sulphide [$(NH_4)_2S$], and sodium thiosulphate ($Na_2S_2O_3$).

The sulphide precipitation process is quite similar to the precipitation of metal hydroxides. To obtain the precipitates, sulphide is administered to the solution and residual precipitates are separated by deposition or gravity filtration. The mechanism of sulphide reaction is described in Equations (11.2) and (11.3).

$$Na_2S + MSO_4 = MS\ (\text{precipitate}) + Na_2SO_4 \qquad (11.2)$$

$$FeS + Cu^{2+} = CuS\ (\text{precipitate}) + FE^{2+} \qquad (11.3)$$

Lower solubility of metal compounds is one of the main advantages of using the sulphide precipitation technique. In addition, the wide range of pH helps to achieve a greater degree of metal reduction in a shorter time. Because of the better settling properties of the precipitated sludge, it becomes easier to thicken and dewater as compared with sludge obtained during hydroxide precipitation. Also, the disadvantages of metal sulphide precipitation include low solubility of metal sulphides, high sensitivity of the process to the dosing of the precipitation agent, and the emission of toxic hydrogen sulphide during the process. Other than the conventional methods, chelating ligands are also used as alternative methods for the precipitation of HMs from aqueous solutions. Three widely used commercial reagents for HM binding include potassium/sodium thiocarbonate (STC), 2,4,6-trimercaptotiazine, trisodium salt nonahydrate (TMT), and sodium dimethyldithiocarbamate (SDTC).

11.4 NANOTECHNOLOGY IN THE REMOVAL OF HMS

Nanotechnology involves the manipulation of matter on an atomic scale. It is the construction and engineering of functional systems at a microscopic level. It deals with structures sized between 1 and 100 nanometres, where 1 nanometre is one billionth of a metre, roughly the width of three or four atoms. The term nanotechnology was coined by Norio Taniguchi in 1974. Nanotechnology is one of the latest advances which makes materials lighter, stronger, faster, smaller, and more durable. According to the National Science Foundation, nanotechnology is the capability to understand, manipulate, and control matter at the level of individual atom and molecules. Nanotechnology is sometimes confused with nano-science. Surface-based adsorption nanomaterials can be used to remove HMs from wastewater. With improvements in nanotechnology, nanomaterials have been delicately designed and fabricated for HM detection and removal, exerting numerous advantages. Various nanomaterials used for removal of toxic HMs, based on their fundamental material, include the following:

Carbon-Based Nanomaterial: Carbon-based nanomaterials like carbon nanotubes, graphene, graphene oxides, and carbon-based nano-composite methods have slowly replaced the activated carbon method.

Carbon Nanotubes: Carbon nanotubes are small-sized cylindrical hollow tubes in which carbon atoms arranged in hexagonal rings have a large surface area, high ground object ratio, good mechanical strength, and high electrical conductivity, which can easily remove HMs from aqueous solution. The physical adsorption, electrostatic attraction, chemical interaction, and sorption-precipitation are various mechanisms by which carbon nanotubes can absorb HM ions from water. Metallic ions like Cd^{2+}, V^{5+}, Cr^{6+}, Co^{2+}, Cu^{2+}, As^{3+}, Ni^{2+}, Pb^{2+}, Cu^{2+}, Zn^{2+}, Sr^{2+}, and Hg^{2+} are removed by carbon nanotubes which have shorter intraparticle diffusion distances and more accessible adsorption sites per unit mass. Single-walled carbon nanotubes and multi-walled carbon nanotubes are two primary allotropes of carbon nanotubes. Single-walled carbon nanotubes undergo a strong redox reaction with Hg^{2+} ions and have high sensitivity towards local dielectric or redox environments due to their band gap, which helps in successful removal of heavy toxic metals. Multi-walled carbon nanotubes have become one of the most active materials in carbon paste electrodes due to their unique structure. Internal sites, interstitial channels, grooves, and outer surfaces are four types of adsorption sites in carbon-based nanotubes. Exposed carbon nanotube bundles have more active sites as compared to capped carbon nanotubes. Adsorbent material is directly exposed to adsorption sites, which results in attaining faster equilibrium on the grooves and outer surfaces as compared to inner sites. Soot, carbon-catalyst particles, and carbon-based impurities stick to active adsorption sites and slow down the rate of adsorption of HMs by carbon nanotubes. Electrostatic interactions of carbon nanotubes via acid treatment and grafting functional groups increase the adsorption capacity of carbon nanotubes. Carboxyl, phenol, and hydroxyl are some important surface functional groups of carbon nanotubes that can absorb metal ions and release H^+ ions which decrease the pH of the aqueous solution. HMs like manganese dioxide, aluminium oxide, and iron oxide are easily removed by metal oxide modification of carbon nanotubes. Cd ions are removed from water using manganese-oxide-coated carbon nanotubes.

Nano-composites: These are eco-friendly nanomaterials made of multiphase and heterogeneous substances. A nano-filter is the first nano-scale phase which is dispersed in the matrix or second phase. Scientists are successful in the artificial synthesis of carbon-layered silicate nano-composite which is an advanced and better choice as compared to traditional carbon nanomaterials. Pb is removed from nano-adsorbent-modified carbon composite which is fully eco-friendly and safe.

Nano-sensors: The latest techniques of green hybrid nano-sensors have been developed with reduced graphene oxide, cellulose nano-whiskers, and polyamide nanofibres which have been used for the detection of HMs in water samples.

Nano-spheres: Colloidal hollow carbon nanospheres are another type of sensory material having large voids, porous shell configuration, spherical morphology, high surface area, low density, and that can be used for the removal of HMs.

Zeolite Nanoparticles: Zeolites are porous, tetrahedral, non-flexible crystalline hydrated alumina-silicate nanomaterials that have a frame of pores similar to a sponge in which small molecules reside. Atoms of aluminium and silicon are surrounded by nascent oxygen atoms to form regular intra-crystalline channels. The removal of toxic metals from wastewater can be efficiently achieved by adsorbing active and large surface areas of the zeolite nanoparticles.

Magnetic Nanomaterials: A magnetic nanomaterial is one of the most advanced nanomaterials that employ nanotechnology and magnetic separation to remove hazardous HMs. These are eco-friendly, dispersible, high surface area, less toxic, recyclable, biocompatible, and have wide applications in environmental remediation. This advance method is cost-effective and reliable in the field of water purification. The presence of different functional groups on magnetic surfaces produces electrostatic interactions, van der Waals interactions, ligand combination interactions, complex formation interactions, and chemical binding interactions due to which interactions between magnetic surfaces and metals are possible. Surface alteration of magnetic nanoparticles can be achieved utilizing numerous modifiers, such as inorganic and organic carbonaceous compounds and polymeric materials. Chemical synthesis produces large numbers of metallic and metal oxide magnetic nanoparticles for environmental applications at low cost. Other types of nonomaterial used for the purpose are:

- Carbon materials
- Magnetic nanoparticles
- Cellulose-based nanomaterials
- Chitosan-based nanomaterials
- Dendrimer-based nanomaterials
- Metal oxides and metal-based nanomaterials
- Polymer-based nanomaterials
- Silica-based nanomaterials

Different types of nanomaterial and their features are listed in Table 11.3.

TABLE 11.3
Types of Nanomaterial Used for the Removal of Heavy Metals

Types of Nanomaterial		Features	Application in removal of HMs
Carbon-based nanomaterials	**Carbon- nanotubes**	• Small-sized cylindrical hollow tubes where carbon atoms are arranged in hexagonal rings • Large surface area, good mechanical strength and electrical conductivity	Cd^{2+}, V^{5+}, Cr^{6+}, Co^{2+}, Cu^{2+}, As^{3+}, Ni^{2+}, Pb^{2+}, Cu^{2+}, Zn^{2+}, Sr^{2+}, Hg^{2+}
	Nano-composite	• Carbon-layered silicate nano-composite	Pb^{2+}
	Graphene	• High dispersity in water • More surface area with functional groups	Cr^{6+}, As^{5+}, Pb^{2+} ions
	Nano-sensors	• Green hybrid nano-sensors	Cu^{2+}
	Nano-spheres	• Large voids and porous shell configuration • Spherical morphology with high surface area and low density	Cd^{2+}, Cu^{2+}, Ni^{2+}, Pb^{2+}, Hg^{2+}
Zeolite nanoparticles	Porous, tetrahedral, non-flexible and crystalline hydrated aluminosilicate		Ni^{2+}, Cu^{2+}, Pb^{2+}, Cd^{2+}
Polymer-based nanomaterials	Highly porous with high gas penetrability		Cu^{2+}

TABLE 11.3 (Continued)
Types of Nanomaterial Used for the Removal of Heavy Metals

Types of Nanomaterial	Features	Application in removal of HMs
Cellulose-based nanomaterials	Excellent synthesized bio-adsorbent	Cu^{2+}, Ca^{2+}, Cr^{6+}, Pb^{2+} ions
Dendrimer-based nanomaterials		
	Radially symmetric and homogeneous	Co^{2+}, Cu^{2+}, Ni^{2+}, Cd^{2+}
Chitosan-based nanomaterials	Non-toxic, biodegradable, compatible, and hydrophilic	Cd^{2+}, Cu^{2+}, Hg^{2+}
Magnetic nanomaterials	Eco-friendly, high surface area, recyclable, and biocompatible	Cd^{2+}, V^{5+}, Cr^{6+}, Co^{2+}, Cu^{2+}, As^{3+}, Ni^{2+}, Pb^{2+}, Cu^{2+}, Zn^{2+}, Sr^{2+}
Carbon material magnetic nanoparticles	Carbonaceous magneto-nano adsorbent	Hg^{2+}, Ag^{+}, Cd^{2+}, Pb^{2+}, Mn^{2+} and Zn^{2+} ions
Silica-based nanomaterials	Non-toxic, eco-friendly with large pore size and surface area	Cu^{2+}, Cd^{2+}, Pb^{2+}
Metal oxides & metal-based nanomaterials	High adsorption capacity and large surface area	Cu^{2+}, As^{3+}, Ni^{2+}, Pb^{2+}, Cu^{2+}, Zn^{2+}, Fe^{2+}

11.5 ADVANTAGES AND DISADVANTAGES OF PHYSICO-CHEMICAL METHODS

On the whole, physico-chemical treatments are more advantageous as they (Table 11.4):

- Occur rapidly
- Are easy to operate and control
- Possess flexibility to temperature change
- Accommodate variable input loads and flow such as seasonal flows and complex discharge
- Can be modified, whenever required
- Have reduced chemical costs and feasible sludge disposal, and also physico-chemical treatments have been found to be the most appropriate treatments for inorganic effluents

TABLE 11.4
Advantages and Disadvantages of Some Physico-Chemical Methods Used for Treating Heavy Metals in Wastewater

Treatment Method	Advantages	Disadvantages
Adsorption with new adsorbents	• Low-cost • Easy operating conditions • Wide pH range • High metal-binding capacities	• Low selectivity • Production of waste products
Chemical Precipitation	• Low capital cost • Simple operation	• Sludge generation • Extra operational cost for sludge disposal
Electro-dialysis	• High separation selectivity	• High operational cost due to membrane fouling • High energy consumption
Membrane Filtration	• Require small space • Low pressure • High separation selectivity	• Owing to membrane fouling, high operational cost
Photo-catalysis	• Concurrent removal of metals and organic pollutant • Produces less harmful by-products	• Long duration time • Limited applications

BIBLIOGRAPHY

Acar, Y. B. and Alshawabkeh, A. N. (1993) Principles of electrokinetic remediation. *Environmental Science & Technology* 27: 2638–2647.

Acar, Y. B., et al. (1995) Electrokinetic remediation – Basics and technology status. *Journal of Hazardous Materials* 40: 117–137.

Amado, F. D. R., et al. (2005) Development of polyurethane/polyaniline membranes for zinc recovery through electrodialysis. *Desalination* 186 (1–3): 199–206.

Arab, F. and Mulligan, C. N. (2018) An eco-friendly method for HM removal from mine tailings. *Environmental Science and Pollution Research* 25: 16202–16216.

Arana, J., et al. (2022) Electrodialysis for metal removal and recovery: A review. *Chemical Engineering Journal* 435(1): 1–20.

Begum, Z. A., et al. (2012) Remediation of toxic metal contaminated soil by washing with biodegradable aminopolycarboxylate chelants. *Chemosphere* 87(10): 1161–1170.

Bisht, R., Agarwal, M., and Singh, K. (2017) Methodologies for removal of HM ions from wastewater: An overview. *Interdisciplinary Environmental Review* 18: 124–142.

Cameselle, C., Gouveia, S., Akretche, D. E., and Belhadj, B. (2013) Advances in electrokinetic remediation for the removal of organic contaminants in soils. In: *Organic Pollutants – Monitoring, Risk and Treatment*, Rashed, M. N. (ed.). London: IntechOpen.

Chen, F. and Fu, M. (2012) Study on remediation of soils polluted by HMs using surfactants. *Sichuan Environment* 31(4): 61–64.

Chen, S. S., et al. (2009) Concentration and purification of chromate from electroplating wastewater by two-stage electrodialysis processes. *Journal of Hazardous Materials* 161 (2–3): 1075–1080.

Chen, X. and Wu, Y. (2018) Remediation mechanism of multi-HM contaminated soil by using different chemical washing reagents. *Chinese Journal of Environmental Engineering* 12(10): 2845–2854.

Cifuentes, L., García, I., Arriagada, P., and Casas, J. M. (2009) Use of electrodialysis for metal separation and water recovery from $CuSO_4$–H_2SO_4–Fe solutions. *Separation and Purification Technology* 68: 105–108.

Dąbrowski, A., Hubicki, Z., Podkościelny, P., and Robens, E. 2004. Selective removal of the HM ions from waters and industrial wastewaters by ion-exchange method. *Chemosphere* 56: 91–106.

Dermont, G., Bergeron, M., Mercier, G., and Richer-Laflèche, M. (2008) Soil washing for metal removal: A review of physical/chemical technologies and field applications. *Journal of Hazardous Materials* 152: 1–31.

El-Gaayda, J., et al. (2021) Natural flocculants for treatment of wastewaters containing dyes or HMs: a state-of-the-art review. *Journal of Environmental Chemical Engineering* 9(5): 106060.

Fang, Q., et al. (2010) Functional mesoporous metal-organic frameworks for the capture of HM ions and size-selective catalysis. *Inorganic Chemistry* 49: 11637–11642.

Fu, F. and Wang, Q. (2011) Removal of HM ions from wastewaters: A review. *Journal of Environmental Management* 92: 407–418.

Gherasim, C., Křivčík, J., and Mikulášek, P. (2014). Investigation of batch electrodialysis process for removal of lead ions from aqueous solutions. *Chemical Engineering Journal* 256: 324–334.

Güvenç, A. and Karabacakolu, B. (2005) Use of electrodialysis to remove silver ions from model solutions and wastewater. *Desalination* 172: 7–17.

Hansen, H. K., Ottosen, L. M., Kliem, B. K., and Villumsen, A. 1997. Electrodialytic remediation of soils polluted with Cu, Cr, Hg, Pb and Zn. *Journal of Chemical Technology & Biotechnology* 70: 67–73.

Ionics Inc. (1984) *Electrodialysis – Electrodialysis Reversal Technology*. Watertown, Massachusetts: Ionics Inc.

Jamil, T. S., et al. (2018) Novel anti-fouling mixed matrix CeO2/Ce7O12 nanofiltration membranes for HM uptake. *Journal of Environmental Chemical Engineering* 6: 3273–3282.

Kanakaraju, D., Rusydahbt-Mohamad, S. N., Lim, Y. C., and Pace, A. (2019) Concurrent removal of Cr (III), Cu (II), and Pb (II) ions from water by multifunctional TiO_2/Alg/FeNPs beads. *Sustainable Chemistry and Pharmacy* 14(8): 100176.

Kim, C. and Ong, S. K. (1999) Recycling of lead-contaminated EDTA wastewater. *Journal of Hazardous Materials* 69: 273–286.

Kobielska, P. A., Howarth, A. J., Farha, O. K., and Nayak, S. (2018) Metal–organic frameworks for HM removal from water. *Coordination Chemistry Reviews* 358: 92–107.

Kumar A. and Pandey G. (2017) A review on the factors affecting the photocatalytic degradation of hazardous materials. *Material Science & Engineering International Journal* 1(3): 106–114.

Leˇstan, D., Luo, C., and Li, X. (2008) Use of chelating agents in remediation of metal-contaminated soils: A review. *Environmental Pollution* 153: 3–13.

Liu, L., Li, W., Song, W., and Guo, M. (2018) Remediation techniques for HM-contaminated soils: Principles and applicability. *Science of the Total Environment* 633: 206–219.

Liu, Y., Yu, L., and Li, D.W. (2012) Study on enhanced electrokinetic removal of HMs in MSWI fly ash assisted by cation exchange membranes. *Research Journal of Chemistry and Environment* 16: 63–66.

Lu, X., et al. (2017) UV-initiated synthesis of a novel chitosan-based flocculant with high flocculation efficiency for algal removal. *Science of The Total Environment* 609: 410–418.

Nath A., Mishra A., and Pande P. P. (2019) A review natural polymeric coagulants in wastewater treatment. *Materials Today: Proceedings* 46: 6113–6117.

Paria, S. (2008) Surfactant-enhanced remediation of organic contaminated soil and water. *Advances in Colloid and Interface Science* 138: 24–58.

Piras, C. C., Fernández-Prieto, S., and De Borggraeve, W. M. (2019) Ball milling: A green technology for the preparation and functionalisation of nanocellulose derivatives. *Nanoscale Advances* 1: 937–947.

Pohl, A. (2020) Removal of HM ions from water and wastewaters by sulfur-containing precipitation agents. *Water, Air, and Soil Pollution* 231(503): 1–17.

Qasem, N., Mohammed, R., and Lawal, D. (2021) Removal of HM ions from wastewater: A comprehensive and critical review. *npj Clean Water* 4(36): 1–15.

Reddy, K. R., Xu, C. Y., and Chinthamreddy, S. (2001) Assessment of electrokinetic removal of HMs from soils by sequential extraction analysis. *Journal of Hazardous Materials* 84: 279–296.

Ren, G., et al. (2021) Recent advances of photocatalytic application in water treatment: A review. *Nanomaterials (Basel)* 11(7): 1804.

Rivas, B. and Eduardo-Pereira, A. M. (2009) Functional water-soluble polymers: polymer–metal ion removal and biocide properties. *Polymer International* 58: 1093–1014.

Saleem, M. and Bachmann, R. T. (2019) A contemporary review on plant-based coagulants for applications in water treatment. *Journal of Industrial and Engineering Chemistry* 72: 281–297.

Seader, J. D. and Henley, E. J. (1998) *Separation Process Principles*. New York: John Wiley & Son.

Shiba, S. and Hirata, Y. (2002) *In-situ* electrokinetic remediation of groundwater contaminated by HM. *Developments in Water Science* 47: 883–890.

Shon, H., et al. (2013) Nanofiltration for water and wastewater treatment – A mini review. *Drinking Water Engineering and Science* 6: 47–53.

Su, X., et al. (2019) Systematic approach to evaluating environmental and ecological technologies for wastewater treatment. *Chemosphere* 218: 778–792.

Wang, L. K., Chen, J. P., Hung, Y.-T. and Shammas, N. K. (2011) *Membrane and Desalination Technologies. Handbook of Environmental Engineering*. Louisville: Humana Press, Springer.

Wasay, S. A., Barrington, S. and Tokunaga, S. (1988) Remediation of soils polluted by HMs using salts of organic acids and chelating agents. *Environmental Technology* 19: 369–380.

Xia, W., et al. (2009) Influence factors analysis of removing HMs from multiple metal-contaminated soils with different extractants. *Journal of Central South University* 16(1): 108–111.

Xuefengm X., Yongjun, S., Jianwen, L., and Huaili, Z. (2021) Flocculation of HM by functionalized starch-based bioflocculants: Characterization and process evaluation. *Separation and Purification Technology* 267.

Yadav, M., Singh, G., and Jadeja, R. N. (2021) Physical and chemical methods for HM removal. *Pollutants and Water Management* 377–397.

Yan, R., et al. (2020) Simultaneous removal of Cu (II) and Cr (VI) ions from wastewater by photoreduction with TiO_2–ZrO_2. *Journal of Water Process Engineering* 33: 101052.

Yang, J. S., Hwang, J. M., Baek, K., and Kwon, M. J. (2013) Soil washing and effluent treatment for contaminated soil with toxic metals. Korean. *Journal of Chemical Engineering* 51(6): 745–754.

Yeung, A. T. and Gu, Y. Y. (2011) A review on techniques to enhance electrochemical remediation of contaminated soils *Journal of Hazardous Materials* 195: 11–29.

Zhang, W., et al. (2016) Process analysis and design suggestion on ex-situ washing technology for HM contaminated soil. *Environmental Engineering* 34(12): 177–182.

Zhang, Y., et al. (2016) Unveiling the adsorption mechanism of zeolitic imidazolate framework-8 with high efficiency for removal of copper ions from aqueous solutions. *Dalton Transactions* 45(32): 12653–12660.

Zimmerman, A. J. and Weindorf, D. C. (2010) HM and trace metal analysis in soil by sequential extraction: A review of procedures. *International Journal of Analytical Chemistry* 2010: 1–7.

12 Bioremediation of Heavy Metals

The Truly Healthy environment is not merely safe but stimulating.

William H. Stewart

After going through this chapter, the reader will learn about microbes used in bioremediation, factors affecting microbial bioremediation, bioremediation techniques, *in situ* and *ex situ* bioremediation techniques, their advantages and disadvantages, and the applications of bioremediation.

FACT FILE

- Bioremediation, a branch of biotechnology, employs living organisms to decontaminate contaminated areas
- Bioremediation is a process where organic wastes are biologically destroyed under controlled conditions to a harmless state, or to levels below concentration limits
- Most bioremediation technologies destroy contaminants in the soil matrix
- Bioremediation technologies are usually meant to reduce toxicity either by annihilation or by converting toxic organic compounds into less toxic ones
- Plants may also be employed to improve biodegradation and stabilize the soil
- Field applications of bioremediation may involve:
  - ✓ Airing of polluted soils
  - ✓ Digging
  - ✓ Removal of fluid
  - ✓ Inoculation of fluid
  - ✓ Introduction of nutrients and substrates
  - ✓ Mixing polluted soils
  - ✓ Soil handling
- Some microscopic "bugs" or microbes living in soil and groundwater enjoy consuming harmful chemicals

DOI: 10.1201/9781003343721-12

- After digesting chemicals, microorganisms change them into H_2O and non-toxic gases such as CO_2
- Bioremediation can be useful only where environmental conditions allow microbial growth and activity

12.1 INTRODUCTION

Bioremediation is a technique where biological organisms are employed to eliminate or deactivate environmental contaminants through a metabolic process. Biological organisms include fungi, algae, and bacteria, and remediation implies addressing a hazardous situation. Microbes grow in varied habitats, including soil, water, plants, animals, deep sea, and even in sub-zero environments. Their number and inclination for a variety of chemicals make them perfect contenders for acting as ecological caretakers. Bioremediation is defined as:

> "…a waste management technique that involves living organisms to get rid of pollutants from a polluted site…"

> "…a treatment technique that makes use of natural organisms to breakdown the detrimental materials into less toxic or non-toxic material..."

Environmental biotechnology is not a recent discipline; composting and wastewater treatment are well-known examples of earlier environmental biotechnology interventions. However, recent advances in molecular biology and ecology offer prospects for more effective biological processes. Remarkable achievements of research include the cleaning-up of polluted water and land areas. Bioremediation as a process has evolved in recent years wherein organic wastes are biologically degraded under controlled conditions to a harmless state. Bioremediation uses living organisms, primarily microorganisms, to degrade environmental contaminants into less toxic forms or to detoxify hazardous substances. Microbes may be indigenous to a polluted area or they may be isolated from elsewhere and brought to the polluted site. Pollutants are changed by living organisms through reactions that take place as a part of their metabolic processes. Biodegradation is a result of the actions of multiple organisms. When microbes are introduced into a contaminated site to enhance degradation, the process is known as bio-augmentation. Bioremediation is generally more cost-effective than traditional methods such as incineration, and some pollutants can be treated on site, thus decreasing exposure risks. Control and optimization of bioremediation processes is governed by the existence of a microbial population capable of degrading the pollutants; availability of pollutants to the microbial population; and environmental factors (type of soil, temperature, pH, and presence of oxygen or other electron acceptors, and nutrients). Bioremediation techniques have extensive uses and continue to expand at an exponential rate. The use of microbial processes for remediation of polluted sites has turned out to be effective and reliable due to its eco-friendly and cost-effective features. New bioremediation technologies have been developed for restoring polluted environments. Both indigenous and non-indigenous microbes have been used in bioremediation. Indigenous microbes present in the environment hold the key to solving the challenges linked with the biodegradation and bioremediation of pollutants. The basic principle of bioremediation is to reduce, detoxify, degrade, mineralize, or transform more toxic pollutants into less toxic ones. The pollutant removal process mainly depends on the nature of the pollutant, which may include:

- Agrochemicals
- Chlorinated compounds
- Dyes
- Greenhouse gases
- Heavy metals (HMs)
- Hydrocarbons
- Nuclear waste
- Organic halogens
- Pesticides
- Plastics
- Sludge
- Xenobiotic compounds

Bioremediation involves:

- Degradation
- Eradication
- Immobilization or detoxification of diverse chemical wastes and physical hazardous materials using all-inclusive microbial action

12.2 MICROBES USED IN BIOREMEDIATION

Microorganisms play an important role in maintaining the biological equilibrium. Bacteria, fungi, algae, and yeasts are involved in bioremediation for removing pollutants. The tolerance of microbes to sub-zero temperatures and extreme heat helps in the removal of hazardous compounds or waste. Moreover, the adaptability of biological systems in diverse environments makes them suitable for the remediation process. Carbon plays an important role in microbial activity. These main microbes used for this purpose are:

- *Achromobacter*
- *Alcaligenes*
- *Arthrobacter*
- *Bacillus*
- *Corynebacterium*
- *Flavobacterium*
- *Mycobacterium*
- *Nitrosomonas*
- *Pseudomonas*
- *Xanthobacter*

12.2.1 Factors Affecting Microbial Bioremediation

Bioremediation involves degrading, eliminating, altering, restraining, or decontaminating various chemicals and physical pollutants from the environment by making use of bacteria, fungi, algae, and plants (see Figure 12.1). The enzymatic metabolic pathways of microbes facilitate the biochemical reactions which eventually degrade pollutants. Microbes act on pollutants which consecutively generate energy and supply nutrients to the microbes for multiplying cells. The efficiency of bioremediation depends on:

Biotic Factors: These factors help in degrading organic compounds by microbes with inadequate carbon sources, and an antagonistic interface among microbes or protozoa and bacteriophages. The rate of contaminant degradation often depends on the concentration of the pollutant and the amount of catalyst present in the biochemical reaction. The major biological factors are:

- Composition
- Enzyme activity
- Growth for biomass production

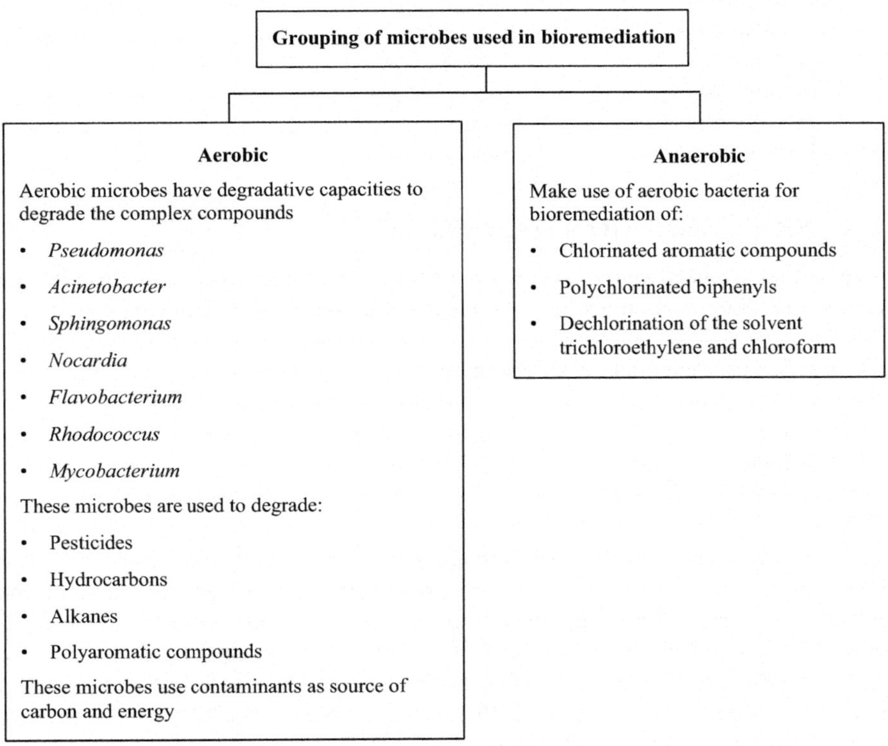

FIGURE 12.1 Grouping of microbes used in bioremediation.

- Horizontal gene transfer
- Interaction (competition, succession, and predation)
- Mutation
- Population size

Abiotic Factors: There is an interaction of the metabolic activity of microbes with environmental pollutants during bioremediation and the success of this interaction depends on the ecological conditions. Microbial growth, degradation kinetics, and activity depend on:

- Chemical structure
- Concentration
- Deficiency of resources
- Moisture content
- Nutrients
- Osmotic pressure
- Oxygen content
- pH
- Physico-chemical bioavailability of pollutants
- Redox potential
- Site conditions
- Soil structure
- Solubility
- Temperature
- Toxicity
- Water solubility

12.3 BIOREMEDIATION TECHNIQUES

Bioremediation can be undertaken *ex situ* or *in situ* (see Figure 12.2). While choosing the bioremediation technique, the following standards are taken into consideration:

- Complexity and quantity of pollution
- Cost-effectiveness
- Location
- Nature of pollutant
- Type of environment

Ex situ bioremediation involves excavating contaminants from polluted sites followed by transportation to another site for treatment. Performance standards regulate the selection of *ex situ* bioremediation techniques. It is divided further into two main types: solid-phase bioremediation and slurry-phase bioremediation. In solid-phase bioremediation, the contaminated soil containing organic waste, industrial waste, and municipal waste is excavated and piled-up. Bacterial growth is initiated through pipes that are dispersed throughout the piles.

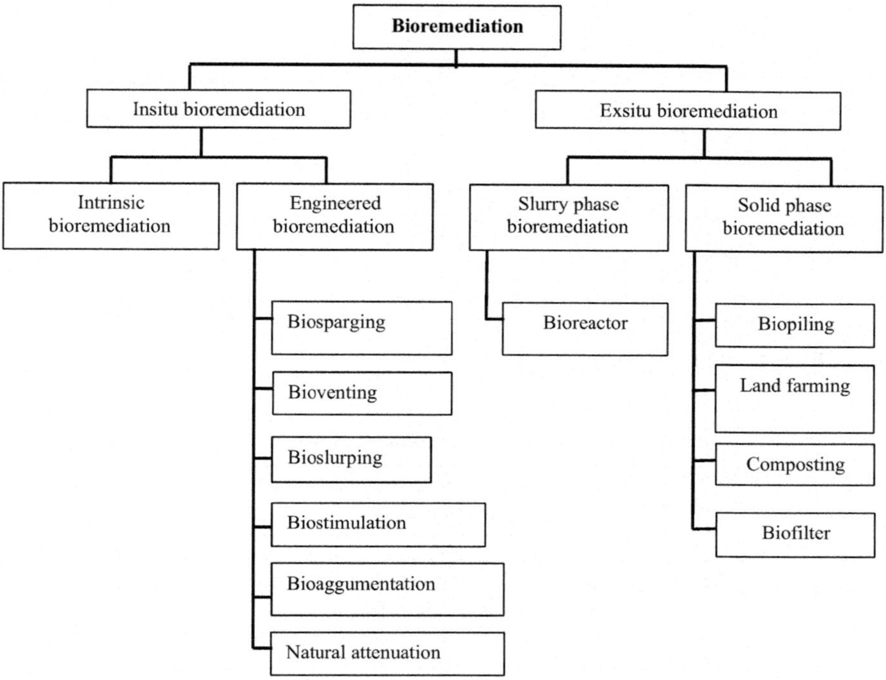

FIGURE 12.2 Different bioremediation techniques.

Air moving through the pipes is essential for ventilation and microbial respiration. A solid-phase system requires a large amount of space, while cleaning requires more time as compared to slurry-phase processes. Solid-phase treatment processes include biopiles, windrows, land farming, composting, etc.

12.3.1 Slurry-Phase Bioremediation Techniques

Slurry-phase bioremediation is a rapid process as compared to other treatment processes. Contaminated soil is mixed with water, nutrients, and oxygen in a bioreactor to create the optimum environment for microbes to decompose the contaminants present in the soil. Processing involves the separation of stones and rubble from the contaminated soil. The concentration of added water depends on the concentration of pollutants, the rate of the biodegradation process, and the physicochemical properties of the soil. After completion of the process, the soil is taken away and dried using vacuum filters, pressure filters, and centrifuges.

12.3.2 Biopile Bioremediation Techniques

This is an *ex situ* technique used for treating volatile low-molecular-weight pollutants. It is a cost-effective technique where the pH, nutrients, temperature, and aeration can

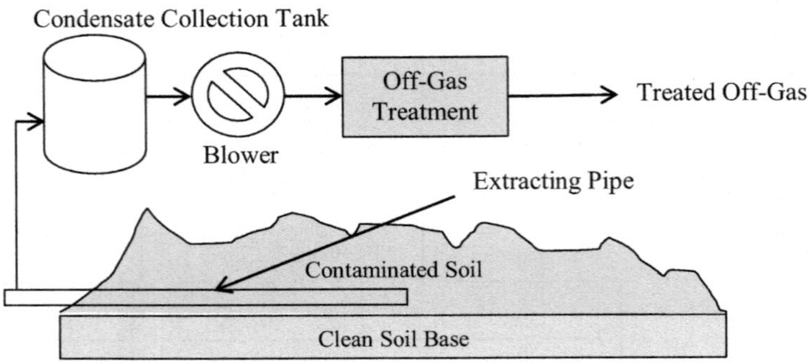

FIGURE 12.3 Biopiling setup.

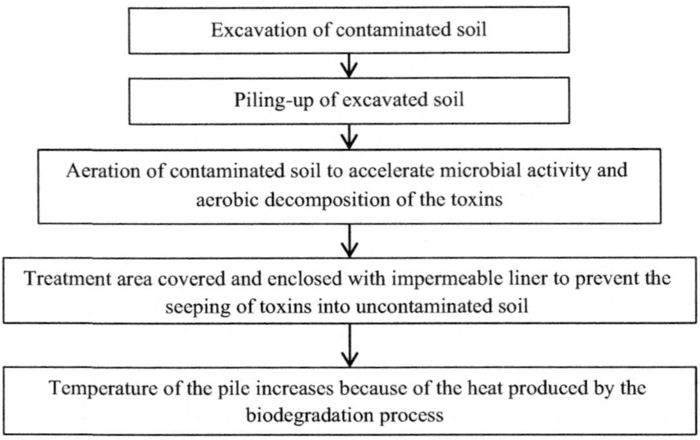

FIGURE 12.4 Steps involved in the biopiling process.

be controlled effectively. Excavated soil or silt is piled over an impervious base or pad with aeration. Pads with a cover have sufficient drainage to manage precipitation exposure, and monitor the temperature, moisture content, and concentration of pollutant. Biopiles are similar to composting and can be modified to accelerate the putrefaction process (see Figures 12.3 and 12.4). To create an appropriate environment for biodegradation, a moisture content of 40–85 per cent of field capacity is maintained through water sprinkling or drip irrigation. To increase penetrability, agents such as straw, bark, or wood chips can be added during construction of the biopile. In order to achieve the success of the bioremediation process through a biopile, it is essential to have proper aeration, which can be achieved through perforated tubing or slotted pipes placed at varying heights throughout the pile. Wind-powered, passive turbine vents can be used at the ventilation piping's protruding ends. Keeping in mind the climatic conditions, an impermeable (plastic) cover can be used to cover the biopile.

Biopiles are employed in the treatment of organic substances which degrade aerobically, such as non-halogenated volatile organic compounds and semi-volatile organic compounds. Pesticides, halogenated VOCs, and semi-volatile organic compounds (SVOCs) can also be treated, although the procedure's efficiency varies.

The cost of maintenance and monitoring of a biopile is minimal, and energy consumption is often low. Moreover, the cost of utilities, such as water and electricity, is often low. Its design and implementation are easy. Under ideal settings, the treatment time ranges from 6 months to 2 years. However, the main disadvantage of this technique is that it is inefficient at removing inorganic pollutants and radionuclides. Substantial heavy metal concentrations of more than 2,500 ppm may inhibit microbial development.

12.3.3 Windrows Bioremediation Techniques

As a modified version or extension to biopiling, the windrows technique involves periodic spinning of piled polluted soil to boost the bioremediation process. Polluted soil is turned by adding water which increases aeration, uniform distribution of pollutants, nutrients, and degradative activity of microbes, thus stepping up the rate of bioremediation, which can be achieved through integration, biotransformation, and mineralization. Research has shown that windrow treatment is more effective as compared to biopiles as it shows a higher rate of hydrocarbon removal. However, because of the periodic turning linked with windrow treatment, it may not be considered the best option in remediating soil polluted with toxic volatiles. The use of windrow treatment is linked with the release of CH_4 because of the anaerobic zone in piled polluted soil, which generally occurs after reduced aeration.

12.3.4 Land Farming or Land Treatment

Due to its cost-effectiveness and lower requirement of equipment for operation, land farming is believed to an uncomplicated, exceptional bioremediation technique. It incorporates liners and other methods to control the leaching of contaminants, which needs digging of contaminated soils, sediments, or sludge. Contaminated media are applied into lined beds and occasionally rotated to ventilate the waste. Soil conditions, such as, moisture content, pH, aeration, etc. are often controlled to enhance the degradation of contaminants. Contaminated media are generally treated in lifts. When the anticipated level of treatment is accomplished, the lift is removed and a new lift is constructed by adding contaminated media to the remaining material and mixing. This inoculates newly added material. Wastes are periodically tilled to aerate the waste. *Ex situ* land farming has been proved most successful in treating petroleum hydrocarbons. The main limitation of this method is that it requires a large space for its operation. Other limiting factors include conditions affecting the biological degradation of contaminants which are mostly uncontrolled, which amplifies the time to complete remediation, and inorganic contaminants are not biodegraded. Volatile contaminants, such as solvents, need to be pre-treated because they volatilize in the atmosphere, causing air pollution.

12.3.5 Bioreactors

The bioreactor technique is an *ex situ* biochemical processing system that uses microbes to remove pollutants from wastewater, pumped groundwater, and solid/

slurry of polluted soil treatment. A bioreactor is a container in which raw material is transformed to specific product(s) after a series of biological reactions. There are diverse operating modes of bioreactor, such as:

- Batch
- Fed-batch
- Sequencing batch
- Continuous
- Multistage

Polluted samples are fed into a bioreactor either as dry matter or slurry (see Figure 12.5). The bioreactor technique can be used to treat soil or water polluted with volatile organic compounds (VOCs) including benzene, toluene, ethylbenzene, and xylenes (BTEX). Since a bioreactor is an enclosed system, genetically modified microorganisms (GEMs) can be used for bio-augmentation, after which the organism can be destroyed before treated soil is returned to the field for landfilling. The bioremediation process in a bioreactor is explained in Figure 12.4. As compared to other processes, bioreaction is an effective and efficient procedure. The treatment rate is better for the reason that slurry bioreactors, being closed systems, speed up and improve the treatment rate by maintaining the environmental factors, including temperature, pH, aeration, nutrients, emissions, and reactions.

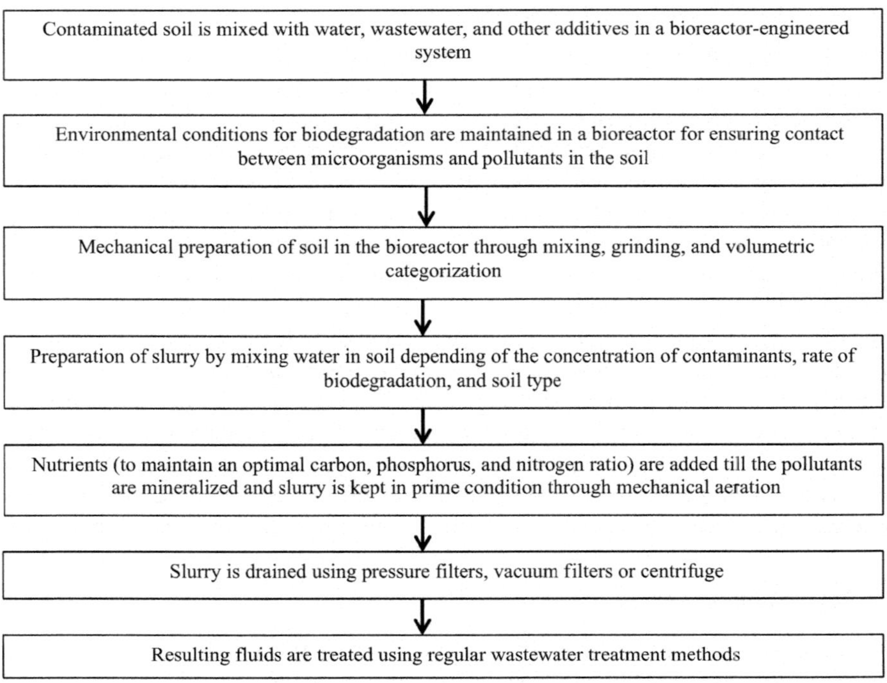

FIGURE 12.5 Steps involved in the bioreactor technique.

Moreover, decontamination with bioreactors is a quick and easy technique. Bioreactors are useful for the treatment of aerobically degradable compounds such as SVOCs, refractory insecticides, explosive substances, aromatic hydrocarbons, chlorinated organic compounds, and PAHs, both at the laboratory and commercial scale. Slurry bioreactors are used as full-scale devices for feasibility studies. This method is used to deal with soils which are difficult to treat using traditional methods.

12.3.6 Composting

Compost bioremediation or composting makes use of bacteria and fungi to isolate or break down contaminants in soil and water. Compost is effective in degrading chlorinated and non-chlorinated hydrocarbons, wood preservatives, solvents, HMs, pesticides, petroleum products, and explosives. Compost rejuvenates contaminated sites to their pre-contamination condition by reclaiming soil and providing nutrients that enable plant growth. It is a self-heating biological process that has been used for millennia as an organic waste management solution. The composting process is categorized into four phases: *mesophilic*, *thermophilic*, *cooling*, and *maturation*.

These phases have different temperature, oxygen demand, microbial community structure, stability, carbon content, nitrogen content, and pH profiles. The mesophilic phase starts after preparing the initial composting mixture. Microbes utilize easily degradable organic matter as a source of nutrition. Consequently, the temperature increases above ambient temperature, depending on the amount of composting, initial material, and composting environment. In the case that the compost mixture contains soluble organic compounds, organic acids are produced during the fermentation of these compounds, resulting in a pH drop to the acidic range. This phase lasts till the temperature reaches 55°C, accompanying the thermophilic phase. This stage has the highest temperature in the entire composting process. The initial stage of the thermophilic phase is the period with highest activity. Less biodegradable and complex organic substances like cellulose and hemicellulose biodegrade during this phase. Ammonia produced from the degradation of nitrogen-containing organic matter increases the pH. The high temperature of this phase destroys most human and animal pathogens such as *E. coli* and *Salmonella* sp. Microbial activity reduces with a depletion of nutrients, decreasing the temperature and initiating the curing phase, which is the cooling and maturation phase. Temperature during the cooling phase is similar to the mesophilic phase, and mesophilic organisms thrive during this stage. The available nutrient source comprises complex organic material. Macrofungi, which consume this complex material, are usually witnessed in the compost, and the pH remains alkaline, drops slightly, and reaches the neutral range. The cooling phase usually takes several weeks and can be mistaken for the maturation phase, which is the last stage of composting when the compost is stable and mature. The end product is a humus-like substance with an earthy smell. At this stage, the temperature of the compost is similar to the ambient temperature, and the pH is neutral or slightly alkaline.

12.4 *In Situ* Bioremediation Techniques

These techniques comprise treatment of polluted substances at the pollution site. They do not require any excavation and there is little or no disturbance in soil consumption. These techniques are cost-effective as compared to *ex situ* bioremediation techniques. Some *in situ* bioremediation techniques like bioventing, biosparging, and phytoremediation may be enhanced, while others may progress without any form of improvement, such as intrinsic bioremediation or natural attenuation. *In situ* bioremediation techniques have been used effectively to treat chlorinated solvents, HMs, dyes, etc.

Types of *in situ* Bioremediation: *In situ* bioremediation is of two types: intrinsic and engineered bioremediation.

12.4.1 Intrinsic Bioremediation

Also known as natural reduction, the *in situ* bioremediation technique is the passive remediation of polluted sites, without any human intervention. It deals with the stimulation of an indigenous or naturally occurring microbial population and is based both on microbial aerobic and anaerobic measures to biodegrade contaminating unmanageable components. Due to the non-involvement of external forces, this method is cost-effective as compared to other *in situ* techniques.

12.4.2 Engineered *in situ* Bioremediation

This involves the participation of microbes at the contamination site. Genetically engineered microbes are used in the bioremediation which speed-up the degradation by mending the physicochemical conditions and boosting microbial growth.

12.4.3 Bioventing

Bioventing is a method of accelerating the natural *in situ* biodegradation of soil contaminants by providing air or oxygen to existing soil microbes. In this method, oxygen is delivered through direct air injection into residual contamination in soil. Besides degradation of adsorbed fuel residuals, volatile compounds are biodegraded as vapours which move gradually through biologically active soil. To increase bioremediation, adjustments are made through the addition of nutrients and moisture. Bioventing finds application in any chemical that can be aerobically biodegraded. It has been used effectively to remediate soil contaminated by petroleum hydrocarbons, non-chlorinated solvents, pesticides, wood preservatives, and other organic chemicals.

12.4.4 Bioslurping

This is a combination of two techniques, bioventing and vacuum-enhanced pumping of free-product to recover free product from the groundwater and soil. Bioslurping involves free-product recovery efficiency without removing groundwater. Bioslurping is intended for product recovery from remediating capillary, light non-aqueous phase liquids (LNAPLs), and unsaturated and saturated zones. It is employed to improve soil polluted with volatile and semi-volatile organic compounds. This method uses a "slurp" that sweeps into the free-product layer and draws liquid from it. A pumping machine transfers LNAPLs to the surface by ascending movement, where it separates air and water. In this technique, soil moisture limits the air permeability and weakens the oxygen transfer rate, which reduces the microbial activities. Though this technique is not appropriate for improving low permeable soil, it is economical owing to the lower amount of groundwater required, while it reduces storage, treatment, and removal costs.

12.4.5 Bio-sparging

This is the process of pumping pressurized air or gas into a polluted area to stimulate *in situ* aerobic biological activity. The technique is similar to bioventing. In this method, air is injected into the soil sub-surface, which improves the microbial activities and stimulates pollutant removal from polluted sites. The technique makes use of indigenous microorganisms to break down organic components in the saturated zone. Petroleum compounds dissolved in groundwater, adsorbed to soil below the water table, can be reduced by bio-sparging. The efficiency of bio-sparging depends on two factors: soil permeability and pollutant biodegradability. This technology targets mineral oils, benzene, toluene, ethylbenzene, xylene, and naphthalene (BTEXN), which can be biodegraded under aerobic conditions. The aim of a bio-sparging system is to increase pollutant biodegradation while reducing the volatile and semi-volatile organic compound volatilization.

12.4.6 Bio-stimulation

Bio-stimulation is employed in *in situ* bioremediation of contaminated water bodies and involves the supply of growth-limiting nutrients, e.g., nitrogen and phosphorus, to enable the interaction of either exogenous or autochthonous microbes in the degradation of polluted environments. Earlier, this strategy was used mostly for the treatment of hydrocarbon-polluted sites to stimulate the growth of indigenous microorganisms and help the rapid breakdown of crude oil. In this technique, the environment is changed to stimulate the existing bacteria capable of bioremediation. This can be achieved by adding rate-limiting nutrients and electron acceptors, such as phosphorus, nitrogen, oxygen, or carbon.

On the other hand, the remediation of halogenated contaminants in anaerobic environments may be promoted by adding electron donors (organic substrates), thereby permitting indigenous microbes to use the halogenated contaminants, such as electron acceptors. Additives are generally added to the subsurface through injection wells, although injection well technology for bio-stimulation is still emerging. While bio-stimulation is generally linked with the remediation of hydrocarbon or high-production-volume chemical spills, it could be beneficial for treating less often encountered contaminant spills, such as pesticides, mainly herbicides. The main advantage of bio-stimulation is that bioremediation can be undertaken by already-present native microbes that are compatible to the subsurface environment, and are well-dispersed spatially within the subsurface. The main disadvantage is that the delivery of additives in a manner that permits the additives to be freely accessible to subsurface microbes is based on the local geology of the sub-surface. Constricted, impervious subsurface lithology makes it difficult to reach additives all over the affected area.

12.4.7 Bio-augmentation

Biological augmentation, a type of bioremediation, involves the addition of archaea or bacterial cultures to accelerate the rate of degradation of contaminants. After identifying that the indigenous bacteria can metabolize the contaminants, more of the indigenous bacterial cultures are added to boost the degradation of contaminants. In the case that the indigenous variety does not have sufficient metabolic ability to execute the remediation process, exogenous varieties are added. The technique is commonly used in municipal wastewater treatment to restart activated sludge bioreactors. Most existing cultures contain microbes such as *B. licheniformis*, *B. thuringiensis*, *P. polymyxa*, *B. sterrothermophilus*, *Penicillium* spp., *Aspergillus* spp., etc. Activated sludge systems are usually based on microbes like bacteria, protozoa, nematodes, rotifers, and fungi, which destroy biodegradable organic matter.

12.4.8 Natural Attenuation

This process naturally converts contaminants to less harmful forms or restrains contaminants so that they threaten the environment to a lesser extent. This includes following processes to effectively reduce contaminant toxicity, mobility, or volume to permissible levels:

- **Biodegradation (Abiotic/Biotic)**: This occurs when microbes consume contaminants and convert them to gases and water during digestion. Microorganisms generally live in soil and groundwater and utilize contaminants for obtaining food and energy.
- **Sorption**: This causes contaminants to adhere to soil particles. It does not damage the contaminants, but keeps them from moving underground or from abandoning the site with groundwater flow.
- **Dispersion/Dilution**: This decreases the concentration of pollutants when they move through and join clean groundwater.
- **Volatilization/Evaporation**: This causes some contaminants, such as gasoline and industrial solvents, to change from liquids to a gaseous state in soil. Once these gases escape to the air at the ground surface, air will attenuate them and sunlight will destroy them.
- **Chemical Reactions**: Chemical reactions with natural substances convert contaminants to less detrimental forms, e.g., in a low-oxygen underground environment, the highly toxic chromium VI can be converted to the less toxic chromium III when it reacts with naturally occurring iron and water.

Engineered applications of the natural attenuation process as a remedial measure are known as monitored natural attenuation (MNA), which involves a monitoring component. MNA works best when the source of contamination has been disconnected, e.g., any waste buried underground must be dug-up and removed suitably, or eradicated using other cleansing methods. When the source is no longer present, natural processes remove the residual contaminants in the soil or groundwater. The site is monitored frequently to ensure that contaminants are decreasing fast enough to meet site-cleansing goals. Degradation is a most incredible process because the pollutants can be transformed into less detrimental products. MNA may take several decades to clean up a site. The actual cleaning time depends on several factors, for example, cleaning will take longer when:

- Concentration of the pollutant is higher
- Contaminated area is large
- Settings of the site (such as temperature, flow of groundwater, type of soil) provide a less favourable environment for biodegradation, sorption, or dilution.

Governing factors may differ from site to site. MNA does not threaten the community or site workers. The method does not involve digging of soil or pumping of groundwater to the surface for treatment; hence, the probability of contacting pollutants is minimal. Continuous, regular monitoring is undertaken to ensure that the contamination does not leave the site and that it is being reduced at a rate that is consistent with the cleaning objectives of the site. The process also ensures that people and the environment are protected during the cleansing process.

12.4.9 Phytoremediation

Phytoremediation is a plant-based approach that makes use of plants to extract and remove elemental pollutants or lower their bioavailability in soil. Plants are capable of absorbing ionic compounds in soil even at low concentrations through their roots. Plants spread their root system in the soil matrix and establish a rhizosphere ecosystem to gather HMs and moderate their bioavailability, thus retrieving the polluted soil and stabilizing soil fertility. Phytoremediation mostly refers to the use of plants and linked soil microbes and reduces the toxic effects of contaminants in the environment.

Being a depolluting technique, it is used for treating contaminated soils, and is based on plant interfaces such as physical, chemical, biological, microbiological, and biochemical in polluted sites to reduce the toxic properties of contaminants. Based on the type of pollutant (elemental or organic), there are various phytoremediation mechanisms. Pollutants like HMs and radionuclides are generally removed by extraction, transformation, and sequestration. Organic pollutant hydrocarbons and chlorinated compounds are generally removed by degradation, rhizoremediation, stabilization, and volatilization, with mineralization being possible when plants such as willow and alfalfa are used.

12.4.10 Phytostabilization

This system is aimed at retaining contaminants in the soil and thus avoiding further scattering. Contaminants can be stabilized in the roots or rhizosphere. Revegetation is a common practice which prevents the further dispersal of contaminants. Mine tailings can be stabilized using commercially accessible varieties of metal-tolerant grasses. The metal tolerance of plants is usually increased by symbiotic, root-colonizing, arbuscular mycorrhizal fungi (AMF), through metal sequestration in the fungal hyphae. Furthermore, elimination of the glycoprotein glomulin by AMF hyphae can complex metals in the soil. AMF adapted to metal-polluted soils helps in phytostabilization.

Soil microbes decrease the toxic effects of pollutants in the soil, e.g., exudates from the bacterium *Pseudomonas putida* can decrease Cd toxicity in plants. Plants can also transform pollutants into less toxic forms, or reduce their bioavailability. Natural chelates released by the roots of some plants can form complexes with metals in the rhizosphere. These include siderophores, organic acids, and phenolics. Enzymes found in the roots of wetland plants can convert highly toxic metals such as toxic Cr (III) to a much less toxic form, viz. Cr (VI). Moreover, plants and related soil microbes can release chemicals which act as biosurfactants in the soil and increase the uptake of contaminants.

Contaminants can become stable in natural and built wetlands through a method called phyto-filtration, where metals are precipitated within the rhizosphere. Iron oxides precipitate along with other metals in the metal plaque. Metal plaque on roots acts as a reservoir for active iron (Fe^{2+}), which thereby increases the resistance of plants to other toxic metals.

12.4.11 Phytodegradation

This mechanism involves direct degradation of organic contaminants, through the release of enzymes from the roots, or metabolic activities in the plant tissues. In this method, hydrophobic organic contaminants are taken up by roots and metabolized in plant tissues to less toxic substances. Poplar trees are used in the phytodegradation of toxic and unmanageable organic compounds. Rhizodegradation involves reducing organic contaminants into less toxic substances in the rhizosphere by way of biodegradation of soil microbes. This process is enabled by organic root exudates that withstand soil microbes. A specific inoculum of bacteria can be included in contaminated soils for improving this process. The bacterial inoculum contains strains that have desirable metabolic activity to damage the targeted contaminants. Inoculating plants with genetically engineered strains of bacteria that reduce a specific contaminant has shown encouraging outcomes. Soil bacteria can be amplified using bio-stimulation, which involves the manipulation of nutrient and pH levels of soil to increase bacterial populations.

12.4.12 Phyto-Volatilization

In this process, plant roots take up contaminants and transform them into a vaporous state, which is released into the atmosphere. The process is driven by the evapotranspiration of plants. Plants that have a high evapotranspiration rate are sought after phytovolatilization. Plants inactively volatilize organic contaminants, mainly the volatile organic compounds (VOCs), e.g., hybrid poplar trees are used to volatilize trichloroethylene (TCE) by converting it into chlorinated acetates and CO_2. Metals such as Se can be volatilized by plants through conversion into dimethylselenide [Se $(CH_3)_2$]. Genetic engineering can be used to allow plants to volatilize specific contaminants.

12.4.13 Phyto-Extraction

This process taps the ability of plants to gather contaminants in the aboveground, harvestable biomass. In this process, repeated harvesting of biomass is done to lower the concentration of soil contaminants. Phyto-extraction can be either a continuous or driven process. Continuous phyto-extraction is based on the ability of certain plants to gradually gather contaminants (mainly metals) into their biomass. Some plants can hyper-accumulate metals without any toxic effects. These plants are acclimatized to naturally occurring, metalliferous soils. More than 400 plant species can hyper-accumulate various metals. However, most plants can only hyper-accumulate one specific metal. Hyperaccumulating plants can contain more than 1 per cent of a metal in their dry biomass. It is possible to extract metals from the harvested biomass in phyto-mining. The underlying mechanism of hyper-accumulation of metals in plants is the overexpression of genes that control cell membrane transporters. These include Cu-transporters and Zn-transporters. The main limitations on the use of hyper-accumulating plants in phyto-extraction are slow growth and low biomass production. The efficacy of phyto-extraction is a function of a plant's biomass production and the content of contaminants in the harvested biomass. Thus, fast-growing crops accumulating metals have great potential in phyto-extraction. The use of crops in phyto-extraction can be enhanced by management of their associated soil microbes. Inoculation of plant growth-promoting bacteria (PGPR) and AMF can increase plant biomass. The AMF–plant association usually brings about reduced gathering of metals in the aboveground biomass of plants. Consequently, overturning AMF activity, by using specific soil fungicides, has amplified metal accumulation in plants. The role of AMF in controlling metal uptake by plants depends on numerous factors, such as:

- AMF populations
- Metal content in the soil
- Nutrient availability
- Plant species

Regulation of AMF is usually metal-specific; where the uptake of essential metals is generally increased, but the uptake of non-essential metals is constrained. Induced phyto-extraction includes the use of fast-growing crops and chemical manipulation of soil. Low bioavailability of metals in the soil is a restrictive factor in phyto-extraction. The bioavailability of metals can be augmented by using synthetic chelates such as EDTA or acidifying chemicals (e.g., NH_4SO_4). The main advantages of phyto-remediation are that the plants involved in the phytoremediation of toxic elements can be harvested, thus eradicating these elements from the polluted site. Moreover, the plants can be easily monitored. It is potentially the least harmful method because it utilizes naturally occurring organisms and sustains the environment in a more natural state. It reclaims the topsoil, upholding the soil fertility and escalates soil health, yield, and plant phytochemicals. While talking about the main disadvantages, it is pertinent to mention that phytoremediation merely shuffles the toxic HMs; it does not remove them from the site. It is restricted to the surface area and depth of roots. With plant-based remediation, it is not possible to entirely prevent leaching of contaminants into the groundwater. Survival of plants is governed by the toxicity of the contaminated land and the soil's general condition.

12.4.14 Permeable Reactive Barrier (PRB)

Owing to its design and mechanism of removing pollutants, the permeable reactive barrier (PRB) technique is generally regarded as a physical method for remediating contaminated groundwater; however, biological mechanisms like precipitation degradation and sorption of pollutant removal are used in the PRB method. Alternative nomenclature, e.g., biological PRB, passive bio-reactive barrier, and bio-enhanced PRB has been proposed to accommodate the bioremediation or biotechnology aspect of this technique. In this technique, the microbes are used for enhancement rather than an independent biotechnology. Generally, PRB is an *in situ* technique used for remediating HMs and chlorinated compounds in groundwater pollution.

In this technique, a permanent or semi-permanent reactive barrier, generally made up of a zero-valent iron, is immersed in the trajectory of polluted groundwater. As polluted water flows through the barrier under its natural gradient, pollutants are entrapped and undergo a series of reactions resulting in clean water. Ideally, barriers are reactive enough to entrap pollutants, penetrable enough to allow the flow of water but not pollutants, passive with little energy input, economical, and instantly accessible. The efficiency of this technique depends mostly on the type of media used, which is governed by:

- Biogeochemical and hydrogeological conditions
- Cost
- Environmental and health influence
- Mechanical stability
- Pollutant type

In recent times, researchers have concentrated on combining PRB and other methods such as electro-kinetics for treating different classes of pollutants. Collective techniques allow polluted soil to sustain suitable environmental conditions (pH, temperature, nutrients) desirable for microbial growth, and have resulted in surfactant biomass dispersal across polluted soil.

12.5 ADVANTAGES AND DISADVANTAGES

All bioremediation techniques have their advantages and disadvantages within their specific applications; however, the general main advantages and disadvantages are listed below.

Advantages

- Harmful chemicals are transformed to water and harmless gases.
- Does not involve the use of any hazardous chemicals for the nutrients mainly fertilizers are added to activate quick microbial growth.
- Helps in complete degradation of the pollutants, where toxic hazardous compounds are converted to less harmful products.
- Cost-effective natural process as compared to other conventional methods used for remediation of toxic hazardous waste.
- Suitable waste treatment process for contaminated materials such as soil.
- Simple, less laborious, and economical, attributable to its role in environmental protection.
- Non-intrusive, probably allowing its continuous use.
- Needs much less effort and can commonly be carried out on site, without disturbing normal microbial activities. It eliminates transportation problems and possible threats to human health and the environment.
- Provides eco-friendly maintainable prospects.
- Treatment products are usually harmless, including cell biomass, water, and carbon dioxide.

Disadvantages

- Techniques are limited to biodegradable compounds.
- Not all compounds undergo complete degradation.
- At times, new products of biodegradation may be more toxic than the initial compounds.
- Contaminants may be solids, liquids, and/or gases, which require a longer treatment time.
- These methods are subject to rapid and complete degradation.
- Biological processes are highly specific but require metabolically competent microbial populations, appropriate environmental conditions, and appropriate nutrients and pollutants.
- Problematic to scale-up bioremediation process from batch and pilot scale to large-scale field operations.
- Advanced research is required to develop contemporary engineering bioremediation technologies suitable for sites with complex amalgamations of pollutants (solids, liquids, and gaseous forms) that are not equitably dispersed in the environment.
- Bioremediation takes a longer time as compared to other treatment techniques, such as quarrying and elimination of soil from polluted sites.

12.6 BIOREMEDIATION APPLICATIONS

- General approaches to bioremediation are:
  - (i) Natural bio-remediation
  - (ii) Bio-simulation (environmental modifications, through nutrient application and aeration)
  - (iii) Bio-augmentation (addition of microbes)

Bioremediation methods can be applied to all states of matter in the environment, such as:

- Solids (soils, sediment, and sludge)
- Liquids (groundwater, surface water, and industrial wastewater)
- Gases (industrial air emissions)
- Sub-surface environments

BIBLIOGRAPHY

Abatenh, E., Gizaw, B., Tsegaye, Z., and Wassie, M. (2017) Role of microorganisms in bioremediation—A review. *Journal of Environmental Biology* 2: 38–46.

Ahsan, Z., et al. (2021) Enzyme-assisted bioremediation approach for synthetic dyes and polycyclic aromatic hydrocarbons degradation. *Journal of Basic Microbiology* 61: 960–981.

Azubuike, C. C., Chikere, C. B., and Okpokwasili, G. C. (2016) Bioremediation techniques–classification based on site of application: Principles, advantages, limitations and prospects. *World Journal of Microbiology and Biotechnology* 32: 180.

Bala, S., et al. (2022) Recent strategies for bioremediation of emerging pollutants: A review for a green and sustainable environment. *Toxics* 10(8): 484.

Bizily, S. P., Rugh, C. L. and Meagher, R. B. 2000. Phytodetoxification of hazardous organomercurials by genetically engineered plants. *Nature Biotechnology* 18: 213–217.

Boopathy, R. (2000) Factors limiting bioremediation technologies. *Bioresource Technology* 74: 63–67.

Boyajian, G. E. and Carreira, L. H. (1997) Phytoremediation: A clean transition from laboratory to marketplace. *Nature Biotechnology* 15: 127–128.

Clemens, S., Palmgren, M. P. and Krämer, U. (2002) A long way ahead: Understanding and engineering plant metal accumulation. *Trends in Plant Science* 7: 309–315.

Garg, S. K., Tripathi, M. and Srinath, T. (2012) Strategies for chromium bioremediation of tannery effluent. *Reviews of Environmental Contamination and Toxicology* 217: 75–140.

Gaur N., Narasimhulu, K. and PydiSetty, Y. (2018) Recent advances in bio-remediation of persistent organic pollutants and its effect on environment. *Journal of Cleaner Production* 198: 1602–1631.

Jaiswal, S. and Shukla, P. (2020) Alternative strategies for microbial remediation of pollutants via synthetic biology. *Frontiers in Microbiology* 11: 808.

Janssen, D. B. and Stucki, G. (2020) Perspectives of genetically engineered microbes for groundwater bioremediation. *Environmental Science: Process and Impacts Journal* 22: 487–499.

Lasat, M. M. (2002) Phytoextraction of toxic metals: A review of biological mechanisms. *Journal of Environmental Quality* 31: 109–120.

Lee, H., Jun, Z., and Zahra, Z. (2021) Phytoremediation: Sustainable strategy for improving indoor and outdoor air quality. *Environments* 8: 118.

Maitra, S. (2018) *In situ* bioremediation—An overview. *Research Journal of Life Sciences, Bioinformatics, Pharmaceutical and Chemical Sciences* 4: 576–598.

Naeem, U. and Qazi, M. A. (2020) Leading edges in bioremediation technologies for removal of petroleum hydrocarbons. *Environmental Science and Pollution Research* 27: 27370–27382.

Pande V., et al. (2020) Bioremediation: An emerging effective approach towards environment restoration. *Environmental Sustainability* 3: 91–103.

Patel, A. K., et al. (2022) Organic wastes bioremediation and its changing prospects. *Science of Total Environment* 824(2): 153889.

Pilon-Smits, E. (2005) Phytoremediation. *Annual Review of Plant Biology* 56: 15–39.

Rayu, S., Karpozas, D. G., and Singh, B. K. (2012) Emerging technologies in bioremediation: Constraints and opportunities. *Biodegradation* 23: 917–926.

Ren, X., et al. (2018) Potential impact on biodegradation of organic pollutants from composting technology for soil remediation. *Waste Management* 72: 138–149.

Salt, D. E., Smith, R. D., and Raskin, I. (1998) Phytoremediation. *Annual Review of Plant Physiology and Plant Molecular Biology* 49: 643–668.

Sharma, B. and Shukla, P. (2020) Designing synthetic microbial communities for effectual bioremediation: A review. *Biocatalysis and Biotransformation* 38: 405–414.

Sharma, J. (2019) Advantages and limitations of *in-situ* methods of bioremediation. *Recent Advances in Biology and Medicine* 5(1): 1–9.

Sharma, K. R., Giri, R., and Sharma, R. K. (2022). Efficient bioremediation of metal containing industrial wastewater using white rot fungi. *International Journal of Environmental Science and Technology* 1(8): 943–950.

Sharma, N., Sodhi, K. K., Kumar, M., and Singh, D. K. (2021) Heavy metal pollution: Insights into chromium eco-toxicity and recent advancement in its remediation. *Environmental Nanotechnology, Monitoring & Management* 15: 100388.

Shukla, P. (2020) Microbial nanotechnology for bioremediation of industrial wastewater. *Frontiers in Microbiology* 11: 1–8.

Sutar, H. and Kumar, D. (2012) A review on bioremediation. *International Journal of Research in Chemistry and Environment* 21: 13–21.

Tarekegn, M. M., Salilih, F. Z., and Ishetu, A. I. (2020) Microbes used a tool for bioremediation of HMs from the environment. *Cogent Food & Agriculture* 6: 1783174.

Wang, L., Rinklebe, J., Tack, F. M., and Hou, D. (2021) A review of green remediation strategies for heavy metal contaminated soil. *Soil Use and Management* 37: 936–963.

Wei, Z., et al. (2021) A review on phytoremediation of contaminants in air, water and soil. *Journal of Hazardous Materials* 403: 123658.

Yaashikaa, P. R., Kumar, P. S., Jeevanantham, S., and Saravanan, R. (2022) A review on bioremediation approach for heavy metal detoxification and accumulation in plants. *Environmental Pollution* 301: 119035.

13 Recent Advances in Eco-Toxicology

All substances are poisons; there is none that is not a poison. The right dose differentiates a poison and a remedy.

Paracelsus

After going through this chapter, the reader will learn about: recent advances in eco-toxicology with respect to the detection of heavy metals, atomic absorption spectroscopy (AAS), inductively coupled plasma optical emission spectrometry (ICP-OES), nuclear magnetic resonance, and green initiatives for environment management.

FACT FILE

- Atomic absorption spectrometry was first discovered in 1802, when the English scientist William Hyde Wollaston observed and described dark lines in the sun's spectrum
- Heavy metal (HMs) trace detection is mainly performed using spectroscopic techniques, such as, atomic absorption spectroscopy, inductively coupled plasma mass spectroscopy (ICP-MS), X-ray fluorescence, and neutron activation analysis
- Their main benefit of using these techniques is their versatility as they are appropriate for a wide selection of elements, their sensitivity, and their limit of detection (LOD) in the femtomolar range
- A great challenge in the area of HM trace detection is the development of electrochemical techniques and devices which are accessible, sturdy, and selective, with low-detection limits and permitting fast examination
- The French scientific academic community has made significant contributions in developing electro-chemical sensors and electro-analytical methods over the last 20 years
- Flame photometry is one of the branches of atomic absorption spectroscopy
- Biomarkers can serve as early warning systems for health

DOI: 10.1201/9781003343721-13

- Meat products have larger carbon footprints per calorie than grains or vegetable because animals like cattle, sheep, and goats produce a lot of methane gas
- A carbon credit is a kind of permit that represents 1 ton of CO_2 removed from the atmosphere
- GRIHA, an acronym for Green Rating for Integrated Habitat Assessment, is a Sanskrit word which means Abode and is a tool of rating habitats that helps people to assess and minimize a building's resource consumption, waste generation, and overall ecological impact within certain permissible limits
- Green energy is a type of energy obtained from natural resources, such as sunlight, wind, or water
- Biodiesel, an alternate biofuel, can be obtained from a range of bio-resources, such as vegetable oil to coffee grounds to exotic plants like *Jatropha*, as long as the plant produces triglycerides (fats and oils)

13.1 INTRODUCTION

Environmental contamination due to heavy metals (HMs) is a global concern. HMs are non-biodegradable elements which get accumulated in living organisms through bio-magnification in the food chain. Inside the body some HM ions are converted to more toxic forms which disturb the metabolism and manifest in the form of disorders causing toxicity in organisms. HM detection from various components of the environment is among the major challenges faced by scientists for decades. Metals which are believed to have an adverse effect on humans include arsenic, lead, cadmium, nickel, copper, zinc, selenium, mercury, and chromium. These metal ions are increasing daily in the natural resources due to anthropogenic activities. Therefore, it is important to develop effective techniques for their detection and removal. Scientists have developed many techniques such as atomic absorption spectroscopy, gamma-ray spectroscopy, X-ray fluorescence, inductively coupled plasma mass spectrometry, etc. for the detection of HMs from various sources. Recent advances in the field of environmental toxicology have resulted in the development of new products, e.g., biomarkers are meeting the emerging need for early warning tools for the detection of exposure and adverse biological responses to environmental pollutants. Ecological restoration, carbon sequestration, carbon footprint, carbon tax, carbon credit, green building, and green energy are some green initiatives which have been introduced globally to reduce environmental pollution.

13.2 ATOMIC ABSORPTION SPECTROSCOPY (AAS)

AAS is a common technique used for the detection of metals and metalloids. It is a reliable and easy-to-use technique. The first atomic absorption spectrometer was developed by Alan Walsh in 1954. It quantifies the absorption of ground-state atoms in the gaseous state and is based on the principle that free atoms generated in an atomizer can absorb radiation at a specific frequency. Atoms absorb UV or visible light and make transitions to higher electronic energy levels. The analyte concentration is determined from the amount of absorption. Atomic absorption methods measure the amount of energy in the form of photons of light that are absorbed by the sample. The detector measures the wavelengths of light transmitted by the sample, and compares them with the wavelength which originally passes through a sample. A signal processor then integrates the changes in the absorbed wavelength, which appear in the readout as peaks of energy absorption at discrete wavelengths.

The energy required for an electron to leave an atom is known as the ionization energy and is specific to each chemical element. The absorption spectrum of an element is in its gaseous atomic state and consists of a series of well-defined and extremely narrow lines arising from electronic transitions of the outer-most electrons. In the case of metals, most of the transitions belong to the visible and UV regions. The most obvious wavelengths at which absorption or emission is observed are associated with the transitions where minimal energy change occurs. For analysis, the sample (element) is converted into atomic vapour and then the absorption, either in the visible or UV region (at a selected wavelength), which is specific for a given element is measured. For converting an element into its gaseous atomic state, the initial step in the procedure of estimation involves spraying a solution of the sample into the flame. This process is accomplished by drawing the solution of the sample as a fine mist into a suitable flame.

AAS has the following basic components (see Figure 13.1):

1. **Radiation Source**: Hollow cathode lamp and electrodeless discharge lamps
2. **Chopper**: Rotating wheel interposed between the radiation source and flame
3. **Atomiser**: Reduces the sample to atomic state
4. **Nebulizer**: Converts liquid sample into tiny droplets
5. **Monochromator**: Selects a given absorbing line from spectral lines emitted from hollow cathode. Example: Prism and grating
6. **Detector**: Photomultiplier tube
7. **Amplifier**: Amplifies the electric current from the detector
8. **Read-Out Device**: Chart recorders are used as read-out devices

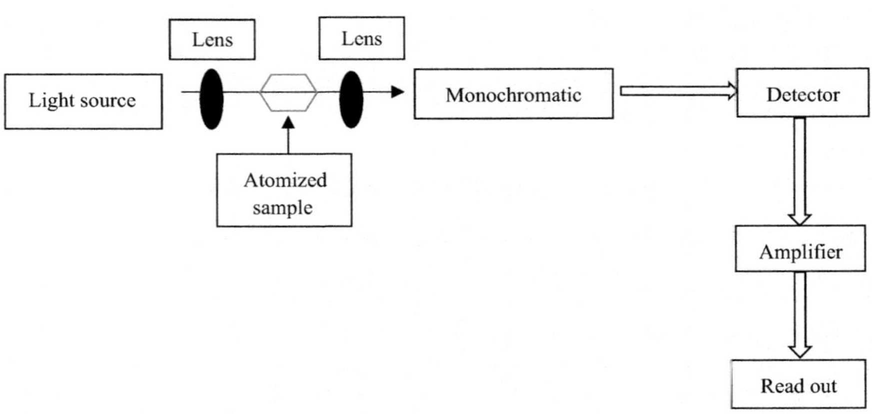

FIGURE 13.1 Diagrammatical representation of atomic absorption spectroscopy (AAS).

AAS for Analysing Water Samples: HMs are usually present in trace amounts in natural water sources, such as ponds, lakes, and underground water sources. Some of the HM ions are very toxic for humans even at low concentrations. Micro-concentrations of HMs in water harmfully affect the environment gradually, accumulating in certain parts of the animal and plant organism, causing a change in their biochemical balance. Metals which are believed to have an adverse effect on humans include arsenic, lead, cadmium, nickel, mercury, chromium, copper, zinc, and selenium. These metal ions are increasing daily in the natural resources as currently the number of industrial complexes near human populations in cities is increasing with the goal of improving the quality of life. Increasing concentrations of contaminants in water bodies are due to their discharge through industrial wastewater, without proper pretreatments. Water sources are constantly polluted by anthropogenic activities such as sewage discharge, agricultural waste, fertilizers, medical waste, domestic waste, and industrial effluents, which adversely affect the quality of water and become major sources of water pollution. AAS is a technique for measuring elements present in water samples and measures the absorbed radiation by the chemical element of interest (see Figure 13.2). This is done by reading the spectra produced when the sample is excited by radiation. Atoms absorb UV or visible light and make transitions to higher energy levels. It measures the amount of energy in the form of photons of light that are absorbed by the sample. When an electron moves from one energy level to another within the atom, a photon is emitted with energy.

AAS determines the elemental composition by its electromagnetic or mass spectrum. In order to detect toxic HMs, the water samples are collected in containers. Polyethylene bottles are used for the collection of water samples because glass bottles absorb metals and will cause inaccuracy in analysis.

The bottles are cleaned by washing with detergent solution and then thoroughly rinsed with distilled water, followed by rinsing with acetone. The container is labelled according to the sample. The sample is weighed and its solution is prepared by suitable dilution. The elements in biological fluids, such as urine and blood, are often measured simply after diluting the original sample. For digestion, 0.75 L of each sample is measured in a clean digestion flask; 9 mL of concentrated nitric acid and 3 mL of concentrated hydrochloric acid are added to the sample in a digestion flask. The entire sample is heated on a hot plate until all the brownish fumes are expelled (nitrogenous compound) which confirms that the sample is digested and the sample is allowed to cool at room temperature. A few millilitres of distilled water are added and the mixture is filtered in 25 mL standard flask and later transferred into a plastic reagent bottle for AAS. A calibration curve is used to determine the unknown concentration of an element in a solution. The instrument is calibrated using several solutions at known concentrations. A calibration curve is produced which is continually rescaled as more concentrated solutions are used. The more concentrated solutions absorb more radiation up to a certain absorbance. The calibration curve shows the concentration against the amount of radiation absorbed. The water sample collected for analysis is characterized using AAS and the conductivity of each water sample is measured using a conductivity meter. The trace number of HMs is obtained using AAS.

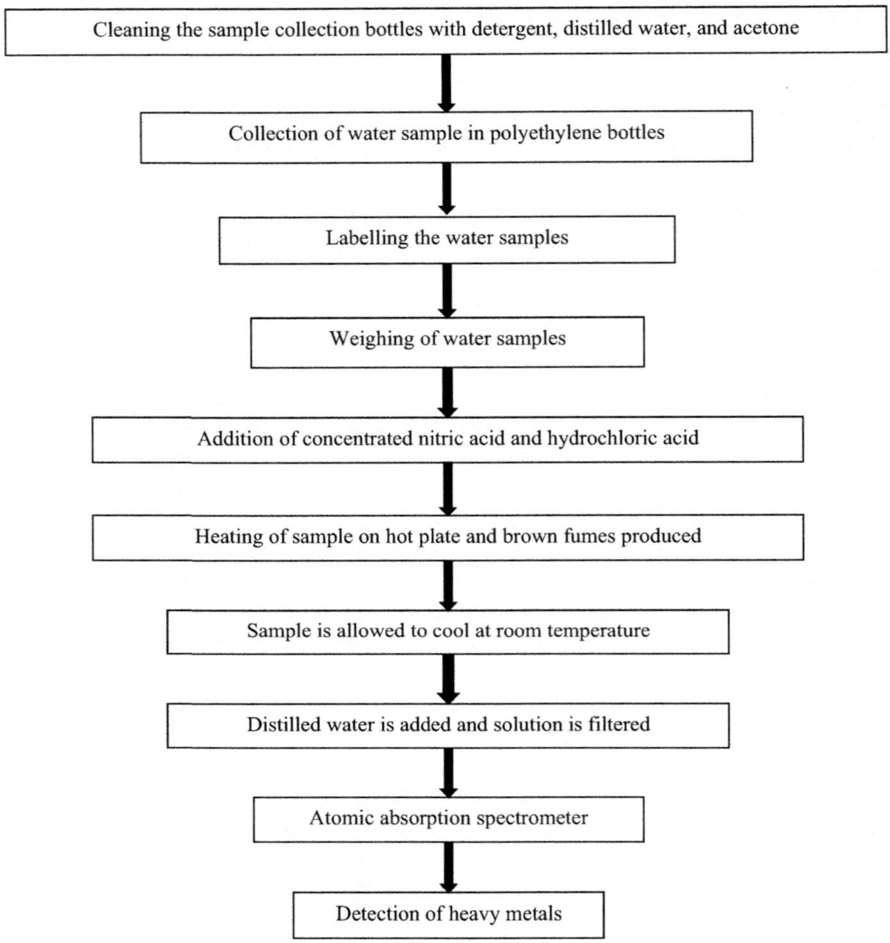

FIGURE 13.2 Steps involved in analysing a sample using atomic absorption spectroscopy (AAS).

13.3 X-RAY FLUORESCENCE

X-ray fluorescence is an elemental analysis technique with broad application in science and technology. It is used for simultaneous determination of elemental composition and film thickness. Modern X-ray fluorescence instruments are capable of analysing solid, liquid, and thin-film samples for both major and trace components. X-ray fluorescence spectrometry (XRF) is an analytical technique that determines the total element concentrations in a variety of materials (see Figure 13.3). The analysis is rapid and sample preparation is minimal. It is based on the principle that when excited by an external energy source, individual atoms emit X-ray photons of characteristic wavelength or energy. By counting the number of photons emitted from a sample, the elements present may be identified and quantitated. When an electron

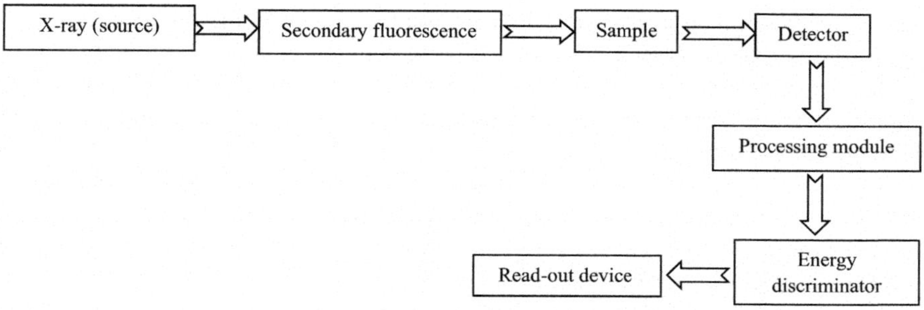

FIGURE 13.3 Diagrammatic representation of X-ray fluorescence.

beam of high energy strikes a material, it results in the emission of photons which have a broad continuum of energies.

X-Ray Fluorescence for Analysis of HMs in Soil: Soil is a complex matrix of soil particles, organic matter, and other inorganic materials that can adsorb pollutants. Soil can be irresponsibly used as a dumping ground by industries which are major sources of toxic HMs. Industries also burn waste on-site, which releases HMs into the atmosphere. These environmentally stable, non-biodegradable pollutants settle in the soil where they accumulate. When surface soils have accumulated HMs, they consequently serve as a transmitter of pollutants to surface water, groundwater, the atmosphere, and food. HMs such as Fe, Mn, Zn, Co, Ca, Cu, Mg, Ni, Mo, and other trace elements are essential for proper functioning of biological systems. Their deficiency or excess can lead to a number of disorders. Excessive accumulation of HMs in soils may not only cause soil pollution, but can also lead to higher HM uptake by plants, and thus affect the food chain, and food quality and safety. Food chain contamination is one of the most important pathways for the entry of toxic pollutants into the human body. Qualitative and quantitative analyses of HMs present in the soil are done using X-ray fluorescence. It determines whether or not the respective concentrations exceed their threshold limit value, and assesses the extent of pollution and its impact on the environment. Stainless-steel core samplers are used to collect samples from the upper level. At each sample site, profiles are laid depending upon the accessibility of bare soil and the number of samples to be taken. At each sampling site, the samples are collected in a grid at 2-metre intervals. The samples are collected from different locations for laboratory characterization and elemental analysis. Samples are air-dried in open air for seven days to ensure proper drainage and to decrease the moisture content below 20 per cent. To achieve this goal, the samples are prepared for drying by breaking them down into aggregates and spreading them evenly on polyethylene sheets or plywood trays in the open air, and ensuring with great circumspection that there is no sample cross-contamination or contamination from any external sources. To reduce soil matrix effects, the soil samples are thoroughly homogenized and sieved to fine particle sizes with Retsch aluminium test-sieves with a vibratory electronic sieve shaker. The fine powdery particles that pass through the sieve are repackaged and carefully stored. Milling or pulverizing of the loose powdered samples further reduces the particle size. Before the milling or mixing procedure, 0.9 g of powdery binder (Hoechst wax) containing cellulose, starch, polyvinyl alcohol, or other organics is weighed into 4 g of each sample using an electronic balance. The resulting mixture (sample and binding material), having a total mass of 4.9 g, is put into deformable aluminium cups (screw-top grinding jars) for the process of thorough milling and homogenization using RETSCH Mixer Mills, which also aid the further pulverization of the sample. By using a SPECAC hydraulic press with a maximum pressure limit of 15 tons, each pulverized sample is manually pressed into pellets of uniform diameter (32 mm) and thickness (3 mm).

The concentration of HMs is determined using polarized energy dispersive X-ray fluorescence (PED-XRF). This is a very sensitive technique; hence great caution is taken to avoid contamination of pellets ahead of the analysis. Poor handling of the samples could seriously affect the results due to the high sensitivity of the instrument, which is sensitive enough to detect fingerprints on the pellet's surface layer. Taking into consideration these precautions, the sample holder disks (sample cups) are filled with the pellets by holding the sample pellets by the edges. Thus, X-ray fluorescence emerges as an important technique for qualitative and quantitative analyses of soil samples.

13.4 INDUCTIVELY COUPLED PLASMA MASS SPECTROMETRY (ICP-MS)

ICP-MS is a highly sensitive analytical technique used for detecting the trace elements in biological fluids, particularly metals of clinical interest. This technique basically determines the metals and non-metals at concentrations as low as parts per billion on non-interfered low-background isotopes in biological fluids. The key features of ICP-MS, such as, high sensitivity, wide linear dynamic range, wide elemental coverage, multi-element capability, and high sample output show the suitability of identifying HMs, particularly in blood samples. Dissolution agents used during the process include nitric acid, perchloric acid, hydrofluoric acid, aqua regia, hydrogen peroxide, or various mixtures of these, which are used for metals in biological samples and soil/sediment samples. In addition, ultrapure reagents and millipore water and argon must be used for preparation and dilution. Before carrying out the analysis, biological samples are usually diluted or thermally digested with the help of diluents, such as dilute acids or alkali. The ICP-MS technique is comprised of two basic conceptual techniques, i.e., plasma emission spectroscopy and mass spectroscopy, while considering inductive coupled plasma as an essential component. It is further divided into different types:

- Plasma emission spectroscopy
- Mass spectrometry (MS)
- Inductively coupled plasma (ICP)

13.5 INDUCTIVELY COUPLED PLASMA OPTICAL EMISSION SPECTROMETRY (ICP-OES)

Also known as ICP-AES (inductively coupled plasma atomic emission spectrometry), this is an analytical technique widely used to identify elements in solid and

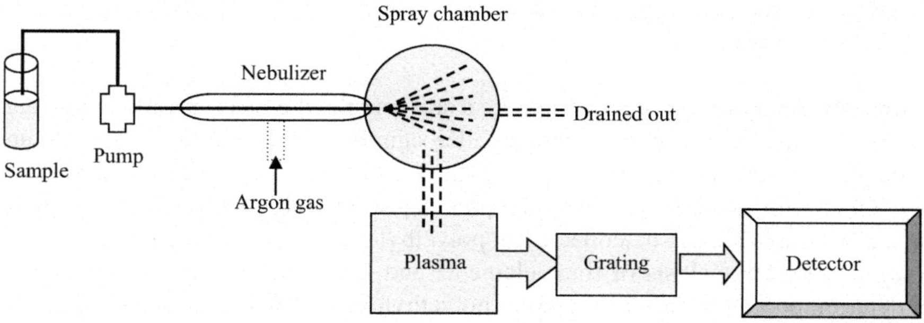

FIGURE 13.4 Diagrammatical representation of ICP-OES.

liquid samples, and is well-suited for the detection of certain HMs. ICP-OES is a technique in which the composition of elements in (mostly water-dissolved) samples can be determined using plasma and a spectrometer. ICP-OES uses extremely hot argon plasma to excite the atoms in a liquid sample from a ground state to an excited state (see Figure 13.4). On returning the element from the excited state to the ground state, light is emitted at specific emission line wavelengths directed to the system's spectrometer for measurement in the detector.

In ICP-OES, the solution introduced for analysis is conducted by a peristaltic pump though a nebulizer into a spray chamber. A high-frequency alternate current flows through a cooled induction coil which generates plasma at the end of a quartz torch. This results in the acceleration of electrons into a circular trajectory due to induced alternative magnetic fields. Collision between the argon atom and the electrons' ionization gives rise to extremely hot stable plasma. However, the atomization and ionization of samples take place in torch desolvation. The ionization of electrons results in emitting different energy levels which are liberated as light or photons. Each element possesses its own characteristic emission spectrum that is measured with a spectrometer.

The ICP-OES instrument consists of three primary components or systems, i.e., the sample introduction system, the inductively coupled plasma (ICP), and the optical emission spectrometer that houses the light-separating optics and detector. The first component, i.e., sample introduction, transports liquid sample to the plasma torch. It consists of a peristaltic pump and tubing, spray chamber with nebulizer, and torch with injector tube where the plasma is formed. The nebulizer is placed inside the spray chamber which converts the liquid sample from the peristaltic pump to a fine aerosol. This sample aerosol is then forwarded to the torch's injector tube where it meets the inductively coupled plasma, the second component. This ICP acts as the source of energy that is used to convert the sample analyte atoms in the aerosol from a ground state to an excited state.

A magnetic field is created through cooled induction copper when alternating current is provided that causes accelerated electrons to move in a circular path. When the electrons fall back to the ground state, they emit light of specific wavelength which is characteristic of the elements of interest. Consequently, the image of the plasma is directed to the optical spectrometer, which is the third significant system of an ICP-OES instrument.

Sample Preparation: In order to acquire a soil sample, the recommended procedure is used. The field is sub-divided into smaller sampling areas where one sample unit is taken from each area along a zigzag path. Sample units are taken by an auger or corer to a uniform depth, usually a plough's depth for arable land but less for grassland. The sample should be collected in polyethylene bags to minimize sample contamination in the analysis of trace elements and HMs. However, soil intended for Hg determination must not be stored in polyethylene containers since any elemental Hg can contaminate other samples stored in the vicinity and invalidates the analysis. Materials used for sampling tools must not contaminate the sample with elements of interest. In most cases, carbon steel is considered, however, aluminium may also be used along with durable polypropylene or polytetrafluoroethylene (PTFE, Teflon). It is also necessary to label the samples outside the sample bag. In the case that the sample is labelled inside the bag, marker pen inks may have their colour affected due to the HM content.

Once the soil sample is taken, it is prepared for drying by breaking down aggregates and spreading the soil on polyethylene sheets in aluminium trays and drying at 25°C. Upon drying, these are passed through an aluminium sieve with a 2 mm mesh, while discarding stones and other aggregates larger than 2 mm. A soil sample of less than 2 mm is stored in polyethylene bags or boxes and is used mainly for determining HMs in various soil extracts. For determination of the total content, a finely ground soil is required. For obtaining a sample from groundwater, water samples are collected in two pre-cleaned containers for duplicate measurement at each sampling location. Initially, the bottles are rinsed thrice with the ground water sample of the particular location and then collected in the final sample to avoid contamination. To prevent the loss of HMs, 0.5 mL of Supra pure grade HNO_3 is added to acidify the samples. All collected groundwater samples are preserved at 4°C using a thermo-coal box with ice packs.

HM Analysis Process: The detection of HM toxicity through analysis by ICP-OES of sample matrices can help in the analysis of drinking water, wastewater, soils/sludge, etc. Soil is often considered for a common location for high concentrations of HMs as by-products of industrial and agricultural pollutants. The following steps are undertaken for performing the detection of HMs:

- Initially, the sample considered for testing is collected from the source
- Samples may be acidified for preservation, depending on the type of sample
- The collected sample is then sent to the laboratory for analysis by ICP-OES within a specified time-span (analytical accuracy will diminish with time)
- The date and time of arrival of the sample are recorded immediately after reaching the laboratory
- Since ICPs analyse samples in liquid form, samples originating from solid matrices require digestion or extraction

- Using special solvents, acids, and laboratory apparatus, the sample is then prepared for analysis
- Appropriate sample introduction components for ICP are selected depending upon the sample type and solvent or acid digestion (peristaltic pump tubing type, type of nebulizer and spray chamber design, and plasma torch injector tube diameter are all carefully chosen to optimize analysis and maintain resistance to the sample matrix)
- Depending on the sample type to be analysed, instrument operating conditions are carefully determined to optimize analysis

Once all these steps have been taken, the sample is analysed for HM content and a report of the analytical data is created.

13.6 NUCLEAR MAGNETIC RESONANCE (NMR)

NMR is a technique used for determining the molecular structure, content, and purity of a sample. This technique uses the magnetic property of nuclei and determines the physical as well as chemical properties of atoms or molecules in which they are contained. NMR is based on the principle that all nuclei are electrically charged and several nuclei spin. On applying an external magnetic field, a transfer of energy can take place between the base energy to a higher level of energy. An NMR spectrometer consists of the following parts (see Figure 13.5):

- **One Radio Frequency Receiver**: To detect weak signals generated by nuclei that are resonating
- **Software**: Used to assist the user in data interpretation produced by the instrument
- **Console**: Controls the pulses of radio frequency and changes the signals detected by the receiver into a digital format

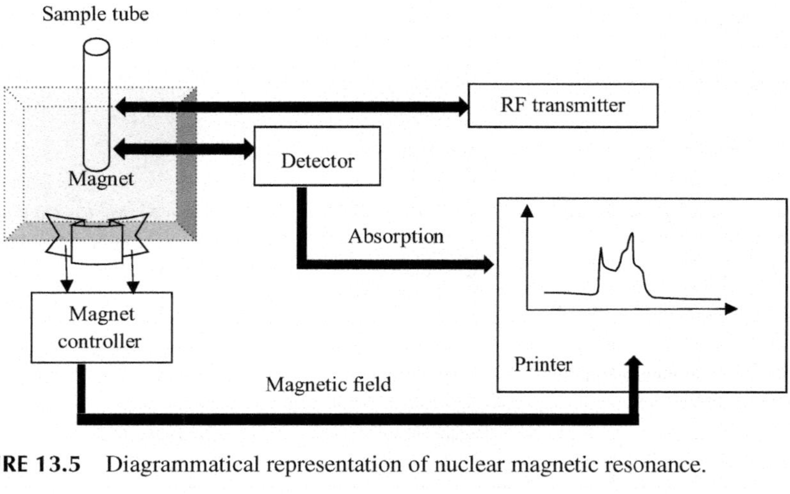

FIGURE 13.5 Diagrammatical representation of nuclear magnetic resonance.

- **Radio Frequency Transmitter**: Used to generate the required electromagnetic radiation
- **Magnet**: Used for generating a uniform magnetic field

Applications: The idea of energy absorption, energy emission, and concepts of NRM resulted in establishing magnetic resonance imaging (MRI). MRI produces images formed on the basis of spatial variations in the phase and frequency of energy of a radio frequency that is being absorbed and further released by the item that is imaged. It also produces an image using the movement of protons within a magnetic field. The movement of paramagnetic ions (Cu^{2+}) through bio-porous membranes can be easily recorded using NRM or simply by performing an MRI scan. A particular strength of MRI is that it is non-invasive, thus any biofilm or tissue analysed could be alive, functioning, and unharmed by the analysis. Magnetic resonance imaging has the potential to examine the transport and fate of a broad range of metals such as Ni^{2+}, Cr^{3+}, Co^{2+}, Ti^{2+}, Nd^{3+}, Th^{4+}, Fe^{3+}, Fe^{2+}, and Mn^{2+} in biofilms.

Table 13.1 summarizes the uses of various techniques in the detection of HMs in different samples.

TABLE 13.1
Heavy Metal Detection Using Analytic Methods in Different Samples

Name of Technique	Type of Sample	Metal Detected
Atomic absorption spectroscopy (AAS)	Water	Cu, Cr, Cd, Hg, As, Pb
	Plant	Mg, Cd, Al
	Soil	Co, Mo
	Blood/urine	Fe
UV/VIS absorbance spectroscopy	Drug/water	Ni, Cr, Cu
	Blood	Fe
X-ray fluorescence	Soil/ash	Fe, Mn, Zn, Co, Ca, Cu, Mg, Ni, Mo
Nuclear magnetic resonance (NMR)	Internal body organs and tissue system	Ni, Cr, Co, Ti, Nd, Th, Fe, Mn, Cu
Gamma-ray spectroscopy	Soil, rock	U, Pu
	Bone	Th
Inductively coupled plasma mass spectrometry (ICP-MS)	Soil, water	Ag, Al, Au, Ca, Cd, Co, Cr, Fe, Mg, Rh, Mo, Mn, Cu.
	Blood/urine	Fe, I
Inductively coupled plasma optical emission spectrometry (ICP-OES)	Soil/water	Pb, Cu, Cd, Zn, Mn, Co
Graphite furnace atomic absorption spectrometry (GFAAS)	Food/water	Pb, Cd, Cu, As, Hg
Flame atomic absorption spectrometry (FAAS)	Water	Cd, Pb, Cu

13.7 GREEN INITIATIVES FOR ENVIRONMENT MANAGEMENT

Contamination caused by HMs has become a global concern for which the conventional treatment methods do not offer a cost-effective and eco-friendly solution. Hence, green initiatives to reduce pollution and HMs present in the environment have become the most acceptable method by researchers. Some of the emerging breakthroughs in this regard are discussed in the following sub-sections.

13.7.1 Ecological Restoration

Ecological restoration – the premeditated or purposeful rejuvenation and restoration of damaged, or demolished ecosystems – has become an important dimension of conservation in the recent years. It has offered an active and positive alternative to both biodiversity protection and justifiable use, targeting the recovery of vanished ecosystem diversity and function. There are three key principles to restoration. The first is upholding ecosystem integrity. The second is species integrity, and the third is the principle of least possible intervention. Ecological restoration is of two main types. The first is passive restoration involving the use of low-intensity techniques aiming at leveraging natural ecological processes for achieving the restoration goals. The focus of passive restoration is on revitalizing soil, water, and natural vegetation cycles. The second type is active restoration, which involves the use of moderate to high intervention techniques which are usually used selectively where passive methods are infeasible or insufficient to realize the restoration goals. As a process of revamping locations in nature whose biological communities and ecosystems have been degraded, ecological restoration is turning out to be an important conservation tool. In many ecosystems, anthropogenic activities have transformed native populations of plants and animals, besides the introduction of invasive species, and transformed natural communities through extractive use, polluted waters, and ruined soil resources. It focuses on repairing the damage caused by human activities and seeks to reinstate them back to an earlier state or a state that is closer to the previous condition.

Restoration specialists relate and use the concepts from the discipline of ecology to repair and restore ecosystems, for example, plant species offer food, important services, such as nutrient cycling, and habitats for various animals in temperate coniferous forests. However, human interference and disturbance, such as deforestation, are so destructive and harmful that they instead impair the normal functioning of the forest and affect the wildlife. To restore a deforested site, restoration ecologists often opt for reversal of ecological changes caused by the elimination of plants and displacement of animals. Ecological succession, long-term evolution of the structure of an ecosystem's biological community that has undergone a disorder, is also a significant component of the restoration process. It starts with planning or assessment of the degraded site, for which the existing situation of the site is properly examined to recognize the need for restoration in the light of possible actions required for restoring it. Restorationists consider the:

- Causes of degradation
- Probability that the degradation can be inverted or decreased
- Ways in which restoration can be achieved

During the planning process, restorationists identify essential resources that affect the feasibility of the project, which needs the support and participation of local communities and government institutions. A critical step in restoration planning is to set realistic goals. Other goals include:

- Erosion control
- Preservation of cultural landscapes
- Protection of wildlife habitats
- Rangeland forage production
- Planning for climate change

Once the goals of the restoration project have been set, the subsequent step is to outline goals and categorize performance indicators.

Planning is followed by implementation, e.g., in some cases, restoration specialists simply remove the source of any issues and allow sites to recover naturally through ecological succession. This process is called passive restoration. In other cases, where the ecosystem has undergone a threshold of degradation, restoration involves the elimination or termination of the cause of disturbance. Thus, in such cases, the restorationists engage in active restoration, which starts or quickens the recovery process or efforts to change the site's ecological succession.

Soil rehabilitation and land stabilization are among the most important components of active restoration. This includes restoration of soil or water to its original physical, chemical, and biological characteristics by making amendments, improving soil pH, curtailing the use of fertilizers, artificial enrichment of soil or water, inoculating soils with beneficial microorganisms, and also undertaking tilling to improve aeration and root penetration. Another constituent of active restoration is reinstating the plant community. After the site is ready, restorationists usually select seeds, seedlings, or cuttings for re-vegetation, that are suitable for local conditions because local plants and animals are likely to be well-adapted to the target ecosystem. On the other hand, restorationists may use plant material from seed companies or nurseries that are usually tailor-made for certain environmental conditions. Many restoration projects try to maximize the genetic diversity of the site by collecting seeds. Increased genetic diversity gives restored plants a chance to survive disturbance events and other ecosystem changes. Such genetic concerns can also be significant for animals, which at times are transported or reinstated to an area.

Monitoring and documentation are important for the restoration process as it guides the adaptive management and determines when and if restoration objectives are met. In some cases only restricted human intervention may be required, while in other cases human intervention may be essential for decades.

HM mobilization due to various anthropogenic activities has been held responsible for the discharge of HMs into the environment posing a threat to ecological health and human wellbeing. Mine waste management practices as per eco-restoration, phytoremediation, or bioremediation methods have shown acceptable results. Hence, the treatment of HM-contaminated soil is urgent and calls for restoration, which can be achieved through remediation methods including physical, chemical, and biological remediation.

13.7.2 Bio-Scrubbers and Bio-Filters

Bio-filters and bio-scrubbers are extremely specific odour treatment methods, which work by filtering odorous air by making use of microbes. Bio-trickling filters usually combine the properties of both bio-filters and bio-scrubbers in that they contain media, water, and odour-treating microorganisms. Unlike bio-filters, they are smaller, enclosed systems where the microbes grow within the bio-trickling filter media. Bio-scrubbers are like bio-trickling filters, being smaller, enclosed biological filtration odour treatment systems, where microorganisms are grown in a separate reservoir and circulated/re-circulated through the media. Bio-filters and bio-scrubbers have different applications, such as VOC treatment, inorganic compounds treatment, or odour reduction. Bio-filtration is a biological process used for the treatment of volatile organic and inorganic compounds that uses microbes which subject contaminants present in the air to biological degradation. During an air treatment process by bio-filtration, polluted air passes through the macro-pores of the filter material. Here, contaminants are destroyed to be later transported to a liquid medium where they are used as carbon and energy sources or as a source of energy. Use of bio-filtration indicates the production of biomass and the partial or total oxidation of the pollutant. Consecutively, biomass, under certain conditions, undergoes oxidation by endogenous respiration. Thus, bio-filtration techniques result in the complete decomposition of the contaminants, creating non-hazardous products. There are three types of bio-filters, as described below.

Fixed-bed bio-filters consist of a synthetic or organic filter material which serves as a support for microorganisms. Porous rock, diatomaceous earth, perlite, ground wood chips, and different types of compost or organic waste are used as filter materials. Fixed-bed bio-filters work by passing a humidified gas stream containing the pollutants through the bed where they are degraded by microbes. The technique is useful for treating pollutants that are poorly soluble in water because of the absence of the aqueous phase.

Drained-bed bio-filters consist of a column with an inert support where biofilm develops. A gaseous stream is fed through the bed that contains the substrate to bio-degrade besides a liquid stream that is continuously recycled through the bed. The liquid stream provides nutrients to the bio-film, as well as removing the degradation products of microorganisms. These systems are suggested for water-soluble compounds. The recirculation of fluids helps in elimination of the reaction products and enables better control over the biological process by pH control and composition of liquid medium.

In bio-scrubbers, the compound to be degraded is first absorbed in the liquid phase, and then placed in an absorption tower filled with liquid. The operation consists of gas flow in a counter current through the liquid where the pollutants and oxygen are absorbed. Afterwards, the liquid is fed to a reactor packed with an inert material covered with biofilm that is responsible for degrading the pollutant. Bio-scrubbers are the most appropriate systems for handling highly water-soluble compounds.

Bio-scrubbers do not accumulate products that have detrimental effects on the microorganisms. In addition, they help in controlling the biological process. However, bio-scrubbers are more costly because they require two systems, one for absorption and another for the biodegradation of pollutant.

Biologically degradable hydrocarbons are converted into H_2O and CO_2 in a bio-scrubber. The non-degradable hydrocarbons remain in the wash water. Components such as H_2S and NH_3 are converted into sulphate and nitrate, respectively. Regular draining is required to keep down the salt content and level of non-degradable hydrocarbons.

Bio-scrubbers are used in the detection of odour in the cigarette industry, and solvent problems when alcohols, ketones, acetates, etc., are used. These are used for removal of odour components, NH_3, H_2S, etc. from gas/odour released from slaughter houses, cattle farms, water purification plants, production of enzymes, production of polymers, aromas, flue gases in the rubber industry, flue gases in the production of methionine, removal of odour, hydrocarbons, and nitrogen components during the processing of paint residues and treatment of flue gases coming from waste sites for dangerous substances.

13.7.3 Biomarkers

Biomarkers are used for any measurement reflecting an interface between a biological system and a potential hazard, which may be physical, chemical, or biological. The measured response may be physiological or biological at the cellular level or during a molecular interaction. The biological marker is simply a measurable indicator of any biological state, which can be something that normally happens in the body or during a disease, or in response to a particular medicine. High BP is an indicator for increased risk of stroke. Likewise, complex genetic changes or mutations in the genes of tumour cells help in identifying a particular type of cancer. A biomarker is an indicator of normal biological processes, pathogenic processes, and pharmacologic responses to a therapeutic intervention. Biomarkers confirm the absorption of toxicants into the human body and measure integrated exposure, and even low-level exposure is detectable. They test and validate exposure models, follow exposure trends, and evaluate public health interventions. Biomarkers have the following characteristics:

- Indicate the risks of exposure to a particular chemical
- Determination can be made over an interval of time
- Determine the presence of xenobiotics in environmental samples and effects in various habitats
- Study the timing and duration of a known exposure
- Determine various routes of exposure
- Help in understanding the exposure–dose relationship
- Provide information on the toxicity of various samples in the environment

Biomarkers are placed in three categories:

1. **Biochemical Indicators**: These are based on the ability of a pollutant to generate a response at the gene level and the activity of specific enzymes involved in the detoxification process.

2. **Immunochemical Indicators**: These are specific reactions between antigens and antibodies that can be used to determine the presence of xenobiotics in environmental samples.
3. **Genetic Indicators**: These are biological indicators identified through gene expression analysis. They include genetically engineered organisms and are used for the study of HM resistance genes.

Biomarkers in Environmental Bio-Monitoring: Pollution biomarkers are defined as quantitative measures of changes in a biological system with respect to its normal status in response to pollutant exposure. Biomarkers offer biologically relevant information on exposure to bioavailable pollutants and on potential impacts of pollutants on the health of exposed organisms (see Figure 13.6). Biomarkers meet the emerging need of early-warning tools for the detection of exposure and adverse biological responses to pollutants. In environmental bio-monitoring, measurements of biomarker response in sentinel species can be used as an early warning of an alteration at population levels with the objective of monitoring environmental quality and evaluating changes in the environment. In an aquatic environment, bivalve mussels and crustaceans are used as sentinel organisms for bio-monitoring.

Their biochemical and cellular responses to pollutant exposure are used as early-warning tools in aquatic environmental monitoring and assessment. Soil invertebrates like earthworms are bio-indicators of soil pollution. They contribute to the mineralization and humification of organic matter by food consumption, respiration, etc. Lysosomal endpoints, oxidative stress, specific responses to metals, neurotoxic pollutants, and geno-toxic substances are the most commonly used biomarkers in environmental bio-monitoring. Lysosomes are involved in cell defence mechanisms and are responsible for the breakdown of cellular constituents and macromolecules, and provide protection against toxic and infectious agents.

Lysosomes react to pollutant exposure through alterations in lysosomal membrane stability, alterations in lysosomal number, and fusion events. Oxidative stress conditions in cells arise from exposure to pollutants. GSH is one of the most commonly used markers of oxidative stress. It neutralizes peroxides to maintain the redox

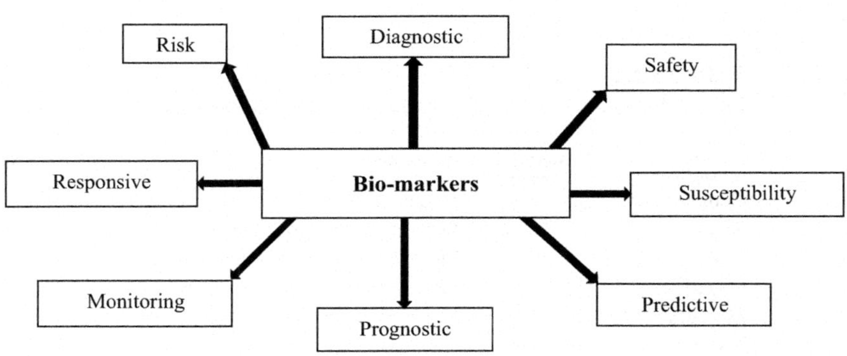

FIGURE 13.6 Categories of biomarkers.

balance of cells by combining with glutathione peroxidase and glutathione reductase. The reduced and oxidized glutathione ratio is a useful tool to assess the oxidative stress in an organism. Malondialdehyde, antioxidant enzymes, such as catalase, superoxide dismutase, and glutathione peroxidase are other biomarkers of oxidative stress suitable for assessing the effects of pollutants in aquatic and terrestrial ecosystems at early stages and with low concentrations. Cysteine-rich metal-binding protein, i.e., metallothionein, is another biomarker which is involved in the detoxification and homeostasis of HMs. Acetylcholinesterase is an important biomarker used in response to neurotoxic compounds. It catalyses the hydrolysis of the neurotransmitter acetylcholine. It is the target site of inhibition by organophosphate and carbamate pesticides. Monitoring of acetylcholinesterase inhibition is widely used as a biomarker of organophosphorus and carbamate exposure either in aquatic or terrestrial ecosystems. Likewise, genotoxicity biomarkers are associated with DNA damage, comet assay with DNA alteration, and micronucleus test with chromosomal damage, which are the most widely applied and validated biomarkers employed in biomonitoring programmes. Some commonly used biomarkers are listed in Table 13.2.

Biomarkers are used to detect the effects of water pollution on aquatic organisms. Benthic macro-invertebrates reside in the sediment on the bottom of streams, where many contaminants settle. These organisms have high exposure to pollution, which makes them good species for experimentation while detecting concentration of a pollutant and its impact on an ecosystem.

There are various biomarkers within an aquatic organism that can be measured, depending on the pollutant. Likewise, there are numerous contaminants in water bodies. To analyse the impact of a pollutant on an organism, the biomarker must respond to a specific pollutant in a specific timeframe or at a certain concentration. Biomarkers used to detect responses of HM pollution in aquatic organisms can be enzymatic or non-enzymatic. Metals lead to oxidative stress in fish by stimulating the generation of reactive oxygen species (ROS), which damages biological components like lipids, proteins, and DNA. Oxidative damage can be determined by measuring lipid peroxidation (LPO), protein carbonylation (PC), and DNA base oxidation. ROS can be detoxified by cellular antioxidant defence systems, both enzymatic and non-enzymatic, while measuring the total antioxidant capacity (TAC), evaluating the overall efficacy of cellular antioxidants against oxy-radicals, and providing a

TABLE 13.2
List of Biomarkers Used for Heavy Metal Responses

Metals	Symbol	Biomarkers
Copper	Cu	Metallothionein
Zinc	Zn	Metallothionein, stress proteins
Cadmium	Cd	Metallothionein, glutathione transferase
Cobalt	Co	Metallothionein, lipid peroxidation
Mercury	Hg	Metallothionein, haem, and porphyrins
Nickel	Ni	Metallothionein
Lead	Pb	Metallothionein, haem, and porphyrins

comprehensive assessment of the receptiveness to oxidative stress. DNA damage biomarkers can be used to assess the geno-toxicity of environmental pollutants and a comet assay is a valuable tool for assessing geno-toxicity in fish in the environment. The induction of stress proteins is the only molecular mechanism used by fish to cope with stress. Overexpression of heat shock proteins (HSPs) and metallothioneins (MTs) can be recognized in response to diverse adverse ecological conditions, including HM exposure. With respect to pollution in general, and HMs in particular, fish condition factor (K) and hepatosomatic index (HSI) are utilized as biomarkers of health and growth. Proper application of biological reactions as biomarkers demands understanding of the seasonal variations in biotic and abiotic variables, which have impacted the basal levels of several biomarkers and their responsiveness to pollution. Biomarkers do not show a single dose–response relationship but a number of dose–response relationships that are the result of all possible interactions of all contaminants present in the ecosystem. Figure 13.7 illustrates the interaction of biomarkers and environmental pollutants in following steps.

Step 1: Presence of Pollutants in the Environment

- Biomarkers are used as early warning systems signalling potentially compromised situations
- Biological change must be caused exclusively by the contaminants

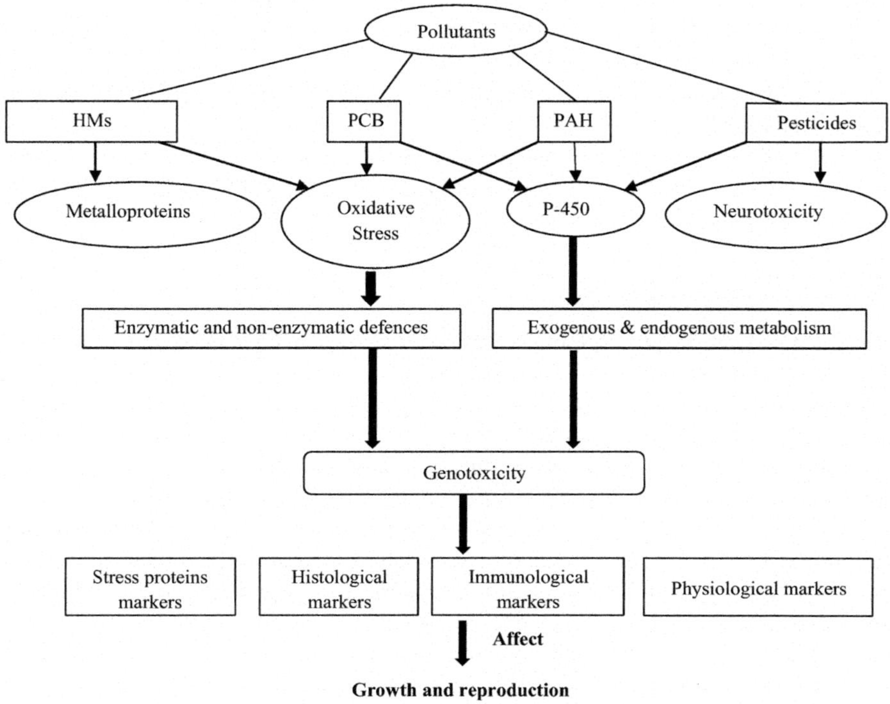

FIGURE 13.7 Interactions between contaminants and biomarkers.

- Various contaminants like HMs, toxic elements, PCBs, PAHs, pesticides, and insecticides are present in the environment
- These pollutants are harmful to the health of the ecosystem

Step 2: Identification and Analysis of Biomarkers

- Lysosomal endpoints, oxidative stress, specific responses to metals, neurotoxic pollutants, and geno-toxic substances are the most commonly assessed biomarkers in environmental bio-monitoring
- GSH, malondialdehyde, catalase, superoxide dismutase, and glutathione peroxidase are the most commonly used markers of oxidative stress condition

Step 3: Interaction of Pollutants with Biomarkers

- Biomarkers are responses correlated both with the exposure to a contaminant and with its toxic effects
- The effect of organophosphorus (OPs) insecticides and carbamates (CBs) triggers inhibition of acetyl cholinesterase (AchE) which causes serious damage to the functioning of the CNS of many organisms
- Oxidative stress, P-450, metalloproteins, and neurotoxicity are specific biochemical markers

Step 4: Effects on Body Metabolism, Growth, and Reproduction

- P-450 affects the endogenous and exogenous metabolism of the body and produces geno-toxic effects in the body
- The oxidative stress affects the enzymatic and non-enzymatic defences of the body
- The less specific biomarkers like stress protein, histological biomarkers, immunological biomarkers, and physiological biomarkers affect the growth and reproduction of organisms.

13.7.4 Carbon Sequestration

CO_2 is one of the most common greenhouse gases (GHG) that entrap heat in the atmosphere as a result of both human and natural activities. Other natural or biological emissions of CO_2 may be due to land-use changes, decomposing organic matter, and forest fires, whereas man-made sources include energy-generating processes, including burning of coal, oil, and natural gas. Carbon sequestration is a process of capturing CO_2 and storing it safely away from the atmosphere. It is a method of reducing the amount of CO_2 in the atmosphere, which aims at reducing global warming and climate change. It aims to stabilize the GHG in the atmosphere and reduce the human carbon footprint. The term carbon sequestration is generally used to describe both natural and anthropogenic processes which cause CO_2 to either get eliminate from the atmosphere or removed from emission sources.

Carbon sequestration is essential to mitigate the challenges of climate change. However, one of the key technologies promoted by policymakers and/or scientists is carbon capture and storage (CSS). This is a geo-engineering process in which CO_2

is initially separated from other gases followed by compressing and transporting the separated CO_2 to a safe location where it is stored for long-term use. The carbon capturing and storage method not only detains CO_2 from its direct source of emissions before releasing it into the atmosphere but also eradicates CO_2 gas from the air. Once the process of storing CO_2 is achieved, it is then transferred to a suitable location, such as deep oceans or deep saline geological formations. These locations favour the process in such a way that CO_2 released can be used effectively as it is used when naturally released to the atmosphere. In other words, carbon sequestration in the ocean practically changes the CO_2, through photosynthesis, into oxygen, by plankton, as is done by trees and plants on land. There are three types of carbon sequestration: terrestrial sequestration, geologic sequestration, and oceanic carbon sequestration.

Terrestrial Sequestration: Terrestrial carbon sequestration or biological sequestration involves the absorption of CO_2 from the atmosphere by means of trees and plants through photosynthesis and storing the carbon in soil and biomass, such as branches, tree trunks, foliage, and roots of plants. As forestry and agricultural activities also release CO_2 into the atmosphere, a carbon sink, where carbon sequestration is greater than carbon releases over a period of time. The term sink refers to cropland, grazing land, and forests, and their ability to sequester carbon. Terrestrial sequestration can also be amplified by effective land use practices which increase the amount of carbon stored in croplands, soils, and forest environments. Methods of storing carbon in cropland ecosystem include crop rotation, reducing summer fallow, planting cover crops like wheat, and vegetation buffers. Other terrestrial sequestration methods include the selection of high-carbon-storage hybrid crops and changing marginal agricultural lands to grasslands or forests. Terrestrial sequestration is accomplished through forest and soil conservation practices. These practices help to enhance carbon storage, such as restoring and establishing new forests, wetlands, and grasslands, and reducing CO_2 emissions.

Carbon Sequestration in Forests: Carbon is stored in woody tissues and organic matter of soil as a forest grows. Generally, the net rate of carbon uptake is high for young trees and gradually reduces with time. This means that old forests can sequester carbon for a long time; however, they contribute no net uptake. When the woody tissue obtained from forests is either burned or processed into other products such as paper, the carbon contained in these tissues will return immediately to the atmosphere. In case that wood is used to produce furniture, a long-lived product, the carbon is retained in the wood and acts as a carbon sink. Therefore, the method of reducing waste and converting most wood into long-lived products appears to be an effective strategy to reduce global atmospheric carbon. However, long-lived wood products still contribute comparatively less sink to carbon; therefore, deforestation eventually reduces the storage of carbon on land.

Carbon Sequestration in Soil: Plants usually assimilate carbon through the process of photosynthesis and return some of the carbon back to the atmosphere through respiration. The carbon left in plant tissue is later consumed either by animals or eliminate through soil decomposition. The fundamental method of storing carbon in the soil is achieved as soil organic matter (SOM). SOM is a complex mixture of carbon compounds and carbon linked with soil minerals collected by the decomposition of plant and animal tissues and microbes, such as protozoa, fungi, and bacteria. This results in capturing carbon stored in soils either for thousand years or be quickly released back into the atmosphere. There are several factors influencing the amount and time of carbon stored in soils such as climatic conditions, natural vegetation, soil texture, and drainage.

Geologic Sequestration: Geologic sequestration of CO_2 involves a three-step process: capturing of CO_2, transportation of CO_2, and storage in deep underground reservoirs. Initially, the sequestration starts with successful capturing of CO_2 from the carbon emission sources. Later, the captured CO_2 is piped below the land surface and injected into porous rock formations. Once the CO_2 is injected underground, it is entrapped beneath an impermeable barrier or seal. Some of the injected CO_2 dissolves gradually in groundwater, while some may be entrapped in the form of carbonate minerals that can be retained for thousand years.

Carbon Sequestration in Oceans: Oceans are long-term sinks for anthropogenic CO_2 emissions and are achieved through various chemical and biological processes which cause chemical reactions between seawater and CO_2 in the atmosphere naturally. In addition, these reactions cause the oceans to achieve a more acidic nature while absorbing atmospheric CO_2. There are many marine organisms that depend on the formation of carbonate skeletons and sediments which are vulnerable to dissolution in acidic waters. Generally, plankton at the ocean surface convert CO_2 into sugars through photosynthesis. These photosynthesizing plankton are consumed by various sea creatures, which causes the carbon-containing sugars to finally sink to the bottom. The deep ocean locks the stored carbon as sediment over millions of years.

13.7.5 Carbon Footprints

A carbon footprint is the total amount of CO_2 emissions produced directly or indirectly as a result of an individual's actions or is gathered over a period of time. It is the sum of all by-products from fossil fuels that are used for manufacturing, building structures, providing food, and travel. The gases produced through these activities, such as CO_2, CH_4, N_2O, etc., are referred to as greenhouse gases and are responsible for entrapping heat in the atmosphere and increasing the Earth's average temperature. The carbon footprint mostly represents the total volume of greenhouse gases resulting from everyday economic and human activity. These environmental indicators measure both direct and indirect emissions of compounds like CH_4, N_2O, HFCs, PFCs and sulphur SF_6, but most importantly CO_2. The term carbon footprint itself is rooted in the literature of the ecological footprint, which estimates the biologically productive land and sea area required to provide the renewable resources for a population to consume and the wastes generated to be absorbed. It basically measures the requirements for productive areas such as croplands, grazing lands, forest areas, marine areas, and most importantly forest land to absorb the CO_2 emissions from energy consumption. Therefore, when the carbon footprint is calculated in the context of the total ecological footprint, tonnes of CO_2 emissions are expressed as the amount of productive land area required to sequester those CO_2 emissions.

Food, consumption, transportation, and household energy are major contributors of carbon footprints. In food, meat in particular is an issue as livestock are accountable for a significant amount of greenhouse gas emissions, with beef considered to be one of the biggest contributors to CO_2. Therefore, reducing the meat intake can help subsequently to lower the carbon footprint. In addition, transportation of foods, pesticide use, and high demand for seasonal food also contributes to carbon footprints. Besides food consumption, home energy consumption also contributes significantly to the carbon footprint. Energy-inefficient homes can cause significant amounts of energy waste through poor insulation, energy-inefficient appliances, and excessive use of water. Of these, poor insulation in homes acts as a lead contributor because heat leaves the house quickly. Moreover, consumption, such as clothing, footwear, household, and personal goods, etc., also tends to increase the individual carbon footprint considerably because all these items have linked emissions from gathering materials, production, and transportation. Reducing or optimal use of consumption can help in lowering the carbon footprint.

13.7.6 Carbon Credit and Tax

Carbon Credits: A carbon credit is a tradable certificate or permit that, per UN standards, entitles the holder to emit one tonne of CO_2 removed, reduced, or sequestered from the atmosphere. A carbon credit is mainly a credit for greenhouse gas emissions by an emission reduction project, which can be potentially used by governments, industry, or private individuals to compensate for their emissions. Thus, generating the credits through greenhouse mitigation projects makes this approach highly susceptible to the use of carbon reduction schemes as finance among trading partners around the world. One carbon credit is equivalent to one metric ton of CO_2, or in some markets, CO_2 equivalent gases ($CO_{2\text{-}}eq$). These credits are bought and sold through online retailers on trading platforms. The projects generally used to sell carbon credits include wind, solar, geothermal, and biomass, which replace fossil-fuel-powered plants. Carbon credits are most often created through agricultural or forestry practices. Carbon credits emerged from the objective of the Kyoto Protocol which was formulated to enable developed nations to assist developing nations economically, impacted by carbon emission constraints and heavily indebted to foreign creditors. Therefore, a carbon credit system was devised to impose national caps on greenhouse gas emissions of developed nations, with each of these countries provided with an allotment number of emission allowances called assigned amount units (AAUs). Other participating countries were required to reduce their emissions below 1990 levels and they were allowed to meet their targets by buying carbon credits. The Kyoto Protocol facilitates three mechanisms that enable countries, or operators in developed countries, to accomplish greenhouse gas reduction credits:

- Under joint implementation (JI), a developed country having high costs of domestic greenhouse reduction can set up a project in another developed country.

- Under the clean development mechanism (CDM), a developed country can sponsor a greenhouse gas reduction project in a developing country in which the developed country would be given credits for meeting its emission reduction targets while the developing country would receive the capital investment and clean technology or beneficial change in land use.
- Under international emissions trading (IET), countries can trade in the international carbon credit market to cover their shortfall in assigned amount units (AAUs).

Carbon offset is another financial approach to reduce greenhouse gas emissions. It is similar to carbon credit as a carbon offset credit is equivalent to a reduction of one metric ton of CO_2 or equivalent greenhouse gas in the atmosphere. Carbon offsetting helps to promote renewable and green energy methods, such as solar and wind energy. By using carbon offsetting for cleaner and renewable energy sources greenhouse emission reduction can more easily be achieved. These systems, carbon credit and carbon offset, encourage organizations or government firms to apply more eco-friendly approach in parallel collaboration with their development.

Carbon Tax: Carbon tax is a form of carbon price on greenhouse gas emissions in which a fixed price is set by government for carbon emissions for certain sectors. It is basically a fee imposed on the burning of carbon-based fuels such as coal, oil, gas, etc. The intention behind implementation of carbon tax is to reduce and eventually eliminate the use of fossil fuels whose combustion is harmful to the environment. The Treasury's Office of Tax Analysis (OTA) is responsible for estimating the revenue, economic, and distributional effects of carbon tax systems tax. The potential carbon tax base is categorized into three categories: fossil fuel emissions, non-fuel emissions, and biomass fuels, such as ethanol, or deforestation, which might ultimately be covered in some way by a carbon tax system.

Fossil Fuel Emissions: Fossil fuel combustion causes the emission of GHGs which could be covered by an excise tax levied on coal, natural gas, and petroleum. This tax is levied at distinct points in the supply chain. Fossil fuel emissions can be taxed using either an upstream approach or a midstream approach. An upstream approach taxes raw fuels, such as crude oil and natural gases, while a midstream approach taxes fuels at a designated point further down the supply chain but before they reach final consumers.

Non-Fuel Emissions: There are a number of GHG emissions that would not be covered by a fuel-based emissions tax, such as emissions that arise during the production of industrial products like cement, lime, glass, ammonia, petrochemicals, etc., in addition to emissions from mining of coal and extraction and refining or processing of oil and natural gas and direct emissions of fluorinated gases.

Biomass Fuels: Biomass fuels such as ethanol, wood, animal waste, etc. are used for transportation, power generation, and heating. There is no way to assess the rationale for taxing these fuels so that these are mandated under non-taxed bio-fuels.

Green Building: Green building is a resource-efficient method of construction that produces healthier buildings which have less impact on the environment and have low-cost maintenance. A green building is a sustainable or high-performance building which:

- Conserves natural resources
- Has enhanced image and marketability
- Has increased user productivity
- Limits waste generation due to recycling and reuse
- Optimizes energy efficiency
- Provides healthier space for occupants as compare to conventional buildings
- Reduces air and water pollution, destruction of natural areas, habitats, and biodiversity, energy consumption without sacrificing comfort levels and pollution loads and soil loss from erosion
- Uses less water

Green Building Rating Systems

LEED (Leadership in Energy and Environmental Design): LEED is the most widely used green building rating system in the world, which was developed by the U.S. Green Building Council and provides a framework for green building design, its construction, operations, and performance to create healthy, highly efficient, and cost-saving green buildings. LEED certification is a globally recognized symbol of sustainability achievement. LEED-certified buildings save money, improve efficiency, lower carbon emissions, and create healthier places for people. They are a critical part of addressing climate change and meeting ESG goals, enhancing resilience, and supporting more equitable communities. To achieve LEED certification, a project earns points by adhering to prerequisites and credits that address carbon, energy, water, waste, transportation, materials, health, and indoor environmental quality. The projects also go through a verification and review process by GBCI and are awarded points that correspond to a level of LEED certification, i.e., certified (40–49 points), Silver (50–59 points), Gold (60–79 points) and Platinum (80+ points).

GRIHA (Green Rating for Integrated Habitat Assessment): Human habitats (buildings) throughout their lifecycles, from construction, operation, to demolition, consume resources in the form of energy, water, materials, etc. and emit wastes either directly in the form of municipal wastes or indirectly as emissions from electricity generation. GRIHA attempts to minimize a building's resource consumption, waste generation, and overall environmental impact to certain nationally acceptable limits. GRIHA attempts to quantify aspects, such as energy consumption, waste generation, and renewable energy adoption, to achieve, regulate, and reduce the same to the best possible extent. It acts as a rating tool that helps people assess the performance of their building against certain nationally acceptable benchmarks. In addition, it estimates the environmental performance of a building holistically over its entire lifecycle, thus offering an ultimate benchmark for what constitutes a "green building". The rating system, based on recognized energy and conservational principles, will seek to strike a balance between the established practices and emerging concepts, both national and international. The system has been developed to help "design and evaluate" new buildings. A building is evaluated based on its projected performance over its entire lifecycle. The phases of the lifecycle that have been recognized for assessment are:

Pre-construction stage (type of soil and land, the flora and fauna on the land before construction activity starts, the natural landscape features)

Building planning and construction stages (issues of resource management including land, water, energy, air, and green cover)

Building operation and maintenance stage (issues of operation and maintenance of building systems and processes, monitoring and recording of energy consumption, and occupant health and comfort)

Some of the breakthroughs of green buildings are:

Pixel Building (Melbourne, Australia): Australia's first carbon-neutral office building, generating all its power and water on site with colourful, eye-catching panels that provide shade and maximize daylight as desired, with wastewater utilization, rainwater harvesting, and wind turbines.

Museum of Tomorrow (Rio de Janeiro, Brazil): With its distinct cantilevered roof, reflective pools, and skeletal structure, Rio's Museum of Tomorrow is a demonstration of future prospects. Its sustainable design features include amendable, fin-like solar panels that add to the building's neo-futurist visual, and a pumping system that takes cold water from the bottom of the nearby Guanabara Bay for use in its air-conditioning system.

Khoo Teck Puat Hospital (Yishun Central, Singapore): An example of a consciously made green environment this complex uses a solar water heating system and other energy-efficient methodologies.

Rajiv Gandhi International Airport (Hyderabad): Commissioned in 2008, this is one of the leading examples of highly environmentally sensitive infrastructure design and management in India.

Green Hydrogen: Hydrogen, which is produced by electrolysis of water using renewable electricity, is referred as green hydrogen. This results in very low or zero carbon emissions that could replace fossil fuel in the sectors which cannot easily switch to electricity. The chief attraction of using hydrogen as a fuel is that the main by-product is water vapour, along with small amounts of nitrogen oxides, making it far less polluting than fossil fuels.

Green Energy: This is a clean, renewable, eco-friendly, waste-free, zero-carbon, replenishable, and sustainable form of energy. It is renewable because it is not depleted easily and is naturally replenished. Green energy comes from natural sources, such as sunlight, hydropower, wind, rain, tides, plants, algae, and geothermal heat. Green energy has zero greenhouse gas emissions, is pollution free, and does not produce waste. Green energy is the best alternative for non-renewable energy. Different types of renewable energy technologies include solar and wind energy, tidal energy, bio-energy, and geothermal energy.

BIBLIOGRAPHY

Brown, R. J. C. and Milton, M. J. T. (2005) Analytical techniques for trace element analysis: An overview. *Trends in Analytical Chemistry* 24(3): 266–274.

Ene, A., Bosneaga, S. A., and Georgescu, L. (2010) Determination of HMs in soils using XRF technique. *Romanian Journal of Physics* 55: 815–820.

Gupta, Y. (2011) Carbon credit: A step towards green environment. *Global Journal of Management and Business Research* 11(5): 16–20.

Kahle, H. (1993) Response of roots of trees to HMs. *Environmental and Experimental Botany* 33: 99–119.

Kalyani, V., Dudy, M., and Pareek, S. (2015) Green energy: The need of the world. *Journal of Management Engineering and Information Technology* 2: 18–26.

Katabata-Pendias, A., and Pendias, H. (1992) *Trace Elements in Soils and Plants*, Second edition, pp. 289–299. Boca Raton: CRC Press.

Khan, S., et al. (2007) Health risks of HMs in contaminated soils and food crops irrigated with wastewater in Beijing, China. *Environmental Pollution* 6: 1–7.

Kodom, K., Preko, K., and Boamah, D. (2012) X-ray fluorescence (XRF) analysis of soil HM pollution from an industrial area in Kumasi, Ghana. *Soil & Sediment Contamination* 21(8): 1006–1021.

Lal, R. (2008) Carbon sequestration. *Philosophical Transactions of the Royal Society B: Biological Sciences* 363: 815–830.

Lam, P. K. and Gray, J. S. (2003) Use of biomarkers in environmental monitoring programmes. *Marine Pollution Bulletin* 46(2): 182–186.

Lionetto, M., Caricato, R., and Giordano, M. (2019) Pollution biomarkers in environmental and human biomonitoring. *The Open Biomarkers Journal* 9: 1–9.

Madden, C., et al. (2022) Portable X-ray fluorescence (pXRF) analysis of HM contamination in church graveyards with contrasting soil types. *Environmental Science and Pollution Research* 29(36): 55278–55292.

Malik, L., Bashir, A., Qureashi, A., and Pandit, A. (2019) Detection and removal of HM ions: A review. *Environmental Chemistry Letters* 17(46): 1–27.

Marchais, T., et al. (2018) Gamma-ray spectroscopy measurements and simulations for uranium mining. *EPJ Web of Conferences* 170(05003): 1–4.

Mijares, M. A., et al. (2013) Determination of HMs contamination using a silicon sensor with extended responsive to the UV. *Journal of Physics: Conference Series* 421.

Mukhopadhyay, G. (2018) Atomic spectroscopy analysis of HMs in plants. *International Journal of Pharmaceutical & Biological Archives* 9(3): 175–179.

Nema, N. K., Maity, N., Sarkar, B. K., and Mukherjee, P. K. (2014) Determination of trace and HMs in some commonly used medicinal herbs in Ayurveda. *Toxicology and Industrial Health* 30(10): 964–968.

Nordberg, G. F. (2009) Historical perspectives on cadmium toxicology. *Toxicology and Applied Pharmacology* 238(3): 192–200.

Nuhu-Aliyu, A. (2022) Atomic absorption spectroscopy analysis of HMs in water at Daura gypsum mining site. *International Journal for Physical and Engineering Sciences* 8(2): 227–234.

Osadchii, S., Petukhov, A., and Dunin, V. (2019) X-ray absorption spectroscopy analysis of HMs by means of a silicon detector. *Journal of Surface Investigation: X-ray, Synchrotron and Neutron Techniques* 13: 683–689.

Phoenix, V. R., Holmes, W.M., and Ramanan, B. (2008) Magnetic resonance imaging (MRI) of heavy-metal transport and fate in an artificial biofilm. *Mineralogical Magazine* 72(1): 483–486.

Radwan, M. A. and Salama, A. K. (2006) Market basket survey for some HMs in Egyptian fruits and vegetables. *Food and Chemical Toxicology* 44: 1273.
Sonawane, M. S. (2018) Types and methods of carbon sequestration: A review. *International Archive of Applied Sciences and Technology* 9(1): 75–81.
Van der Oost, R., Beyer, J., and Vermeulen, N. P. (2003) Fish bioaccumulation and biomarkers in environmental risk assessment: A review. *Environmental Toxicology and Pharmacology* 13(2): 57–149.
WHO. (1995) *Lead Environmental Health Criteria*. Geneva: World Health Organization.
Wiedmann, T. and Minx, J. (2008) A definition of carbon footprint. In: *Ecological Economics Research Trends: Chapter 1*, C. C. Pertsova (ed.), pp. 1–11. Hauppauge NY, USA: Nova Science.

Index

E

F

G

H

I

L

M

N